D0816632

NOTABLE SCIENTISTS

A TO Z OF BIOLOGISTS

LISA YOUNT

Facts On File, Inc.

For Alec
Sometimes the "outsiders" turned out to be right!

A TO Z OF BIOLOGISTS

Notable Scientists

Facts On File, Inc.
132 West 31st Street
New York NY 10001

Library of Congress Cataloging-in-Publication Data

Yount, Lisa.
A to Z of biologists / Lisa Yount.
p. cm. — (Notable scientists)
Includes bibliographical references (p.).
ISBN 0-8160-4541-0
1. Biologists—Biography. I. Title. II. Series.
QH26.Y68 2003
570'.92'2—dc21 2002013816

Facts On File books are available at special discounts when purchased in bulk quantities for businesses, associations, institutions, or sales promotions. Please call our Special Sales Department in New York at 212/967-8800 or 800/322-8755.

You can find Facts On File on the World Wide Web at http://www.factsonfile.com

Text design by Joan M. Toro
Cover design by Cathy Rincon

Printed in the United States of America

VB Hermitage 10 9 8 7 6 5 4 3 2 1

This book is printed on acid-free paper.

CONTENTS

LIST OF ENTRIES

ACKNOWLEDGMENTS

I wish to thank all the scientists who read their entries, and their assistants, for patience in answering questions and, in some cases, providing photographs.

Thanks also to the National Library of Medicine for the availability of their wonderful image archive, which reaches far beyond the boundaries of medicine, and to Infotrac and the Internet in general for saving many tedious trips to the library.

My gratitude to Frank K. Darmstadt, my editor at Facts On File, for his patience and cheerful attitude. And, as always, thanks and love to my husband, Harry Henderson, for unending support and for being the most ideal life-partner imaginable.

INTRODUCTION

Although the term *biology* was not introduced until the 19th century, humans have been informal students of that science since the beginning of their existence. Indeed, biology—the study of all the living things on the planet, including humans themselves—was a life-and-death matter. Early people had to know the qualities and behavior of animals and plants in order to hunt and gather, to protect and clothe themselves, and, later, to farm and raise domestic animals. They had to learn about their own bodies in order to maintain health and treat injury and disease.

The study of biology is still a life-and-death matter. Some modern biologists explore life in the wild, trying to grasp the incredibly complex interactions between living things and their environment in the hope of preserving both from destruction—a destruction that could take humanity with it. Others delve into mysteries of cells, genes, and molecules that earlier biologists could hardly have imagined, attempting to create new kinds of food that might assuage world hunger, to root out cancer or inherited diseases at their source, or even perhaps to take control of human evolution. Today, as never before, biologists are shaping as well as studying the living world.

Like other sciences, biology has attracted a diverse mix of intelligent and sometimes eccentric men and women. Some were drawn to nature in childhood as they took long walks through the countryside or collected "miniature zoos" of local animals. Others were brought there by sudden changes in their lives: Time spent outdoors while convalescing from illness inspired Jean-Baptiste Lamarck and John Ray, and wartime experiences turned Maurice Wilkins from physics to biology and J. Craig Venter from surfing to medical research. Some, especially the women, had to struggle hard to reach their goals. Nettie Maria Stevens spent much of her life as a teacher and librarian before she found a chance to obtain an advanced degree and enter research. Florence Wambugu's mother sold the family's only cow in order to start her daughter on the road to education. Once in science, all showed the combination of inspiration, careful observation, and hard, dedicated work that lies behind any scientific achievement.

THE SCIENTISTS

A to Z of Biologists presents the stories of 184 biologists from a wide range of countries, time periods, backgrounds, and fields. They include medical researchers as well as students of plants and animals. A few scientists who were not biologists but had major impacts on biology, such as geologist Charles Lyell (whose work helped to pave the way for Charles Robert Darwin's theory of evolution) and physicist Wilhelm Conrad Röntgen (whose discovery of X rays revolutionized medicine), appear here as well.

The entrants were chosen chiefly because of their importance to science, as judged by reference works and scientific publications, and many of their names will be familiar to readers with a background in biology. Some are more obscure but seem to deserve wider recognition or represent interesting examples of work in various fields. Inevitably, many other scientists could have been included with equal justification, but for various reasons (such as lack of easily obtainable biographical information or simple oversight on my part) were not. I offer my apologies to them or their descendants; no disrespect of their work was intended by their omission.

THE ENTRIES

Entries are arranged alphabetically by surname, with each entry given under the name by which the entrant is most commonly known. The heading for each entry provides the entrant's complete name, birth and death dates, countries where the entrant was born and (if different) where the entrant lived at the time of his or her chief scientific achievement, and field of work.

To make terminology more accessible, I have kept the number of field designations fairly small; thus, for example, biologists who studied the brain and nervous system are all listed as neurobiologists rather than being divided into neuroanatomists, neurophysiologists, and so on. In some cases, I have used field names that would not have been in existence at the time a scientist worked because they seem to give the clearest picture of what the scientist did. Gregor Mendel, for instance, would have had no idea of what the term *geneticist* meant, but it is hard to think of anyone who better deserves that designation. Conversely, I have kept the old-fashioned term *naturalist*—one who studies all of nature—for those scientists of earlier times who, unencumbered by today's insistence on specialization, let their curiosity range throughout biology and, sometimes, into physical sciences, such as geology and meteorology, as well.

The text of the entries ranges from about 500 to 1,500 words, with most running around 1,000 words. They include the usual biographical information: date and place of birth and death, family information, educational background, places worked and positions held, prizes awarded, and so on. The bulk of the entries describes the scientists' work, explained in terms that I hope are relatively simple and clear. (A glossary at the back of the book provides definitions of the biological terms most frequently used in the entries; others are defined in the text.) In addition, often using quotes from the scientists themselves or from those who knew them or commented on their work, I have attempted to convey some feeling for the personalities of these fascinating men and women and for their impact on biology and society. Names in small caps within the essays indicate cross-references to other scientists described in the book. For those who wish to learn more about a particular scientist, a short list of further reading, including both print and Internet resources, is provided at the end of each entry.

The book concludes with several appendices that may aid readers seeking particular types of information. In addition to the glossary, the appendices include a list of scientists arranged by country of birth; a list by country of major scientific activity; a list by year of birth; and a chronological chart showing the chief discoveries of the scientists in the book arranged by approximate date. This chronology provides a sort of capsule history of the science—one might say, "Biology's Greatest Hits."

I hope that readers will find the entries interesting as well as informative. I certainly was often thrilled or amused as well as enlightened while researching and writing this book. I pictured myself walking down a sort of endless receiving line of these fascinating people, regretting only that I had such a short time to stop and chat with each one before I had to move on to the next. I thank all of them, living and dead, for their company.

Adrian, Edgar Douglas

(1889–1977)
British
Neurobiologist

Edgar Adrian made important discoveries about the way nerves send messages and interact with muscles and sense organs, such as the eyes and ears. He was born in London on November 30, 1889, the second son of Alfred Adrian, an attorney, and his wife, Flora. Adrian studied physiology and medicine at Cambridge University and St. Bartholomew's Hospital in London, earning his M.D. in 1915. During World War I, he served in the Royal Army Medical Corps, treating soldiers suffering from nervous system or psychological injuries.

Trinity College, part of Cambridge, was Adrian's home for most of his adult life. He began teaching and doing research there as a staff lecturer in 1919. He was professor of physiology at the college from 1937 to 1951, then Master of Trinity College until 1965. He was Chancellor of Cambridge University from 1968 to 1975.

One of Adrian's specialties was applying and adapting new technology. Other scientists had learned that nerves send messages by means of tiny discharges of electricity. In the late 1920s, Adrian developed a way to use a new device called a thermionic valve to amplify nerve signals as much as 5,000 times. He combined this tool with delicate surgical techniques to record electrical signals from individual fibers within a nerve, something scientists had not been able to do before.

Keith Lucas, one of Adrian's professors, had shown that a stimulus (incoming signal, such as pressure on the skin or light reaching the eye) had to reach a certain strength before it made a nerve cell fire an electric pulse in response. Adrian extended this work by showing that all electric signals from a nerve have the same strength. If a stimulus becomes stronger, the nerve sends out pulses more often, but the signals do not become more powerful or travel more quickly along the nerve. "Stimulating a nerve may be compared to firing a gun," Adrian told a *New York Times* reporter in 1934. "We may pull too feebly on the trigger, but if we pull hard enough to fire the bullet no amount of extra pulling will make it travel any faster." The frequency with which a nerve sends signals tells the brain how strong a stimulus is.

Adrian also mapped the areas of the cerebral cortex (the part of the brain devoted to conscious thinking) that receive messages from different parts of the body. He found that the maps are different in different animals, depending on the animals' needs. In humans and monkeys, for

Edgar Douglas Adrian showed how nerves send electrical signals in response to stimuli. *(National Library of Medicine)*

instance, the largest area of the cortex is devoted to messages from the hands and face. In ponies, on the other hand, the area taking input from the nostrils is as big as the area receiving messages from all the rest of the body.

In the 1930s and early 1940s, Adrian experimented with a newly invented machine called the electroencephalograph, a device that detects and records electrical patterns, called waves, made by the firing of thousands of nerve cells in the brain. "The search for the mechanisms of the brain . . . may lead us to a new understanding of human behavior," he said in 1944. In his final years as a researcher, from about 1937 to 1959, Adrian studied the sense of smell. During his career he wrote several books, including *The Mechanism of Nervous Action* (1932) and *The Physical Background of Perception* (1947).

Adrian's studies of nerve cells earned a portion of the Nobel Prize in physiology or medicine in 1932. (He shared the prize with CHARLES SCOTT SHERRINGTON, who also did research on the nervous system.) He also received awards such as the Copley Medal of the Royal Society (1946) and the Gold Medal of the Royal Society of Medicine (1950). Queen

Elizabeth made Adrian the first Baron of Cambridge in 1955.

Adrian was a popular figure at Cambridge, famous for riding his bicycle rapidly through the streets and even along the underground corridors of Cambridge's physiology laboratory. He enjoyed fencing, mountain climbing, and sailing. He married Hester Pinsent in 1923 and had a son and two daughters. He died on August 4, 1977.

Further Reading

"Adrian, E(dgar) D(ouglas)." *Current Biography Yearbook 1955*. New York: H. W. Wilson, 1955.

"Edgar Douglas Adrian—Biography." *Nobel Lectures: Physiology or Medicine 1922–1941*. Available online. URL: www.nobel.se/medicine/laureates/1932/adrian-bio.html. Last modified 2001.

Anderson, W. French

(1936–)
American
Geneticist, Physician

Reporters have often called William French Anderson "the father of gene therapy." He was the leader of the group that, in 1990, first successfully used altered genes to treat a human illness.

Born in Tulsa, Oklahoma, on December 31, 1936, to Daniel French Anderson, a civil engineer, and LaVere Anderson, a book editor for a local newspaper, "Bill" Anderson, as he was known at the time, grew into what he later called "a rather weird little boy." By the time he was eight years old, he was reading college science texts. A few years later, he decided to become a physician.

The summer before he was to enter Harvard University, Anderson first heard about genes and DNA. He learned that some diseases are caused by defects in DNA, which produce mistakes in the coded information that genes carry. He soon made himself two promises: "I was going to be in the Olympics, and I was going to cure defective molecules." The notion of repairing or replacing damaged genes seemed so far-fetched, however, that one of Anderson's Harvard professors called it daydreaming. Scientists were just starting to learn how genes worked in the 1950s, and they had no idea how to change them.

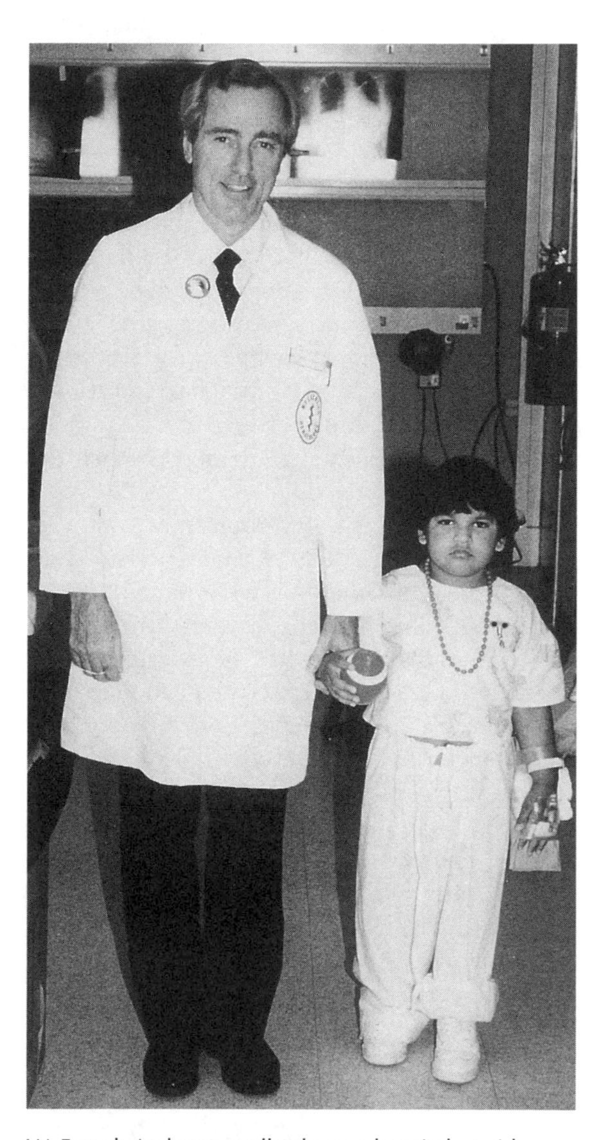

W. French Anderson walks down a hospital corridor with Ashanthi DeSilva, the girl to whom he gave the world's first human gene therapy in 1990. *(W. French Anderson)*

Anderson graduated from Harvard in 1958 and went on to earn a master's degree from Cambridge University in England in 1960 and an M.D. from Harvard Medical School in 1963. In 1965, he joined the National Institutes of Health (NIH) in Bethesda, Maryland, where he helped a well-known NIH scientist, MARSHALL NIRENBERG, work out the details of the genetic code. He then helped to devise a treatment for a rare, inherited blood disease called thalassemia, but he was frustrated because he could not find a way to modify the defective gene that caused the disease.

In the early 1970s, researchers learned how to insert a gene from one kind of living thing into the genome of another and make it function in its new location. Anderson and a few other farsighted scientists began to hope that they could use this technique to substitute healthy genes for diseased ones in humans. At first, however, this genetic engineering could be done only in bacteria.

Transferring genes into human cells proved nearly impossible until 1984, when Richard Mulligan of the Massachusetts Institute of Technology developed a way to use certain viruses to do this job. Viruses are nature's genetic engineers. They reproduce by inserting their genomes into cells and making the cells copy the viruses' genetic material along with their own. Mulligan took out the genes that let the viruses reproduce and substituted the genes that he wanted to transfer. The viruses inserted the new genes into the cells with the rest of their genomes.

Excited by this advance, French Anderson began planning how to use it to treat an inherited disease. He knew he had to choose a disease caused by a defect in a single gene that had been identified. Because genetic engineering was still very difficult, the disease also had to be treatable by putting a healthy copy of the gene into a small number of cells that could be easily reached. Only a few diseases of blood cells met both requirements.

Anderson finally focused on a rare condition called ADA deficiency. People with this disease lack a working gene that carries the instructions for making a chemical called adenosine deaminase (ADA), which blood cells in the immune system require. Children born without the ADA gene essentially had no immune systems. Like people with AIDS, they suffered from one infection after another, and they usually died at an early age.

Anderson's wife, Kathy, a pediatric surgeon, introduced him to another NIH scientist, Michael Blaese, who was an expert on immune system diseases in children. In 1984, Anderson and Blaese began experimenting with inserting the gene for ADA into human blood cells in the laboratory. They encountered many difficulties before Don Kohn, a member of their research team, finally succeeded in 1987. For example, they were unable to transfer the genes into stem cells, long-lived cells that make all other types of blood cells.

Gaining government permission to test the treatment in humans was equally hard. The group had to win the approval of both the federal Food and Drug Administration (FDA), which oversees all human tests of new medical treatments, and the Recombinant DNA Advisory Committee (RAC), a committee that NIH had established in the mid-1970s to oversee genetic engineering experiments. To earn this permission, they had to convince the committees that their proposed gene therapy would be both effective and safe in human patients.

They succeeded only after teaming up with another NIH scientist, STEVEN A. ROSENBERG, and using engineered genes as part of an experimental cancer treatment. The committees let the group proceed with this treatment because the patients on whom it was being tested were already dying of cancer and because the altered genes were not expected to have any effect on their health. The team utilized the genes only as markers to track the immune system cells used in the treatment after they entered the patients' bodies. When the cancer patients developed no problems that could be traced to the altered genes, the committees decided that the gene treatment for ADA deficiency could also go ahead.

The child whom the group chose to treat first was a solemn-faced girl named Ashanthi, or Ashi, DeSilva. At the beginning of September 1990, they took immune system cells from Ashi's blood and engineered them in the laboratory with viruses that carried healthy ADA genes. Then, on September 14, just after Ashi's fourth birthday, they injected the cells back into her arm. Afterward, Anderson told reporters that the treatment was not only a scientific but "a cultural breakthrough . . . an event that changes the way that we as a society think about ourselves." He meant that from then on, people would know that their genetic heritage did not prescribe an unavoidable fate.

Ashi had gene treatments every other month until she had received 11 in all. Tests showed that her immune system improved steadily, and Anderson wrote that by 1995 she was "a healthy, vibrant nine-year-old who loves life and does *everything*." Cynthia Cutshall, another ADA-deficient child on whom the treatments were tried, also benefited from them. Neither girl was cured, but they became able to lead basically normal lives. They were still doing well in 2002.

The success of Anderson's treatment created excitement among scientists and the media and opened the door for other researchers to begin trying gene therapy. Today, gene treatments for more than 500 diseases are being tested in the United States alone, including not only ones for inherited illnesses like ADA deficiency but also treatments aimed at more common conditions, such as cancer and heart disease, that are influenced by genetic defects or can be helped by added genes. Few of these treatments have worked as well as the one given to Ashi DeSilva, and questions about the safety of gene therapy have continued to arise. Still, Anderson is confident that this new form of treatment will eventually revolutionize medicine.

French Anderson is still exploring frontiers in gene therapy, such as improvements in the use of viruses to deliver genes. He moved to the University of Southern California in Los Angeles in 1992, and he now heads the gene therapy laboratories at the university's Keck School of Medicine. He is also a professor of biochemistry and pediatrics there and a founder of a Maryland biotechnology company called Genetic Therapy, Inc.

Among Anderson's most controversial plans is a proposal to inject healthy genes into a child with an inherited disease before the child is born. This could prevent irreversible damage that the disease would otherwise cause before birth. Some scientists oppose such treatment because it might alter genes in the baby's sex cells (sperm or eggs), which would be passed on to future generations. All gene treatments so far have affected only the individuals to whom they were given. Some people feel that germ-line, or inheritable, genes should never be changed because a mistake in such a treatment could affect many generations.

Although some people find his work disturbing, French Anderson has received many honors, including a Distinguished Service Award from the U.S. Department of Health and Human Services (1992) and the National Biotechnology Award (1995). In 1994, *Time* magazine named him a runner-up for its Person of the Year, and in 1997 it named him one of its "Heroes of Medicine." He was inducted into the Oklahoma Hall of Fame in 1998.

In a sense, Anderson has even achieved his ambition to take part in the Olympics. He became an expert in sports medicine and in the Korean martial art called tae kwon do, and he was the official physician for the United States tae kwon do team in the 1988 Olympics in Seoul, Korea. Practicing tae kwon do, in which he holds a fifth degree black belt, is still one of his favorite activities.

Further Reading

"Anderson, W. French." *Current Biography Yearbook 1994*. New York: H. W. Wilson, 1994.

Lyon, Jeff, and Peter Gorner. *Altered Fates*. New York: W. W. Norton, 1996.

Thompson, Larry. *Correcting the Code*. New York: Simon & Schuster, 1994.

Yount, Lisa. *Milestones in Discovery and Invention: Genetics and Genetic Engineering*. New York: Facts On File, 1997.

Aristotle

(384–322 B.C.)
Greek
Philosopher of Science, Naturalist, Taxonomist

The ancient Greek philosopher Aristotle is one of the founders of science. He originated several basic ideas in biology, such as the idea of classifying living things according to features they had in common. Respected modern evolutionary biologist ERNST MAYR has written, "No one prior to Darwin has made a greater contribution to our understanding of the living world than Aristotle."

Aristotle was born in 384 B.C. in Stagira, a colony on the coast of northern Greece that at the time belonged to Macedonia. His father, Nicomachus, was the Macedonian king's court physician. Aristotle's parents both died when he was young.

In 367 B.C., when Aristotle was about 17 years old, he traveled to Athens, the center of the Greek intellectual world. He lived there for 20 years, studying under the famous philosopher Plato and later teaching in Plato's academy. After Plato died in 347 B.C., Aristotle began traveling around Greece. He spent three years in Assos, where he married Pythias, the niece of the local ruler. They had one daughter, but Pythias soon died, perhaps in childbirth. Aristotle also lived on the island of Lesbos for several years and probably studied marine animals there.

Around 342 B.C., Philip, the king of Macedonia, asked Aristotle to tutor the ruler's teenaged son, Alexander. Aristotle did so for about three years, until Philip died and Alexander took over the throne. Aristotle then returned to Stagira, where he married a woman named Herpyllis and had a son.

Aristotle went back to Athens in 335 B.C. and set up his own philosophical academy, the Lyceum. He lectured his advanced students in the mornings and gave general talks to larger audiences in the afternoons or evenings. Most of his approximately 400 surviving books date from this time in his life. Many seem to be collections of lecture notes.

Aristotle tried to assemble and classify all existing knowledge. He made important contributions to logic, politics, ethics, and literature, as well as biology. He also attempted to describe physics and astronomy, though much less successfully. He stressed similarities and relationships among all parts of nature. He also emphasized the importance of observing the natural world, writing that "more trust should be put in the evidence of sense perception than in theories."

A basic idea in Aristotle's philosophy was that everything in nature has a purpose. He tried to find out and describe what those purposes were. He also held that each part of a living thing has a purpose or function, and that function explains why the part has the features it does. For instance, he pointed out, hawks and other birds that kill birds or animals for food (raptors) have hooked beaks and sharp, curved talons "to obtain mastery over their prey, that being suited better for deeds of violence than any other form." Biologists still study carefully the relationships between form and function in living things.

Aristotle believed that every living thing has a built-in urge to fulfill its purpose and develop itself as completely as possible. The purpose of an acorn was to become an oak tree, he said, and the acorn somehow contained both the instructions and the drive to carry out this purpose. These ideas foreshadowed modern scientists' understanding of the "program" for the development of each kind of living thing that is carried in its genes.

Three of Aristotle's books—*The History of Animals*, *The Parts of Animals*, and *The Generation [Reproduction] of Animals*—were concerned almost entirely with biology. They described more than 500 types of animals in detail. Aristotle's descriptions drew on his personal experience, reports and specimens from students and friends, and accounts by other writers. Some descriptions contained serious errors or even referred to creatures that never existed, but others were extremely accurate.

In *The History of Animals*, Aristotle introduced the idea of classifying animals according to features that they have in common. "Animals may be characterized according to their way of living, their actions, their habits, and their bodily parts," he wrote. "It is by resemblance of the shapes of their parts, or of their whole body, that the groups are marked off from each other." Aristotle's comparisons laid the foundation for comparative anatomy, which has proved very useful in understanding relationships among living things. His classification system included both an animal's specific type, such as tiger, and the group of similar creatures into which that type of animal might be placed, such as mammals. CAROLUS LINNAEUS developed a similar, though more systematic, method of classification some 2,300 years later.

The ancient Greek philosopher Aristotle provided one of the earliest systematic descriptions and classifications of animals. *(National Library of Medicine)*

Aristotle attempted to describe how animals' bodies worked as well as what they looked like. His ideas about physiology were much less accurate than those about anatomy, however. He believed, for instance, that the heart was the seat of intelligence, whereas the brain had no function except to cool the blood. He made errors in descriptions of human anatomy and physiology because he assumed that human bodies were just like the bodies of animals he had studied. Belief in some of Aristotle's mistaken ideas held biology and medicine back for centuries.

Aristotle devoted a whole book to considering how animals reproduce. He recognized that the male and the female are equally necessary for reproduction, but, like many thinkers of his time, he downplayed the female's role. He believed that the male provided the form and energy that created the offspring, while the female provided only the substance from which it was made. He also mistakenly believed that small creatures such as insects could arise spontaneously from nonliving matter. In spite of such errors, Aristotle's writings about reproduction are one of the foundations of embryology.

Aristotle believed that complex living things were "higher," or more valuable, than simpler ones because they were more organized. (For the same reason, he held that all living things were "higher" than nonliving things.) He did not imagine that lower forms had changed into higher ones, however, as later supporters of evolution, such as CHARLES ROBERT DARWIN, did. Aristotle thought that each kind of animal was exactly the same as it had been throughout time.

Athens, which had been under Macedonian control, became free once more after Alexander died in 323 B.C., and the Athenians turned against everyone connected with that kingdom. The Athenian government had executed Plato's teacher, the philosopher Socrates, about 75 years earlier because of his "dangerous" ideas, and Aristotle feared that the same thing might happen to him if he remained in the city. He therefore retired to Chalcis, on the island of Euboea, where he owned some property. He died there of a stomach ailment in 322 B.C.

Christian thinkers of the Middle Ages adapted some of Aristotle's ideas to their own ways of thinking but rejected others, causing them to be all but forgotten in most of Europe. Thinkers of the Renaissance, a period of new interest in nature and science that began around 1350, rediscovered the writings of Aristotle and other ancient Greek philosophers and tended to accept them uncritically. Later scientists, discovering Aristotle's errors, just as blindly rejected his ideas, including many that make sense. Most modern science historians agree that, although he misinterpreted many things, Aristotle was one of the greatest thinkers in all of history.

Further Reading

The Internet Encyclopedia of Philosophy. "Aristotle." University of Tennessee at Martin. Available online. URL: www.utm.edu/research/iep/a/aristotl.htm. Posted 2001.

Lennox, James G. *Aristotle's Philosophy of Biology: Studies in the Origin of Life Science*. New York: Cambridge University Press, 2000.

Parker, Steve. *Aristotle and Scientific Thought*. Broomall, Penn.: Chelsea House, 1995.

Taylor, Alfred E. *Aristotle*. Mineola, N.Y.: Dover, 1955.

Avery, Oswald Theodore
(1877–1955)
Canadian/American
Bacteriologist

Oswald Avery's research helped to convince other biologists that DNA carries inheritable information. Avery was born in Halifax, Nova Scotia, on October 21, 1877, but his father, a Baptist minister, was invited to take over a church in New York City and brought his family there when Avery was about 10 years old. Avery earned a bachelor's degree from Colgate University in 1900 and an M.D. from Columbia University in 1904.

After working as a physician for several years, Avery turned to research. He worked first at the Hoagland Laboratory in Brooklyn, then moved to the hospital of the Rockefeller Institute for Medical Research (now Rockefeller University) in 1913 and remained there for the rest of his career. He became a United States citizen in 1918.

Avery's special interest was bacteria that cause pneumonia, a serious lung disease, especially a species called *Diplococcus pneumoniae*. Some types, or strains, of this bacterium produced the disease in animals such as mice, while others were harmless. His most important research, done during the early 1940s, built on an experiment that a British scientist, Frederick Griffith, had done on this same kind of bacteria in 1928. Griffith had mixed living bacteria of a strain that could not cause disease with killed bacteria of a strain that could cause pneumonia and injected the mixture into mice. The mice came down with pneumonia, and Griffith showed that their blood swarmed with living bacteria of the disease-causing type. Most important of all, these "transformed" bacteria produced offspring that also belonged to the disease-causing strain.

At first, Avery had trouble believing Griffith's results. The British scientist had done something that seemed as impossible as, say, changing a dog into a fox. Other researchers confirmed Griffith's work, however, and showed that the transformation could be made to occur in laboratory dishes, where the changed bacteria were identified by their ability to make shiny shells called capsules. Working with Colin MacLeod and Maclyn McCarty, Avery then began doing chemical tests to find out what substance caused the transformation. The tests eventually eliminated all possibilities except nucleic acids, most likely the kind of nucleic acid called DNA. Avery published this conclusion in 1944.

Avery, a cautious man, did not state directly that DNA carries inherited information—but that was the clear implication of his research. Many scientists, feeling that DNA was not a complicated enough molecule to do this job, were as suspicious of his results as he had been of Griffith's. Nonetheless, later experiments, especially those of ALFRED DAY HERSHEY and Martha Chase, confirmed and expanded Avery's findings by showing that viruses injected nucleic acid, but not protein, into bacterial cells when they infected the bacteria. Nobel Prize-winning geneticist JOSHUA LEDERBERG has said that Avery's work was "the historical platform of modern DNA research."

Avery continued to work in his laboratory until 1948. He received major honors, including the British Royal Society's Copley Medal (1945), the Albert Lasker Medical Research Award from the American Public Health Association (1947), and election to the U.S. National Academy of Sciences. After his retirement, he moved to Nashville, Tennessee, where he died on February 20, 1955.

Further Reading

Dubos, René. *The Professor, the Institute, and DNA*. New York: Rockefeller University Press, 1976.

McCarty, Maclyn, and Bradie Metheny. "Dr. McCarty Recalls Dr. Avery as a Man." September 3, 1998. Lasker Foundation. Available online. URL: www.laskerfoundation.org/awards/library/1947b_int_oamm.shtml. Accessed 2003.

"The Oswald T. Avery Collection: Biographical Information." National Library of Medicine Profiles in Science. Available online. URL: www.profiles.nlm.nih.gov/CC/Views/Exhibit/narrative/biographical.html. Accessed 2001.

Baer, Karl Ernst von

(1792–1876)
Estonian/Russian
Embryologist, Naturalist

Karl Ernst von Baer was the first person to see the ovum, or egg, of a mammal. This and other discoveries of von Baer's helped to found modern embryology.

Karl von Baer was born in Piep (now Piibe), Estonia (later a part of the Soviet Union, now an independent republic), on February 29, 1792, to a noble family of Prussian (German) descent. Magnus von Baer, his father, was wealthy and owned a large estate. He and his wife, Juliane, had a large family as well; Karl was one of 10 children. Karl studied medicine at the university in Dorpat (now Tartu) and obtained his M.D. in 1814. He then took advanced training in Vienna, Austria, and Würzburg, Germany.

While von Baer was at Würzburg, one of his professors interested him in embryology. Von Baer began to do research in this field as well as teach at Königsburg (later Kaliningrad) University in 1817. Three years later, he married Auguste von Medem, and they had six children.

Von Baer's first studies extended the work of a friend, Christian Pander. Pander found in 1817 that soon after the fertilized egg cell that would become a chick began dividing, it formed three layers of tissue—a sort of sandwich. All the body parts of the chick grew out of these three layers. Von Baer showed that all vertebrates (animals with backbones) began development with these same three layers.

Scientists had long known that insects, reptiles, and birds developed from eggs. They believed that female mammals, including humans, must also have eggs. Because mammals' eggs are very tiny and most microscopes were of poor quality until the late 19th century, no one had ever seen a mammal's egg. A Dutch researcher, Regnier de Graaf, had found egglike structures in organs called ovaries in the abdomens of female mammals in 1673, but no one was sure whether these were really eggs. In 1826, von Baer cut open the body of a fellow scientist's pet dog and found tiny yellow spots in de Graaf's structures that proved to be the actual eggs. He went on to find eggs in the ovaries of other mammals as well, including human females. He announced his discovery in a book called *On the Mammalian Egg and the Origin of Man*, published in 1827.

At the time von Baer did his research, scientists' understanding of the way animals develop before birth was still primitive. Some believed that the eggs or the sperm (male sex cells) contained tiny versions of the animals, already completely formed and that the animals matured simply by growing larger. Others thought that animal embryos started out with very little form and then acquired their organs and other features in a series of steps as they grew. Von Baer proved that this second theory, called epigenesis, was correct.

Von Baer went on to compare stages of development in the embryos of different animals. He disproved a commonly held idea that vertebrate embryos went through stages during which they looked like adults of other species. He found that the earlier in development the embryos were, the more they looked alike, even if they would grow into creatures as different as a dog and a snake. Unique, complex structures such as legs or wings formed late in the development process. The idea that generalized features developed before specialized ones came to be known as the biogenetic law. Von Baer published his descriptions of embryos in a two-volume work called *On the Development of Animals* (1828 and 1837).

Von Baer moved to St. Petersburg (later Leningrad), Russia, in 1837. He worked for the Russian Academy of Sciences and founded several research societies. Erki Tammiksaar of the Baer Museum in Tartu says that von Baer "had an immense role in the organization and directing of natural scientific research in Russia."

Giving up his research on embryology, von Baer spent the rest of his life studying completely different areas of science. In 1837, for instance, he headed an expedition to a peninsula called Novaya Zemlya in the Russian Arctic. He became the first biologist to bring back plant and animal specimens from this region and, Tamiksaar claims, "laid the basis of ecological research in Russia." He studied Russian fisheries and fish biology extensively in the 1850s, and his work led to the passage of the country's first fish protection law in 1859.

Between 1858 and 1862, von Baer turned his attention to anthropology and gathered, measured, and compared an extensive collection of human skulls. In 1859, the same year that CHARLES ROBERT DARWIN's *On the Origin of Species* appeared, von Baer published a book about his skulls in which he speculated that all humans might have developed from a single ancestor. This idea was similar to some of Darwin's, but von Baer disagreed with Darwin's beliefs about evolution and spent most of his last years writing articles that opposed him. Von Baer returned to Dorpat in 1867 and died there on November 28, 1876.

In spite of his great energy and contributions to many scientific fields, von Baer suffered from depression, which sometimes kept him from working. Once, he did not leave his house for a whole year. Traveling seemed to ease his sadness, which may explain why he was such an eager explorer.

Further Reading

The Autobiography of Dr. Karl Ernst von Baer. Translated by Jane M. Oppenheimer. Canton, Mass.: Watson Publishing International/Science History Publications, 1986.

Tammiksaar, Erki. "Homepage of Karl Ernst von Baer, 1792–1876." Baer Museum. Available online. URL: www.zbi.ee/baer/. Updated November 27, 2001.

Baltimore, David

(1938–)
American
Virologist

David Baltimore earned a share of a Nobel Prize in 1975, when he was only 37 years old, for his discovery of an enzyme that allows certain viruses to copy themselves "backward" into the genetic material of cells. He was born in New York City to Richard and Gertrude Baltimore on March 7, 1938. His mother, a psychologist, stirred his interest in science. He turned toward biology in high school after spending a summer at the Jackson Memorial Laboratory in Bar Harbor, Maine, a famous center for the study of mammalian genetics. One of the other students at Bar Harbor that summer was HOWARD MARTIN TEMIN, who later would independently make the same Nobel Prize–winning discovery as Baltimore.

Baltimore earned a bachelor's degree in chemistry from Swarthmore College in Pennsylvania in 1960. His graduate studies were at the Massachusetts Institute of Technology (MIT) and the Rockefeller Institute (later Rockefeller University) in New York, and he received his Ph.D. from Rockefeller in 1964. After postdoctoral work at several institutions, he returned to MIT in 1968, first as an associate professor and then, in 1972, as a full professor.

In the late 1960s, Baltimore began to focus on certain viruses, such as some that cause cancer in animals, that have genes made of RNA rather than DNA, the closely related substance that carries the genetic information of most living things. FRANCIS CRICK and others had stated as a central dogma (basic belief) of genetics that DNA always copies itself into RNA and then into protein; this order could never be reversed. How, then, Baltimore wondered, did RNA viruses make cells copy the viruses' genetic material so that the viruses could reproduce?

Working with his wife, microbiologist Alice S. Huang, whom he had married in 1968, Baltimore discovered in 1970 that an RNA virus that causes cancer in mice possesses an enzyme that he called reverse transcriptase. This enzyme proved able to do what Crick had said could not be done: make the process of copying DNA into RNA run backward, so that infected cells copied the viruses' RNA genes into their own DNA genomes. When the cells then copied their DNA again into RNA in the more usual way, they also reproduced the viruses' genes. The added virus genes made the cells reproduce endlessly, forming cancers.

Baltimore's old Bar Harbor classmate, Howard Temin, found reverse transcriptase in another cancer-causing virus at almost the same time. They and another researcher, Renato Dulbecco, shared the Nobel Prize in physiology or medicine in 1975 for their discoveries about what came to be called retroviruses ("backward viruses"). The discovery of reverse transcriptase forced the central dogma to be revised and also showed how some viruses cause cancer. It later shed light on AIDS as well, because HIV, the virus that causes this disease, is a retrovirus.

Baltimore had his first encounter with political controversy in the mid-1970s, just after scientists became able to combine and alter genes for the first time. He and some other researchers, such as PAUL BERG, feared that genetic engineering experiments might create new kinds of bacteria or viruses that would prove dangerous to humans. Baltimore helped to organize a meeting at the Asilomar Conference Center in Pacific Grove, California, in February 1975, at which gene scientists worked out safety standards for their experiments. He later served on the Recombinant DNA Advisory Committee (RAC), a group that the National Institutes of Health (NIH) in Bethesda, Maryland, created in 1976 to oversee this kind of research.

David Baltimore became the first director of the MIT-sponsored Whitehead Institute for Biomedical Research in 1982. He also continued his own work, which came to focus on the immune system. He became embroiled in controversy once again when Thereza Imanishi-Kari, a Brazilian-born scientist he had hired, published a paper on the effects of altering certain genes in mouse immune system cells in 1986, and a younger researcher in her laboratory, Margot O'Toole, claimed that the paper contained faked results. Baltimore's name became tarnished along with Imanishi-Kari's because, as the laboratory's most senior scientist, he had been in charge of her work (he was listed as an author on her paper) and because he defended it against O'Toole's claims. Those claims launched a fraud investigation that went on for 10 years and eventually involved the NIH, the media, a House of Representatives subcommittee, and even the Secret Service. The researchers were finally cleared of all fraud charges in June 1996.

Baltimore left the Whitehead Institute to become president of Rockefeller University in 1990, but he was forced to resign after a year, partly because of the cloud cast by the fraud accusations. He continued as a professor at the university until the mid-1990s. In 1997, he became the seventh president of the California Institute of Technology (Caltech) in Pasadena, a position he still holds. He has also continued to do research, most recently on AIDS vaccines.

Baltimore's research on viruses has won many awards in addition to the Nobel Prize, including an American Cancer Society professorship (1973), election to the U.S. National Academy of Sciences (1974), and the National Medal of Science (1999). He has said, "I work [in science] because I want to understand."

Further Reading

"Baltimore, David." *Current Biography Yearbook 1983*. New York: H. W. Wilson, 1983.

Bernstein, Jeremy. "Science, Fraud and the Baltimore Case." *Commentary*, December 1998.

Crotty, Shane. *Ahead of the Curve: David Baltimore's Life in Science*. Berkeley, Calif.: University of California Press, 2001.

Kevles, Daniel J. *The Baltimore Case: A Trial of Politics, Science, and Character*. New York: W. W. Norton, 2000.

Banting, Frederick Grant

(1891–1941)
Canadian
Physiologist

Millions of people with diabetes owe their lives to Frederick Banting. Before Banting's 1921 discovery of insulin, the hormone that controls the body's use of sugar, diabetics, who lack this substance, faced certain death.

Frederick Grant Banting was born on November 14, 1891, on his family's farm near Alliston, in the Canadian province of Ontario. He was the youngest of William Thompson Banting and Margaret Grant Banting's five children. He went to the University of Toronto with the intention of becoming a minister, but he soon changed his mind and began medical training. He obtained his medical degree in 1916 and immediately joined the medical corps of the Canadian army. He served in England and France during World War I and was awarded a Military Cross in 1919.

After the war, Banting worked briefly as a physician in London, Ontario, and taught part time at the University of Western Ontario. He returned to the University of Toronto in 1921. By this time, he had decided to do research on diabetes. This disease, in which sugar builds up in the body, had been known since ancient times, but until the 19th century, no one had any idea what caused it.

In 1889, scientists had found that dogs quickly developed diabetes if an abdominal organ called the pancreas was removed. This organ was known to make chemicals that play a part in digestion. A scientist named Paul Langerhans had shown that it contains clumps, or islands, of cells that look different from the rest. As long as these "islets of Langerhans" remained healthy, animals did not develop diabetes. A British researcher, Edward Sharpey-Schäfer, had predicted that the islets of Langerhans would be found to make a substance that controls the body's use of sugar. He gave the chemical a name, insulin (from the Latin word for "island"), even though no one had yet proved that it existed.

Banting wanted to find Sharpey-Schäfer's mystery substance in the hope that it could help diabetics. Doctors had tried feeding them animal pancreases, but when the organs were ground up, the digestive chemicals in them destroyed the insulin. If insulin could be given by itself, this problem might be avoided.

Banting had read a scientific paper saying that if surgeons closed off the opening that led from a dog's pancreas to its digestive system, the part of the pancreas that made the digestive chemicals was destroyed, leaving only the islets. He believed he could use this procedure to separate the islets from the other tissue and then extract insulin from them. He asked John J. R. Macleod, head of the University of Toronto's physiology department, to help him obtain the resources for such an experiment. Macleod did not think much of Banting's ideas, but he was planning a summer vacation in Scotland, and he gave Banting permission to use his laboratory, including its experimental dogs, while he was away. He also helped Banting pay an assistant, a medical student named Charles Best.

As Banting had predicted, he and Best succeeded in obtaining islet cells from Macleod's dogs and extracting a tiny amount of chemical from them. They injected this substance into another dog that was dying of diabetes, and within a few hours the dog sat up and wagged its tail. They found that the amount of sugar in its blood had dropped considerably. They then began to work with pancreases from unborn

Canadian physician Frederick Grant Banting, left, and his assistant, Charles Best, right, purified the hormone insulin in 1921 and tested it on dogs before giving it to humans as a treatment for diabetes. *(National Library of Medicine)*

calves, which they obtained from a nearby slaughterhouse, because these contained more islet tissue than adult pancreases. After much difficulty, with the help of biochemist James B. Collip, they were able to make a small quantity of fairly pure insulin from these organs.

Banting and Best injected a little of their insulin extract into themselves to verify that it did not harm human beings. Then, on January 11, 1922, they tried it on Leonard Thompson, a 14-year-old boy in the Toronto General Hospital who was dying of diabetes. The amount of sugar in the boy's blood began falling almost at once, and within weeks he was able to return home. The story of this success made international headlines.

Other scientists developed ways to extract insulin from the pancreases of pigs on a large scale, and the substance became commercially available in 1923. Unfortunately, insulin could not be taken as a pill because chemicals in the stomach destroyed it. Diabetics had to give it to themselves in daily injections. Still, if they did this and kept most sugar-containing compounds out of their diet, they usually found that they could live essentially normal lives.

For his lifesaving work, Frederick Banting, then only 32 years old, received a share of the Nobel Prize in physiology or medicine in 1923. He was the first Canadian to win a Nobel Prize. Macleod, as the laboratory's supervisor, was also awarded part of the prize, even though he had been out of the country when the research was done. Best, on the other hand, was not included. Banting, angry at this oversight, gave half his prize money to Best. Macleod, in turn, shared his portion with Collip.

The University of Toronto and the Canadian and British governments also showed their gratitude to Banting. The university made Banting a professor in 1923 and later named several medical research facilities after him. In 1923, too, the Canadian Parliament voted to give Banting a yearly payment of $7,500 for the rest of his life. Britain's King George V, himself a diabetic, made Banting a knight in 1934.

During the late 1920s and 1930s, Banting did research on a range of subjects, including cancer, silicosis (a lung disease that affected miners), and drowning. In 1937, he married for the second time, to Henrietta Ball, a technician in his laboratory. He had previously married Marion Robertson in 1924, and the couple had had a son, William, in 1928. They were divorced in 1932.

When World War II broke out in 1939, Banting began studying medical problems that affected pilots, such as blackouts from lack of oxygen at high altitudes. He also was a liaison officer between the North American and British medical services. He was killed when a plane taking him to a meeting in England crashed in Newfoundland on February 21, 1941.

In 2000, the Canadian popular magazine *Maclean's* called Banting a "genuine Canadian hero." It noted that a survey of the magazine's readers in 1927 had chosen him as "the greatest living Canadian."

Further Reading

Bliss, Michael. *Banting: A Biography*. Toronto, Canada: University of Toronto Press, 1992.

"Frederick Grant Banting—Biography." *Nobel Lectures, Physiology or Medicine, 1922–1941*. Available online. URL: www.nobel.se/medicine/laureates/1923/banting-bio.html. Last updated 2001.

Hume, Stephen Eaton. *Frederick Banting: Hero, Healer and Artist*. Montreal, Quebec, Canada: XYZ Publishing, 2001.

Bateson, William

(1861–1926)
British
Geneticist

William Bateson was a man of contradictions. He believed in evolution, but he strongly disagreed with CHARLES ROBERT DARWIN's ideas about how and why evolution took place. He translated GREGOR MENDEL's paper about patterns of inheritance into English and coined the term *genetics* but remained unconvinced of the physical reality of genes. Many of his ideas have been disproved, yet he is considered a founder of modern genetic science.

Bateson was born in Whitby, England, on August 8, 1861. His father, William Henry Bateson, was a minister, classical scholar, and eventually master of St. John's College, part of Cambridge University. His mother, Anna Aiken Bateson, was one of the first British women to work for women's right to vote.

The contradictions in Bateson's life began early. The headmaster at Rugby School, which he attended as a child, wrote his father that he doubted whether "so vague and aimless a boy will profit by University life." When Bateson went on to St. John's College to study zoology, however, he achieved high honors.

After his graduation in 1883, Bateson went to the United States and studied for two years at Johns Hopkins University in Baltimore, Maryland. There, he became interested in evolution and embryology. He showed that a wormlike sea creature called *Balanoglossus* had features of both echinoderms (a family of animals that includes starfish, or sea stars) and chordates (animals that have a nerve cord in their backs, including humans). This was the first evidence that chordates had evolved from echinoderms.

Bateson returned to Cambridge for additional studies in 1885. He wanted to learn how differences in environment change inherited characteristics, a subject that few other researchers cared about at the time. A fellow scientist wrote later that this interest made other scientists label Bateson "a renegade." However, he added, "Bateson was never deterred by other men's opinions."

Bateson met opposition in his personal life as well. In 1889, when he returned to England from two years of investigating variations in shellfish (molluscs) in salty lakes in Russia, Egypt, and Europe, he and a young woman named Beatrice Durham fell in love. Her parents distrusted him, however, and forbade her to go on seeing him. Only seven years later, when an acquaintance showed him a story that Durham had written about a woman who longed to be reunited with her lost lover, did he dare to contact her again. By that time her parents had died, so the couple could resume their relationship. They were married in 1896 and had three sons.

Meanwhile, Bateson went on studying heredity and being a renegade. Darwin had maintained that living things evolve by means of slow, gradual change, but Bateson's studies of molluscs had convinced him that evolutionary changes often occurred as sudden, large "jumps" with no intermediate stages. He wrote a book describing his ideas in 1894, but many scientists rejected his theory because he could not explain how or why these abrupt changes occurred.

A train journey changed Bateson's life in April 1900. On his way to speak at a meeting of the Royal Horticultural (plant study) Society, he read a paper that a scientific acquaintance, Dutch botanist HUGO DE VRIES, had sent him. The paper, written in 1865 by Gregor Mendel, an obscure monk living in what is now the Czech Republic, had remained all but forgotten until De Vries rediscovered it earlier in the year. Bateson immediately realized that Mendel's paper, which described mathematical rules governing the way living things inherit characteristics from their parents, provided the missing theory to explain his own ideas. He abandoned the speech he had planned and, instead, told the Horticultural Society about Mendel's work. Soon afterward, he translated Mendel's paper into English, giving it a much wider readership.

Bateson now focused his research on Mendel's theories. Through breeding experiments with chickens, he showed that Mendel's rules, which had been worked out in studies of pea plants, also applied to animals. Furthermore, he pointed out, physician Archibald Garrod's recent investigation of the pattern of inheritance of a rare illness showed that human traits could be inherited according to Mendelian rules as well.

On the other hand, Bateson's research proved some of Mendel's ideas wrong. Mendel had believed that each characteristic of a plant or animal was inherited separately, so that possessing one trait did not make a creature any more or less likely to possess another. Around 1904, however, Bateson found during breeding experiments with sweet peas that some pairs of characteristics were almost always inherited together, a phenomenon that came to be called linkage.

Realizing that a name was needed for the developing field of science that focused on biological inheritance, Bateson suggested in 1906 that it be called *genetics*. Other scientists of his time were beginning to believe that threadlike bodies called chromosomes, found in the nucleus of most cells, played some part in transmitting traits to new generations, but Bateson never accepted this theory. To him, a gene (a term coined by another scientist in 1909) was a biological concept, not a physical entity.

Bateson won awards for his work, including the Darwin Medal in 1904. Cambridge did not give him the advancement he felt he deserved for many years, however. Even after he was finally made a professor—Britain's first professor of genetics—in 1908, he was poorly paid. He therefore left the university in 1910 to become the director of the new John Innes Horticultural Institute, a position with a higher salary. He held that post until his death from a heart attack on February 8, 1926.

Further Reading

Bateson, William. *Mendel's Principles of Heredity*. Reprint. Placitas, N.M.: Genetics Heritage Press, 1996.

"Historic Figures: William Bateson, 1861–1926." BBC. Available online. URL: www.bbc.co.uk/history/historic_figures/bateson_william.shtml. Accessed 2001.

Beadle, George Wells

(1903–1989)
[illegible]
Geneticist

George Beadle began life as a Nebraska farm boy, and he never strayed too far from his roots. According to an article in *Scarlet*, a magazine published by the University of Nebraska at Lincoln, when Beadle was dressing for the ceremony in which he would receive a share of the Nobel Prize in physiology or medicine in 1958, he said to his wife, Muriel, "Honey, I wish we were home making compost!"

Beadle was born on October 22, 1903, on the farm of his father, Chauncey Elmer Beadle, near Wahoo, Nebraska. His mother, Hattie, died when he was a child. A high school teacher interested him in science and persuaded his father to let him attend college. He went to the University of Nebraska at Lincoln and earned his B.A. in 1926. He followed this with a master of science degree a year later.

Beadle did graduate work at Cornell University in Ithaca, New York. He became interested in genetics there and did his Ph.D. project on the inheritance of traits in maize, or Indian corn. He obtained his degree in 1931. He then went to the California Institute of Technology (Caltech) in Pasadena, where he studied the genetics of fruit flies in the laboratory of THOMAS HUNT MORGAN, a Nobel Prize–winning expert on the subject. Beadle remained at Caltech from 1931 to 1935.

In research done during a six-month stay in Paris, Beadle showed that genes appeared to affect the color of fruit flies' eyes by controlling production of pigment, or coloring chemical, in the eyes. Before this time, scientists had known that genes determined a living thing's characteristics, or traits, but they had not known how this happened. Beadle's research suggested that the answer lay in alterations that the genes made in cells' ability to manufacture chemicals.

The presence or absence of these chemicals, in turn, determined which traits would appear.

After returning to the United States in 1936, Beadle spent a year at Harvard. He then moved to Stanford University in California in 1937 and remained there as a professor of genetics for nine years. After several years, he gave up his work on fruit flies and began experimenting with a simpler kind of living thing, a reddish bread mold called *Neurospora crassa*. Beadle and another researcher, Edward L. Tatum, bombarded the mold with radiation to cause random changes in its genes. They then studied how these mutations affected the molds' ability to make different nutrients. They showed that particular mutations dependably made the molds lose the power to make particular nutrients.

Beadle and Tatum's research confirmed the idea that genes express themselves through chemistry. In particular, they seemed to control the making of proteins, especially the kinds called enzymes, which control most chemical reactions in the cell. During the early part of the century, a gene had been defined as the unit of inherited information that determines one trait. Beadle and Tatum's work eventually changed this definition to the unit of inherited information that causes a cell to make one protein or, in particular, one enzyme. Although later research showed that this was an oversimplification—some genes control other genes rather than making proteins, for instance—the work provided one of the first important clues about how genes do their job. It won both men a share of the 1958 Nobel Prize and created the new specialty of biochemical genetics.

Beadle returned to Caltech in 1946 and took up Morgan's old position as chairman of the division of biology. In 1961, he moved to the University of Chicago, where he was chancellor and then president of the university. After his retirement in 1968, he returned to genetic research on his first subject, corn. He grew some of his corn plants in his front yard.

Beadle's research won many awards and prizes. In addition to the Nobel Prize, these included the American Public Health Association's Albert Lasker Medical Research Award (1950), the Emil Christian Hansen Prize from Denmark (1953), the Albert Einstein Commemorative Award in Science (1958), and the Kimber Genetics Award of the U.S. National Academy of Sciences (1956). Beadle was president of the American Association for the Advancement of Science in 1956 and was also elected to the National Academy of Sciences.

Beadle married twice. He had a son by his first wife, Marion Hill. He married Muriel McClure, a writer, in 1953. He died on June 9, 1989.

Further Reading

"Beadle, G(eorge) W(ells)." *Current Biography Yearbook 1956*. New York: H. W. Wilson, 1956.

Cranford, Andrea Wood. "George Beadle: A Renaissance Man Spun from Cornsilk." *Scarlet*, September 15, 1995. Available online. URL: www.unl.edu/scarlet/v5n23/v5n23special.html#beadle. Accessed 2003.

"George Wells Beadle—Biography." *Nobel Lectures, physiology or medicine, 1942–1962*. Available online. URL: www.nobel.se/medicine/laureates/1958/beadle-bio.html. Last updated 2001.

⊠ Beaumont, William

(1785–1853)
American
Physiologist, Physician

One of William Beaumont's surgical patients gave him a unique opportunity to learn how the human stomach digests food. Beaumont's discoveries were the first important contribution of a United States scientist to the field of physiology.

Beaumont was born on November 21, 1785, on his family's farm near Lebanon, Connecticut, the second of Samuel and Lucretia Beaumont's nine children. In 1807, he began teaching school in Champlain, New York, but two years later he decided to become a doctor instead. Like many other physicians of the time, Beaumont learned medicine by assisting an older doctor. The state of Vermont awarded him a license to practice medicine in 1812.

The War of 1812 was under way, so Beaumont enlisted in the army as a surgeon's assistant. After the war ended, he began private practice in Plattsburgh, New York. He rejoined the army in 1819, this time as a surgeon, and was assigned to Fort Mackinac, on an island in Lake Michigan near the Canadian border. In 1821, he married Deborah Platt, whom he had met in Plattsburgh, and brought her back to the fort.

On June 6, 1822, Beaumont was called to treat a young French Canadian fur trader named Alexis St. Martin. An accidental gunshot had torn open St. Martin's abdomen, leaving a wound that Beaumont described as "more than the size of the palm of a man's hand." St. Martin eventually recovered, but an opening remained between his stomach and the outside of his body.

St. Martin could no longer do his old job, so in 1823 he moved in with Beaumont's family and worked for them as a handyman. Two years later, Beaumont, realizing that the young man's wound offered a chance to observe human digestion in a way never possible before, added "guinea pig" to St. Martin's duties. He tied small pieces of foods such as beef, pork, and cabbage to silk threads and put them into St. Martin's stomach through the wound opening. He removed the food after different amounts of time, then weighed and examined it to see how much had been digested. Sometimes, he removed gastric juice, the fluid in the stomach that digests food, and compared its action in a test tube with its action on the same kind of food in St. Martin's stomach.

At first, St. Martin cooperated, but some of Beaumont's procedures gave him indigestion, and he became irritated at the time and discomfort they required. (Beaumont noted that anger slowed down the digestive process.) After about a month, the "guinea pig" had had enough and went back to Canada, where he married and had children.

U.S. Army surgeon William Beaumont learned about digestion in the late 1820s and 1830s by experimenting on a man whom a shooting accident had left with a permanent opening from his stomach to the outside of his abdomen. *(National Library of Medicine)*

Beaumont went on to serve in army posts in New York and what later became Wisconsin. Perhaps needing money for his growing family, St. Martin rejoined him at Fort Crawford in Prairie du Chien in 1829, and Beaumont made additional experiments on him from December 1829 to March 1830. St. Martin left again in 1831, but he and Beaumont did a final round of experiments in Washington, D.C., in late 1832 and early 1833.

Beaumont had become interested in possible relationships between weather and disease, so in some of these later experiments he tested

St. Martin's digestion on different days and at different times to see whether weather conditions affected digestion. He found that damp weather lowered the temperature in St. Martin's stomach, whereas dry weather raised it. He observed the muscular movements of the stomach and did many further experiments on gastric juice, showing that this liquid dissolved food by a chemical process and that it needed heat to do so. The process worked well at 100°F, the temperature inside St. Martin's stomach, but in a test tube at a colder temperature it took place slowly or not at all. Beaumont published the results of his experiments—about 238 of them—in 1833 in a book titled *Experiments and Observations on the Gastric Juice and the Physiology of Digestion*. This was the first detailed, accurate description of human digestion.

Beaumont was transferred to Jefferson Barracks, near St. Louis, Missouri, in July 1834. He tried to persuade St. Martin to join him there, but he did not offer enough money to let the trader bring his large family as well, so the French Canadian refused. Beaumont himself liked St. Louis, and when the army tried to move him to Florida in 1839, he resigned rather than leave. He remained in the city for the rest of his life, teaching at St. Louis University and maintaining a profitable medical practice. He died on April 25, 1853, after slipping on an icy step and hitting his head. His former "guinea pig" outlived him by 27 years, finally dying in 1880 at the age of 86.

Further Reading

Beaumont, William. *Experiments and Observations on the Gastric Juice and the Physiology of Digestion*. 1833. Reprint, Mineola, N.Y.: Dover Publications, 1996.

Epstein, Samuel, Beryl Williams Epstein, and Joseph Scrofani. *Dr. Beaumont and the Man with a Hole in His Stomach*. New York: Coward, McCann & Geoghegan, 1978.

Horsman, Reginald. *Frontier Doctor: William Beaumont, America's First Great Medical Scientist*. Columbia, Mo.: University of Missouri Press, 1996.

"Life of Dr. William Beaumont, 1785–1853, 'Father of Gastric Physiology.'" James.com. Available online. URL: www.james.com/beaumont/dr_life.htm. Last updated 2001.

Behring, Emil von

(1854–1917)
Prussian/German
Bacteriologist, Immunologist

In a dramatic demonstration in a German hospital on Christmas 1891, Emil Adolf von Behring showed that he had found a new weapon against diphtheria—one of the greatest killers of children in his time—and a new way of fighting other diseases caused by bacteria as well. He was born in Hansdorf, Prussia, on March 15, 1854, the first child of August Behring, a teacher, and his second wife, Auguste. The Behrings had no money to spend on higher education, but in 1874 Emil won a scholarship to the Friedrich-Wilhelm Institute of Medicine and Surgery in Berlin. The German government paid for his medical training there in return for several years of work as an army physician. He graduated in 1878.

While doing his military service in what is now Poland, Behring began to study the toxins, or poisons, that certain disease-causing bacteria make. The government, impressed with his research, sent him to famed bacteriologist ROBERT KOCH's Hygiene Institute in Berlin in 1888. There, Behring began working with a longtime friend, Erich Wernicke, on diphtheria and with Japanese bacteriologist SHIBASABURO KITASATO on tetanus in 1890. The symptoms of both diseases were caused primarily by bacterial toxins.

By injecting weakened diphtheria or tetanus bacteria into rabbits, rats, and guinea pigs, Behring and his coworkers spurred the animals' immune systems to develop resistance to the bacteria and their poisons. They then extracted blood from immunized animals and separated the liquid serum from the blood cells. When they injected the serum into other animals that had been given a dose of bacteria strong enough to kill them, those animals, too, survived. In other words, unlike the vaccines developed by EDWARD JENNER and LOUIS PASTEUR, which had to be given before a person or animal was exposed to full-strength microbes, Behring's antiserums, or antitoxins, could cure people or animals that already had diphtheria or tetanus.

Behring and Kitasato developed an antiserum for tetanus in 1890, and the antiserum for diphtheria followed in 1891. Behring's Christmas demonstration showed that diphtheria antiserum could save a child's life, but the treatment was undependable at first because the amount of antitoxin in the serum varied from batch to batch. PAUL EHRLICH, another researcher in Koch's laboratory, solved this problem around 1894, and the antitoxin then came into wide use. The rate of deaths from diphtheria dropped by more than 50 percent during the following 10 years, largely as a result of this treatment.

Behring left Koch's institute in 1894 and joined the University of Marburg in 1895. He became a full professor and director of the university's Institute of Hygiene the following year, a post he kept until his retirement in 1916. In 1896, he married Else Spinola, daughter of the administrative director of Berlin University's medical clinic. They later had six sons.

Behring's diphtheria treatment made him famous. When the first Nobel Prizes were given in 1901, he won the prize in physiology or medicine because of it. In that same year, the German government made him a noble, with the right to add "von" before his name. He was elected to the Privy Council, a high government post, in 1903. Other countries also awarded him honors and medals; France, for instance, made him an officer of the Legion of Honor.

In Marburg, Behring developed a diphtheria vaccine that contained a mixture of his serum antitoxin and the diphtheria toxin. The small amount of toxin stimulated the vaccinated person's immune system to make its own antitoxin and gave long-term protection against the disease. He introduced this "toxin-antitoxin" vaccine in 1913. He failed, however, in repeated attempts to develop a treatment or vaccine for tuberculosis, which depressed him so severely that he had to leave his work and rest in a sanatorium (convalescent home) between 1907 and 1910.

Von Behring's other major codiscovery, the tetanus antitoxin, brought him new fame during World War I. Many soldiers died of tetanus in the early months of the war because tetanus bacteria in the soil of the battlefields were driven into their bodies when they were wounded. The German government was reluctant to send tetanus antitoxin to the front at first because the treatment had not been widely tested in humans (it had been used primarily to treat farm animals exposed to the disease). Von Behring helped to refine the antitoxin and procedures for using it, however, and it went into widespread use in April 1915. For this work, the German emperor called him the "savior of the German soldiers" and awarded him the Iron Cross, an honor seldom given to civilians. Von Behring died in Marburg on March 31, 1917.

Further Reading

Grundmann, Kornelia. "Emil von Behring: The Founder of Serum Therapy." Nobel Foundation. Available online. URL: www.nobel.se./medicine/articles/behring. Last updated 2002.

Beijerinck, Martinus Willem

(1851–1931)
Dutch
Botanist, Microbiologist

Martinus Beijerinck was the first to recognize the existence of viruses and describe some of their features, even though he could not see them. He was born in Amsterdam on March 16, 1851.

Although he was chiefly interested in botany, the degree he took in 1872 from the Delft Polytechnic School (now the Delft Institute of Technology) was in chemical engineering. He taught botany and other subjects to support himself while working for his doctor of science degree, which he earned in 1877 in Leiden.

Beijerinck worked for the Netherland Yeast and Alcohol Factory in Delft from 1887 to 1893, heading the first industrial microbiology laboratory in the Netherlands. He then returned to the Delft Polytechnic School, where he founded a new microbiology laboratory in 1897 and researched and taught until his retirement in 1921. He died on New Year's Day, 1931, near Gorssel.

Beijerinck spent most of his life studying microorganisms in the soil and in crop plants. In 1888, he became the first to identify an important kind of bacteria that live in the roots of a family of plants called legumes, which includes peanuts and peas. These bacteria combine nitrogen gas from the air with other substances to form nutrients essential to plant growth, a process called nitrogen fixing. When the legumes die, these compounds enter the soil, enriching it and spurring the growth of other plants.

Beijerinck is chiefly remembered, however, for his research on the then-unidentified agent that causes a sickness of tobacco plants called tobacco mosaic disease. Plants with this disease are shorter than normal and have spotted leaves. He showed that liquid squeezed from infected plants could produce the disease when injected into other plants, even after it was strained through a filter with pores small enough to remove bacteria. This meant that whatever caused tobacco mosaic disease must be smaller than any kind of microbe known at the time.

Beijerinck called the mystery agent a virus, the Latin word for poison. To find out whether it was a chemical, he infected a plant, withdrew fluid from this plant, and used it to infect another, continuing this process through many plants. A chemical would have been diluted by being mixed with other substances in the plant each time he did this. The fluid from the successive plants, therefore, should have contained less and less of the original poison and should have become less able to cause disease. This weakening did not occur, however. Beijerinck therefore concluded that the "poison" was a living thing that could maintain its power by reproducing. He called it a "contagious living fluid."

Beijerinck published his conclusions about tobacco mosaic virus in 1898. His research won Denmark's Hansen Prize in 1922; he was the first person from the Netherlands to receive this award. Building on his work, later researchers showed that other diseases, including some deadly ones that affect humans, are caused by agents like the one he had described. Scientists finally became able to see viruses in the 1930s, when electron microscopes were invented. Martinus Beijerinck's work set the stage for the discovery of this important group of microorganisms.

Further Reading

Bardell, David. "The Tobacco Mosaic Virus." *Science Teacher*, December 1997.

Iterson, Gerrit van, et al. *Martinus Willem Beijerinck: His Life and His Work*. The Hague, Netherlands: Martinus Nijhoff, 1940.

Johnson, James. "Martinus Willem Beijerinck, 1851–1931." *The Plant Health Instructor*, January 29, 2001. American Phytopathological Society. Available online. URL: www.apsnet.org/education/feature/TMV/intro.html. Accessed 2003.

Berg, Paul

(1926–)
American
Microbiologist, Biochemist

Paul Berg has been called "the father of genetic engineering." In 1972, he became the first person to combine pieces of DNA, the chemical that carries genetic information, from two different kinds of living things. For this and other work in what came to be known as recombinant DNA technology or genetic engineering, he was awarded a share of the Nobel Prize in chemistry in 1980.

Berg was born in Brooklyn, part of New York City, on June 30, 1926. He was one of three sons of Harry Berg, who manufactured clothing, and his wife, Sarah. Berg began studying biochemistry at Pennsylvania State College in 1943 but left a year later to fight in World War II. After three years in the navy, he returned to the university, finally earning his bachelor's degree in 1948. He took his Ph.D. from Western Reserve University (now Case Western Reserve) in Cleveland, Ohio, in 1952.

Berg did postdoctoral research on cancer in Denmark and at Washington University in St. Louis, Missouri. He taught biochemistry in that university's school of medicine from 1955 to 1959. He then joined the faculty of Stanford University in California, where he has spent the rest of his career. He was head of the department of biochemistry from 1969 to 1974. He became the Willson Professor of Biochemistry in 1970 and the director of the Beckman Center for Molecular and Genetic Medicine in 1985. He is also the Cahill Professor of Cancer Research. Berg married Mildred Levy in 1947, and the couple has one son, John Alexander.

Berg did his first major work in the late 1950s. It built on several earlier discoveries, beginning with GEORGE WELLS BEADLE and Edward Tatum's finding in the early 1940s that genes determine characteristics by telling cells how to make proteins. Genes proved to be made of DNA, and FRANCIS CRICK, the codiscoverer of DNA's structure, had suggested that the order in which small molecules called bases are arranged within the large DNA molecule specifies the order in which the cell connects other small molecules called amino acids to make a particular protein.

Crick proposed in 1955 that what he called "adaptor molecules," following the instructions originally specified in the cell's DNA, attach themselves to different kinds of amino acid molecules and tow them into place as a protein is being assembled. A year later, Berg isolated the first of these molecules, which came to be called transfer RNAs (RNA is a nucleic acid related to DNA). Berg showed that this molecule always attached itself to a type of amino acid called methionine. His discovery helped to prove that Crick's theory of protein manufacture was basically correct.

The work that Berg did in his Stanford laboratory in the early 1970s, however, was even more important. It grew out of Crick and JAMES WATSON's discovery that DNA has a structure like a twisted ladder, with pairs of bases as the steps. There are four kinds of bases, and the kind called adenine (A) always pairs with another one called thymine (T). Similarly, the base cytosine (C) always pairs with guanine (G). When DNA reproduces, it splits apart lengthwise into two strands, and each strand attracts its complementary bases—the other half of the pairs—from chemicals floating free in the cell. In other words, a single-stranded DNA fragment with the base sequence C-A-A-T-G would attract bases that assemble themselves in the order G-T-T-A-C. The result is two identical pieces of double-stranded DNA.

Berg noticed that if two complementary single strands of DNA touched each other, they would stick together. In 1972, he used chemical means to attach such "sticky ends" to a DNA fragment from a common bacterium called *Escherichia coli* (*E. coli*, for short) and one from a monkey virus called SV40, which causes cancer in some animals and can infect humans. He then inserted a piece of DNA from *E. coli* into the virus DNA, creating the first recombinant DNA—a single piece of DNA containing genetic material from two different types of living things. This was the start of genetic engineering, which today is revolutionizing medicine, agriculture, and many other fields.

Recombinant DNA experiments presented potential dangers as well as benefits, and Paul Berg was one of the first people to recognize this. He had intended to continue his experiments by inserting SV40 DNA into the genomes of living *E. coli* cells. When a microbiologist named Robert Pollack heard about these plans, however, he warned Berg that putting genes from a cancer-causing virus into a bacterium that infects humans, as *E. coli* does, might be risky. What if the engineered bacteria escaped from Berg's laboratory and proved able to cause cancer in people?

Berg could not disprove Pollack's concerns. He therefore decided to abandon his own experiment and also began suggesting to other scientists that they hold off doing similar ones until ways could be developed to make sure that recombined organisms could not escape. Eventually, he organized a three-day meeting of 100 researchers from 16 countries, which took place at the Asilomar Conference Center in Pacific Grove, California, beginning on February 27, 1975. The scientists at the conference worked out safety guidelines for different kinds of recombinant DNA experiments. A year later, the National Institutes of Health used the Asilomar guidelines as a basis for drafting safety regulations that all laboratories using U. S. government funding had to follow.

Some scientists, both during the conference and later, felt that Berg and others overstated the risks of genetic engineering. Berg, however, believes that his concern was justified at the time. He also feels that the fact that scientists themselves first created the rules—and agreed to temporarily halt certain types of experiments until safe facilities could be built—was very important. In 2001, 25 years after the historic conference, he wrote in *Perspectives in Biology and Medicine* that the scientists' actions at Asilomar "gained the public's trust" and were "widely acclaimed as laudable [praiseworthy] and ethical behavior."

Paul Berg's development of genetic engineering and other research won many awards in addition to the Nobel Prize, including the Eli Lilly Prize in Biochemistry (1959), the Albert Lasker Medical Research Award (1980), and the National Medal of Science (1985). Berg was named California Scientist of the Year in 1963 and was elected to the U.S. National Academy of Sciences in 1966.

As an emeritus professor at Stanford, Berg continues his study of combining genes, investigating ways in which genes in certain microorganisms recombine naturally. He is also working on ways to change genes in special cells called stem cells, which can transform themselves into any type of cell in the body. Finally, his laboratory is studying how HIV, the virus that causes AIDS, invades and destroys certain cells in the immune system.

Further Reading

Berg, Paul. "Reflections on Asilomar 2 and Asilomar 3: Twenty-five Years Later." *Perspectives in Biology and Medicine*, Spring 2001.

———. "A Stanford Professor's Career in Biotechnology, Science Politics, and the Biotechnology Industry." University of California Programs in the History of the Biosciences and Biotechnology. Available online. URL: http://www.sunsite.berkeley.edu:2020/dynaweb/teiproj/oh/science/berg/@Generic_BookView. Accessed 2000.

———, and Maxine Singer. *Dealing with Genes: The Language of Heredity*. Herndon, Va.: University Science Books, 1992.

Bernard, Claude

(1813–1878)
French
Physiologist

Claude Bernard helped to establish physiology as a modern laboratory science and made several major discoveries about how parts of the body work. His parents, Pierre and Jeanne, worked in the fields of France's famous Beaujolais wine-growing region. Bernard was born there near the village of St. Julien on July 12, 1813. A local minister encouraged him to seek an education, but his family had no money to pay for it. At age 19, therefore, Bernard went to work for a pharmacist in Vaise, a suburb of Lyons.

Bernard dreamed of becoming a playwright; indeed, he spent so much time writing that the pharmacist finally fired him. When he finally reached Paris and showed one of his plays to the professor of literature at the renowned Sorbonne, however, the professor told him dismissively, "You have done some pharmaceutical work. Study medicine. You have not the temperament of a dramatist."

Crushed, Bernard took the professor's advice. He studied at the medical school in Paris and at the Collège de France, where he met a more encouraging professor, physiologist François Magendie. Magendie was rude to most people, but he recognized Bernard's brilliance. In 1841, he hired the young man to prepare material for his class lectures. After seeing several of Bernard's preparations, he reportedly muttered, "You're better than I am," and stalked out of the room.

Bernard earned his medical degree in 1843. He continued doing research with Magendie, but he failed to win a teaching position. Probably out of financial desperation, he married Marie Martin, the daughter of a wealthy doctor, in 1845. Her dowry, a sum of money that her father gave her when she married, eased Bernard's monetary problems, but the marriage was a miserable one. Marie, a very religious woman, strongly objected to experimentation on living animals, which became Bernard's specialty. The couple raised two daughters, but in 1869 they decided to live apart.

Most French medical researchers of the time emphasized the study of human patients in hospitals, but Bernard preferred to work in a laboratory, where he could carry out carefully controlled experiments like those used in physics and chemistry. Since experimenting on humans would be unethical, he used dogs, rabbits, and other animals as "living laboratories," feeling sure that what he learned from them would apply to humans as well.

Many of Bernard's experiments concerned digestion. In 1849, he showed that the pancreas, an organ in the abdomen whose function had been unknown, produces chemicals that help to break down fats and carbohydrates. This research earned him the ribbon of the Legion of Honor from the French government. Bernard also proved that most digestion of food takes place in the small intestine, not in the stomach as had been thought.

Later, through experiments on dogs, Bernard showed that sugar in the blood of animals comes from the liver. (The Sorbonne awarded him a doctorate in zoology for his research on the liver

Claude Bernard, an eminent 19th-century French physiologist, showed that birds and mammals can control their body temperature and create a stable internal environment. *(National Library of Medicine)*

in 1853.) In 1857, he purified a starchy compound that he named glycogen. He showed that the liver makes this substance from sugar and stores it. When the body needs extra energy, the liver breaks down some of its glycogen into sugar once more and releases it into the blood. Body cells then use the sugar as fuel. Scientists had not known before this time that the bodies of animals could create as well as break down complex substances.

Bernard also investigated nerves and their relationship with blood vessels. He found that one group of nerves makes blood vessels expand, while another makes them contract. Together, these nerves, along with certain chemicals that Bernard also identified, help to control blood pressure and body temperature.

The fact that the bodies of birds and mammals (warm-blooded animals) can control their temperature and other aspects of their internal environment, and indeed require such a controlled environment, was probably Bernard's most important discovery. This ability to create and maintain a constant internal environment, which was later called homeostasis, lets the animals be fairly independent of the conditions around them. Mammals, for instance, can remain active even in cold weather, when insects and other creatures that lack homeostasis (cold-blooded animals) are paralyzed because their body temperature falls along with the air temperature.

In contrast to his frustrating early years, Bernard was very successful in middle age. He took over Magendie's faculty position at the Collège de France when his old mentor died in 1855. In 1869, he was elected to the French Academy of Sciences, an organization of the country's top scientists, and he also became a senator.

When failing health forced Bernard out of his laboratory in his later years, he turned his attention to the philosophy of science. In *Introduction to the Study of Experimental Medicine*, published in 1865, he described the classical approach to learning about nature that scientists still use. In this method, a scientist observes some natural phenomenon, then uses knowledge and creativity to devise a tentative theory, or hypothesis, that explains why this phenomenon occurs. The scientist tests the hypothesis by performing experiments and observing whether the results match those that the hypothesis would predict. Bernard once wrote that a good scientist "see[s] what everybody has seen, and think[s] what nobody has thought." He died in Paris on February 10, 1878.

Further Reading

Bernard, Claude. *Introduction to the Study of Experimental Medicine*. 1865. Reprint, Mineola, N.Y.: Dover Publications, 1957.

Holmes, Frederick L. *Claude Bernard and Animal Chemistry*. Boston: Harvard University Press, 1974.

Bishop, J. Michael

(1936–)
American
Molecular Biologist, Geneticist, Virologist

When John Michael Bishop was growing up, no one—least of all Bishop himself—would have predicted that he would become a Nobel Prize–winning researcher, uncovering genetic secrets of both cancer and normal cell growth. Born on February 22, 1936, in York, Pennsylvania, he spent his childhood in a rural home near the Susquehanna River. His father was a Lutheran minister, and exposure to church liturgy provided Bishop with a lifelong love of music. His education in a two-room schoolhouse stirred an interest in history. A family physician brought medicine and biology to his attention. When he entered Pennsylvania's Gettysburg College, he planned to go to medical school, but he also considered becoming a philosopher, a historian, or a novelist. He graduated with a B.A. in chemistry in 1957.

Only after Bishop entered Harvard Medical School did he realize that he wanted to do biomedical research. He began to focus on molecular biology and the study of viruses that infect animals. During his medical school years, he married Kathryn Putman, whom he had met at Gettysburg College. They later had two sons. Bishop earned his M.D. in 1962 and did postdoctoral research at the National Institutes of Health (NIH) and in Germany.

In 1968, Bishop joined the University of California at San Francisco (UCSF), where he has remained ever since. Two years later, he met HAROLD E. VARMUS, and the two decided to work together. They wanted to find out how certain viruses cause cancer in animals. One of these viruses existed in both a common form, which caused cancer in chickens, and an unusual form, which could not cause the disease. Other researchers found that the two forms differed in only a single gene, which the cancer-causing form possessed and the harmless form lacked. They named this key gene *src*, short for "sarcoma," the type of cancer that the virus caused.

Scientists went on to find similar genes in other cancer-causing viruses. Robert Huebner and George Todaro of the National Cancer Institute, part of NIH, called them *oncogenes*, from the Greek word for cancer. Huebner and Todaro believed that these viruses inserted oncogenes into the genomes of the cells they infected. Sometimes, however, instead of causing cancer, the genes were passed down through generations of cells in a form that remained harmless unless a chance mutation returned them to their cancer-causing form.

In 1976, working with Dominique Stehelin and Deborah Spector, Bishop and Varmus turned Huebner and Todaro's idea on its head. They not only found a gene resembling *src* in normal chicken cells but proved that it was a chicken gene rather than a virus one. They theorized that, instead of having inserted the *src* gene into chicken cells in the distant past, the viruses had picked up the gene from the cells. The gene had then become modified in a way that made it cause cancer when the viruses later reintroduced it. Spector went on to find similar genes, which Bishop and Varmus called proto-oncogenes, or cellular oncogenes, in normal cells from fish, birds, and mammals, including humans. "Cancer may be part of the genetic dowry [inheritance] of every living cell," Bishop has said.

The discovery of cellular oncogenes, which won the Nobel Prize in physiology or medicine for Bishop and Varmus in 1989, completely changed scientists' understanding of cancer. For one thing, it showed that all cancer, whether inherited or triggered by something in the environment such as a virus or radiation, is ultimately due to changes in genes. Furthermore, the fact that cellular oncogenes were so widespread in nature suggested that, in their normal, harmless form, they had a very important function. Since cancer results from uncontrolled reproduction of cells, Bishop and Varmus suspected that the genes played roles in cell reproduction and growth.

Bishop, Varmus, and other researchers confirmed this idea during the early 1980s. They found that oncogenes direct cells to make various chemicals that cause the cells to reproduce. Normally, these genes are active only at certain times in a living thing's existence—before birth, for instance, or when new cells are needed to heal a wound. Sometimes, however, they are damaged in a way that activates them at the wrong time or leaves them "turned on" constantly, causing uncontrolled reproduction and cancer. Researchers also discovered a second group of genes, called tumor suppressor genes, that normally control and limit cell reproduction. Cancer can result when these genes are inactivated or destroyed.

Bishop continues to try to learn what oncogenes and tumor suppressor genes do and how they do it in both normal and cancerous cells. His goal, he says, is to learn how normal cells control their growth and reproduction and why cancer cells fail to do so. However, he has had less time for his research since 1998, when he became chancellor of UCSF. He also heads the G. W. Hooper Research Foundation, is a member of the Markey Program in Biological Sciences, and is a University Professor, the highest faculty level in the University of California. He has been showered with awards, including the Albert Lasker Medical Research Award (1982), the Armand Hammer Cancer Research Prize (1984), and the Trustees' Medal of Massachusetts General Hospital (1992).

Much as Bishop loves research, he has said that "one life-time as a scientist is enough." Near the end of the autobiography he submitted to the Nobel Foundation in 1989, he wrote, "If offered reincarnation, I would choose the career of a performing musician with exceptional talent, preferably, in a string quartet."

Further Reading

Bishop, J. Michael. "J. Michael Bishop—Autobiography." *Les Prix Nobel 1989*. Available online. URL: www.nobel.se/medicine/laureates/1989/bishop-autobio.html. Last updated 2002.

———. "Oncogenes." *Scientific American*, March 1982.

Wallis, Claudia. "Advances in the War on Cancer." *Time*, November 8, 1982.

Yount, Lisa. *Milestones in Discovery and Invention: Genetics and Genetic Engineering*. New York: Facts On File, 1997.

Black, James Whyte

(1924–)
British
Physiologist, Pharmacologist

James Whyte Black created two drugs that improved the health of millions. Even more important, his use of basic research on the ways cells communicate as a starting point for inventing drugs helped to make drug design more rational and effective than ever before. "He has designed drugs that behave like a rifle bullet, instead of a shotgun," a coworker has said.

Black was born in Addingston, Scotland, in 1924, the second-youngest of five boys. His father was a mining engineer and manager of a coal mine. Black earned a residential scholarship to St. Andrew's University in Fife, where he studied medicine, earning his M.D. in 1946. While still an undergraduate, he met a young woman named Hilary Vaughan at a dance, and they were married as soon as he graduated. They had a daughter, Stephanie, in 1951.

Black went into medical research, specializing in physiology. He did postdoctoral work at St. Andrew's for a year and then, needing money to pay off medical school debts, taught physiology in Malaya (now Malaysia) for three years. On his return to Britain, he joined the veterinary school at the University of Glasgow, where he set up a new physiology department. He remained there from 1950 to 1958.

During this time, Black became interested in heart disease, especially a type called angina. Angina causes chest pain, especially when a person exercises. Some of the millions of middle-aged and elderly people who suffer from this condition cannot even climb a flight of stairs. Fatty deposits block or narrow the arteries that carry blood to the hearts of people with angina, limiting blood flow. Their hearts are thus starved for oxygen, which is carried in the blood. At the time Black started his research, the chief treatment for angina was a drug called nitroglycerin, which makes arteries expand and therefore allows more blood to flow through them. Nitroglycerin was not always effective, however, and it produced side effects such as headaches.

Most people looking for better drugs to treat angina were trying to find other chemicals that dilated blood vessels. James Black, however, has what has been called "an incurable mental habit of inverting every proposition" he encounters. Therefore, instead of trying to bring more oxygen to the heart, he decided to look for ways to reduce the heart's need for oxygen.

Scientists knew that whenever the body undergoes physical or psychological stress, certain nerves signal the adrenal glands, tiny organs just above the kidneys, to release a hormone called adrenalin. Among other effects, adrenalin speeds up the heartbeat. The faster the heart beats, the more oxygen it needs. Black reasoned that if he could find a chemical that blocked this action of adrenalin, a heart under stress would need less oxygen and might therefore be able to survive on the limited supply available to it. At the same time, he did not necessarily want to stop other effects of adrenalin, which are useful to the body.

In the 1950s, researchers were just beginning to realize that hormone molecules interact with other molecules, called receptors, on the surface of cells. Each hormone has a different kind of receptor, which appears only on the types of cells that that hormone is meant to affect. The hormone molecule fits into the receptor like a key into a lock, sending a signal into the cell.

Scientists were puzzled by the fact that some hormones have different effects on different kinds—or sometimes even the same kind—of cells. Some suspected that this occurred because these hormones could "lock into" more than one kind of receptor. Interaction with one type of receptor produced one set of effects, whereas interaction with the other type produced a different set. Some cells might show both sets of effects because they had both types of receptors.

This same theory could also explain why certain chemicals block some effects of a hormone but not others. The molecules of chemicals that block a hormone's effects are usually shaped enough like the hormone molecules to let them fit into the same receptors, yet they are different enough to keep them from sending a signal into the cells. Once a molecule of such a substance attaches to a receptor, it acts like a bent coin in a parking meter, preventing hormone molecules from attaching to the receptor and sending their signal. If a chemical blocked only one of several types of receptors, it would prevent only the effects produced when the hormone interacted with that type of receptor. Effects produced by the hormone's interaction with other types of receptors would not be changed.

In the late 1950s, Black read an article by a scientist named James Ahlquist that suggested that adrenalin might have two kinds of receptors. Ahlquist called them alpha and beta. On further investigation, Black found that adrenalin apparently produced changes in heart rate by interacting with beta receptors. He hoped, therefore, that a drug that blocked only the beta receptors could be used to treat angina. In 1956, he proposed his idea to the drug division of Imperial Chemical Industries (I.C.I.), a British firm, and the company agreed to provide a laboratory for his research. Black worked for I.C.I. from 1958 to 1964.

With the help of chemist John Stephenson, Black studied the chemical structure of adrenalin and other substances known to attach to beta receptors. They then modified these structures to create new compounds. One of these compounds, propranolol, proved able to block beta receptors and prevent the changes in heartbeat that adrenalin caused without stopping other effects of adrenalin or causing side effects of its own.

First marketed in Britain under the name Inderal in 1964, propranolol became the chief medication used to treat angina. It has reduced or ended chest pain for millions of people with the condition, allowing them to lead more active lives. It also substantially reduces deaths from heart attacks among angina sufferers and people who have had earlier heart attacks. Indeed, propranolol has proved to have more uses than Black had ever dreamed. It lowers high blood pressure, a dangerous condition that can lead to heart attacks or strokes. It also helps people with certain kinds of kidney disease. It even reduces the psychological effects of stress, which are also caused by adrenalin. Black has said that some concert pianists and public speakers—including himself, on occasion—take it before appearances to reduce stage fright.

By the time propranolol became a best-selling drug, however, James Black was no longer associated with it. He likes to explore the basic biology involved in designing drugs and leave drug development—the long process of preparing a drug for market—to others. He changed companies, going to work for Smith, Kline and French (now GlaxoSmithKline) in 1964, and set to work on a new problem. It involved another hormone, histamine, which, like adrenalin, proved to have two different kinds of receptors. Black wanted to block the kind that controls histamine's action in the stomach.

In the early 1970s, using methods similar to those that had worked so well in making propranolol, Black invented a second billion-dollar drug, cimetidine. It keeps histamine from increasing the stomach's production of the acid liquid called gastric juice. Gastric juice is needed to digest food, but too much of it can eat painful and sometimes life-threatening sores called ulcers in the walls of the stomach and nearby parts of the digestive system. Sold under the trade name Tagamet, cimetidine became as popular for treating ulcers, heartburn, and other stomach problems as propranolol was for treating heart disease.

Black moved again in 1973. He headed the pharmacology department at University College, London, until 1977, then worked at another drug company, the Wellcome Foundation, until 1984. In 1988, working with King's College (part of the University of London) and supported by money from the drug company Johnson & Johnson, Black opened his own research institution, the James Black Foundation, in Dulwich, England. Because he feels that the best research is done by small groups, he specified that the organization never employ more than 20 people. He also heads the department of analytical pharmacology at the King's College School of Medicine and Dentistry. Black was knighted in 1981 and shared the 1988 Nobel Prize in physiology or medicine with two other pioneer drug designers, GERTRUDE BELLE ELION and GEORGE HERBERT HITCHINGS. According to science reporter Thomas A. Bass, "Black is widely acknowledged to be the founder of the modern scientific approach to pharmacology."

Further Reading

Bass, Thomas A. *Reinventing the Future*. Reading, Mass.: Addison-Wesley, 1994.

Black, James. "Sir James W. Black—Autobiography." *Les Prix Nobel 1988*. Available online. URL: www.nobel.se/medicine/laureates/1988/black-autobio.html. Last updated 2002.

Stapleton, Melanie Patricia. "Sir James Black and Propranolol: The Role of the Basic Sciences in the History of Cardiovascular Pharmacology." *Texas Heart Institute Journal*, 1997. Available online. URL: www.tmc.edu/thi/sirjames.html.

Boussingault, Jean-Baptiste

(1802–1887)
French
Chemist

Jean-Baptiste-Joseph Dieudonné Boussingault made important discoveries about the way plants and animals interact with the nonliving environment, especially chemicals in the air and soil. He was a pioneer of agricultural chemistry.

Boussingault was born in Paris in 1802. As a young man, he studied the science of mining and supervised mines in Alsace. A meeting with famous scientist-explorer ALEXANDER VON HUMBOLDT stirred his desire to visit distant lands, and he found a chance to fulfill his wish when Simón Bolívar, who had freed several South American countries from Spanish control, asked Humboldt for help in finding scientists to staff a research institution and Humboldt gave him Boussingault's name. Boussingault spent 10 years in Venezuela, Colombia, and Ecuador, studying earthquakes, volcanoes, mineral deposits, and climate.

Boussingault returned to France in 1832 and began teaching chemistry at the University of Lyons. In 1838, he married the daughter of an expert in agronomy, the science of farm management, and became interested in his father-in-law's profession. He devoted the rest of his life to this study, teaching and doing research at the Conservatoire des Arts et Métiers in Paris from 1839 until his death in 1887.

Scientists knew that one of the nutrients plants need is nitrogen, an element that, as a gas, makes up 78 percent of the air. Researchers of the time believed that plants absorbed nitrogen from the air, but Boussingault showed that most plants died if raised in soil that lacked nitrogen-containing compounds. His experiments suggested that these compounds, which are present in most soils, are the plants' true source of nitrogen. He found that the only plants that could survive without nitrogen compounds belong to a family called legumes, which includes peas and beans. MARTINUS WILLEM BEIJERINCK later discovered that certain bacteria living in the roots of these plants convert nitrogen from the air into compounds that plants can use.

On his farm in Alsace, Boussingault did elaborate experiments that compared plants and animals in terms of the nutrients they need and the substances that their bodies return to the environment. He showed that, in many respects, the two groups are opposite. Plants need nitrogen as a nutrient, for example, whereas animals give off nitrogen-containing compounds in their waste. Animals exhale carbon dioxide gas in their breath, and the leaves of plants absorb this same gas from the air. Conversely, animals take in oxygen when they breathe, while plants give off this gas.

The result of these oppositions, Boussingault showed, is a series of cycles in which substances such as nitrogen and carbon dioxide pass repeatedly through plants, animals, and the nonliving environment. He is especially credited with discovering the nitrogen cycle. Understanding of these cycles proved very important to agriculture, for instance, in showing why adding animal wastes to soil increases plant growth. Boussingault's cycles also underlined the fact that all parts of nature depend on one another.

Further Reading

Hays, H. R. *Birds, Beasts, and Men: A Humanist History of Zoology*. New York: G. P. Putnam's Sons, 1972.

Boyer, Herbert Wayne

(1936–)
American
Biochemist

Herbert Boyer and STANLEY N. COHEN were the first to transfer a gene from one species of living thing into another species and show that the gene could function in its new location, thus making genetic engineering practical. Boyer also cofounded one of the first businesses based on the new technology.

Boyer, born in Derry, Pennsylvania, in 1936, was a football star in high school. He studied biology and chemistry at St. Vincent College in Pennsylvania, from which he obtained a bachelor's degree in 1958. He earned his master's degree in 1960 and his Ph.D. in 1963 from the University of Pittsburgh. In 1966, he began to do research at the University of California in San Francisco (UCSF). He was made a professor of biochemistry and biophysics there in 1976.

In the early 1970s, a graduate student in Boyer's laboratory had discovered chemicals called restriction enzymes, which some bacteria use to protect themselves against attacking viruses. Restriction enzymes act like molecular scissors, snipping the viruses' genetic material into pieces so that the viruses cannot reproduce. The enzymes cut DNA at any point where they find a certain sequence of bases within the DNA molecule.

Boyer gave a talk about his work with restriction enzymes at a scientific meeting in Honolulu, Hawaii, in November 1972. Afterward, he took an evening walk with Stanley Cohen that ended with a chat over corned beef sandwiches at a nearby delicatessen. The two, who worked within a short drive of each other in California, discovered that their research interests fitted together as neatly as the pairs of bases in a DNA molecule. Each had expertise that the other could use.

Boyer told Cohen that he had learned that his molecular scissors did not snip cleanly. When they sliced through a double-stranded DNA molecule, they left a short single strand of bases at each end of the cut piece. The bases on these single strands attracted and attached to bases that made up the other half of their natural pairs, just as they do when the strands of DNA split apart and duplicate their missing halves before a cell divides. Any two pieces of DNA cut with the same restriction enzyme had single-stranded ends with the same sequences. In theory, therefore, two such pieces could be joined together, even if they came from different kinds of living things.

Cohen told Boyer that he had been doing a different kind of research with bacteria. In addition to their main genome, some bacteria carry small, ring-shaped pieces of genetic material called plasmids. Each plasmid contains just one or a few genes. Bacteria in nature sometimes exchange plasmids, and Cohen had developed a way to imitate this process, removing plasmids from bacteria and making other bacteria take them up on demand.

Back in California in the spring of 1973, Boyer and Cohen began experiments that combined the techniques they had developed. They used a restriction enzyme from Boyer's laboratory to cut open some of Cohen's plasmids. The "sticky ends" produced by the process allowed them to join the two plasmids together to make a single larger one, even though the plasmids had come from two different strains of bacteria. They then used Cohen's technique to make other bacteria take up the new plasmids. They showed that the bacteria copied the plasmids along with their own genes each time they reproduced. Furthermore, they proved that the genes in the combined plasmid, which made bacteria resistant to two types of antibiotic, functioned in their new location.

Boyer and Cohen went on to transfer plasmids from one species of bacteria to another and then, more daringly, to insert genes from a toad into bacteria. These genes also functioned in their new location and were passed on when the bacteria reproduced. Boyer and Cohen's experiments provided the first proof that "gene splicing" could be used dependably in living things. After hearing an account of their work in June 1973, one researcher exclaimed, "Now we can put together any DNA we want to."

Boyer was one of the first people to realize that the technique he and Cohen had developed might have commercial value. Bacteria divide very rapidly, so millions can be made from one in a single day. If a gene carrying the instructions for making a medically useful substance was put into a bacterium, Boyer reasoned, the bacterium's many descendants should all become microscopic "factories" churning out this substance. The material could then be harvested and sold.

In 1976, Boyer and Robert Swanson, a young venture capitalist, founded a company that they named Genentech, short for GENetic ENgineering TECHnology. They planned to use genetically altered bacteria to make insulin, a vital hormone that people with diabetes lack. Insulin could be obtained from slaughtered cattle and pigs, but some diabetics were allergic to it because animal insulin is not quite the same as human insulin. If the human gene for insulin was inserted into bacteria, however, Boyer and Swanson expected the bacteria to make insulin exactly like that in the human body. Genentech researchers showed that bacteria could make human insulin in 1978, and the company began selling its recombinant insulin in 1982. In the years that followed, Genentech and other companies made many other recombinant products, such as human growth hormone, greatly benefiting medicine, agriculture, and other industries and sometimes making a considerable profit for the companies.

Boyer has been a director of Genentech from the time of its founding, and he was also the company's vice president from 1976 to 1990. He was elected to the National Academy of Sciences and the National Inventors Hall of Fame (2001). His numerous awards, most of which were shared with Stanley Cohen, include the Albert Lasker Medical Research Award (1980), the Swiss Helmut Horten Research Award (1993), and the Lemelson-MIT Prize for inventors (1996). When awarding Boyer and Cohen the Lemelson-MIT Prize, Charles M. Vest, the president of MIT (Massachusetts Institute of Technology), said, "Boyer and Cohen's ingenuity has revolutionized the way all of us live our lives."

Further Reading

Boyer, Herbert W. "Recombinant DNA Research at UCSF and Commercial Applications at Genentech." University of California Programs in the History of the Biosciences and Biotechnology. Available online. URL: http://sunsite.berkeley.edu:2020/dynaweb/teiproj/oh/science/boyer/@Generic_BookView. Posted 1994.

Yount, Lisa. *Genetics and Genetic Engineering*. New York: Facts On File, 1997.

Buffon, Georges-Louis, comte de
(1707–1788)
French
Naturalist

Although he did not believe in evolution, Georges-Louis Leclerc de Buffon, usually referred to as Comte de Buffon, raised many of the questions about the history of the Earth and changes in living things over time that supporters of evolution such as CHARLES ROBERT DARWIN would later discuss. In *The Growth of Biological Thought*, ERNST MAYR writes, "Except for ARISTOTLE and Darwin, there has been no other student of organisms who has had as far-reaching an influence" as Buffon.

Buffon was born into a well-to-do family in Montbard, part of the Burgundy region of France, on September 7, 1707. As a young man he studied law and politics at a Jesuit college in Dijon and medicine in Angiers. Friendship with a young Englishman, the duke of Kingston, introduced him to science. While visiting the duke in England in the 1730s, Buffon studied mathematics and physics and became very impressed with the ideas of Isaac Newton. He translated Newton's *Fluxions* into French.

In 1739, influential friends helped the 32-year-old Buffon become director of the Jardin du Roi (King's Garden), a botanical garden and natural history museum in Paris (now the Jardin des Plantes). His work in cataloging the museum's collections inspired him to begin writing an encyclopedia, *Histoire naturelle* (Natural history), which attempted to sum up all scientific knowledge. Thirty-five volumes were published during his lifetime, beginning in 1749, and nine more after his death. Other people helped him gather material for this monumental work, but he rewrote all their contributions in his own elegant style. Buffon's encyclopedia became a best-seller throughout Europe.

Buffon was born in the same year as CAROLUS LINNAEUS, the great Swedish classifier of plants and animals, but at first the two had opposite views of biology. Linnaeus was chiefly interested in biological relationships among living things, whereas Buffon preferred to describe plants and animals as individuals. The detailed accounts in *Natural History* included not only the anatomical features that Linnaeus stressed but also accounts of behavior and geographical distribution. Linnaeus divided plants and animals into different groups and subgroups, but Buffon saw nature as unified and continuous.

Buffon's views grew closer to Linnaeus's as the two men aged, however. For instance, in the first volume of *Natural History*, Buffon wrote that he did not believe in the existence of species, or specific types of living things that could be separated from others. Later, however, he held that individuals could be considered to belong to different species if they could not mate with each other and produce offspring that in turn were able to bear young.

Powerful Catholic theologians criticized Buffon's books because he questioned church-sanctioned views of Earth's past. For example, church leaders claimed that Earth was only 6,000 years old, but Buffon held that the correct figure was 75,000 years. He wrote that different kinds of rocks seemed to have come into existence at different times, rather than having appeared all at once as the biblical account of the Creation suggested. To avoid trouble, Buffon had to retract some of his ideas in print, but geologists such as CHARLES LYELL later built on them.

Buffon was unusual for his time in considering humans to be animals. "We are superior to the animal only by a few characteristics granted us by the tongue and the hand" and the ability to think or reason, he wrote. He was among the first to suggest that similar kinds of creatures, such as humans and apes, might have descended from a common ancestor, although he himself did not believe this theory. He did not think that species of living things had changed, or evolved, over time, although he was willing to accept the possibility that some species had died out and new ones had arisen.

Buffon was highly honored for his work. In 1739, he was elected to the Academy of Sciences, France's top scientific organization, and also became a member of the Royal Society, a comparable organization in Britain. King Louis XV awarded him the title of Comte in 1771. He married Marie de Saint Belin-Malin in 1752, and the couple had one son. Buffon died on April 16, 1788, in Paris after a long battle with kidney disease.

Further Reading

Buffon, Comte de. *368 Animal Illustrations from Buffon's "Natural History."* Reprint, Mineola, N.Y.: Dover Publications, 1993.

Lyon, John, and Philip Slan. *From Natural History to the History of Nature: Readings from Buffon and His Critics*. Notre Dame, Ind.: University of Notre Dame Press, 1981.

Roger, Jacques. *Buffon: A Life in Natural History.* Translated by Sarah Lucille Bonnefoi. Ithaca, N.Y.: Cornell University Press, 1997.

Burkholder, JoAnn Marie

(1953–)
American
Ecologist

Like RACHEL LOUISE CARSON and THEO E. COLBORN, JoAnn Burkholder has caused controversy by warning of dangers to wildlife and human health that result from pollution. Burkholder was born in 1953 in Rockford, Illinois. She earned a bachelor's degree from Iowa State University in 1975, a master's from the University of Rhode Island in 1981, and a Ph.D. from Michigan State University in 1986. Later that year, she joined the botany department of North Carolina State University in Raleigh as an assistant professor. She became an associate professor in 1992 and a full professor in 1997. She currently directs the

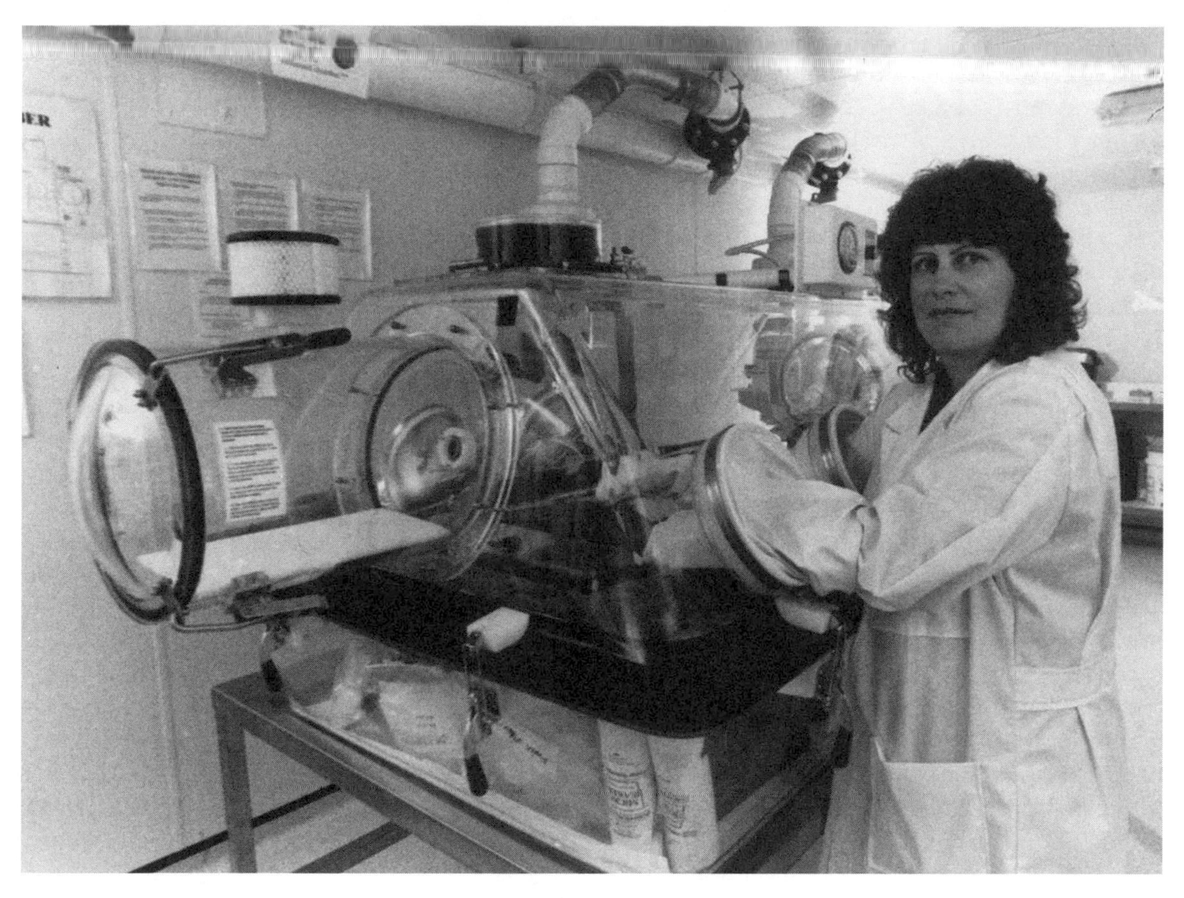

Mysterious deaths in a fish tank led JoAnn Marie Burkholder of North Carolina State University, Raleigh, to discover that a microorganism named *Pfiesteria* was the cause of massive fish kills in several states. *(News Service, NCSU)*

university's Center for Applied Aquatic Ecology. Her specialty is the interaction of living things in lakes, rivers, estuaries, and the seacoast. Beginning in the 1990s, she has concentrated on estuaries, places where fresh water from rivers and salt water from the sea mingle.

In 1989, Burkholder became curious when E. J. Noga, a coworker at the university, told her that fish in his laboratory were dying mysteriously. She learned that millions of fish at a time had also been dying in waters along the North Carolina coast. By 1991, she had identified a newly discovered type of microorganism that was killing fish in estuaries. With the help of K. A. Steidinger, she named it *Pfiesteria piscicida*—the fish killer. It is a dinoflagellate, a one-celled creature related to those that can cause poisonous "red tides."

Burkholder's research team showed that *Pfiesteria piscicida* could kill fish and shellfish, such as scallops, oysters, and crabs, in the laboratory. They linked it and a second newly discovered, closely related species of *Pfiesteria*, *P. shumwayae*, with the mass fish deaths in rivers and estuaries. When no fish are nearby, they discovered, these organisms usually lie quietly at the bottom of the waterways or float in the water, consuming other microbes and organic debris. If large schools of migrating menhaden or other fish enter the water, however, the *Pfiesteria* sense chemicals given off by the fishes' bodies and turn into predator forms that swim up to and attack the fish. Poisoned by the microorganisms' toxin, the fish begin to swim weakly and erratically. Within hours to a few days, they develop large, bleeding sores and often die. Once the supply of fish runs out, the fish-killers disappear into the river-bottom ooze once more or change back to their harmless, floating form.

During the early to mid-1990s, the North Carolina fish kills were a major ecological problem because they occurred almost every summer in an estuary system that is the most important fish nursery on the Atlantic coast. Furthermore, Burkholder's team discovered, *Pfiesteria* toxin can affect human health. Scientists exposed to water containing active *Pfiesteria* or air above water in which fish kills were in progress, including Burkholder herself, have suffered problems ranging from burning skin and difficulty in breathing to memory loss so severe that for several weeks they could not remember their own names or addresses. Fishers who handled infected fish and shellfish or people who swam in waters where *Pfiesteria* was active have reported similar signs of illness.

Burkholder's work became controversial in the early 1990s, when her group linked the *Pfiesteria* outbreaks to high amounts of nitrates, phosphates, and other nutrients in the water. These nutrients usually come from runoff (wastewater) from cities, hog or chicken ranches, or farms that use animal waste as fertilizer. Burkholder maintained that limiting such pollution was necessary to control *Pfiesteria*, but North Carolina water officials were slow to act because some supporters of development, the seafood industry, the swine industry, and scientists who received funding from them claimed that she had exaggerated the problem and tried to discredit her work. She even received anonymous death threats.

When a *Pfiesteria* outbreak in Chesapeake Bay attracted wide media attention and criticism in 1997, however, the Maryland state government responded very differently. In 1998, it approved what Burkholder calls "the strongest regulations for nonpoint pollution control in the nation." Since then, two national panels of scientists have verified Burkholder's conclusions, and both federal agencies and those in several states, including North Carolina, have set up programs to reduce nutrient pollution, detect *Pfiesteria* outbreaks quickly, and evaluate their effects on human health.

Burkholder has received several awards for her research, including an Admiral of the Chesapeake Award from Maryland's governor and the Scientific Freedom and Responsibility Award from the American Association for the Advancement of Science. Her research team has purified a potent *Pfiesteria* neurotoxin (nerve poison), and they are studying how the toxin affects fish and human health. They also continue to examine the effects of pollution on estuaries and freshwaters. Burkholder hopes that, as she wrote in the October 2001 issue of *Bioscience*, "the *Pfiesteria* issue has helped many people realize that water quality, fish health, and human health are strongly linked."

Further Reading

Barker, Rodney. *And the Waters Turned to Blood: The Ultimate Biological Threat*. New York: Touchstone Books, 1998.

Burkholder, JoAnn M., and Howard B. Glasgow. "History of Toxic *Pfiesteria* in North Carolina Estuaries from 1991 to the Present." *Bioscience*, October 2001.

Burnet, Frank Macfarlane

(1899–1985)
Australian
Virologist, Immunologist

Frank Macfarlane Burnet made discoveries about viruses and the workings of the immune system that earned a share of the 1960 Nobel

Prize in physiology or medicine. He was born in Traralgon, part of the Australian state of Victoria, on September 3, 1899, the second of Frank Burnet and the former Hadassah McKay's seven children. Burnet grew up in Terang, Victoria, where his father, a bank manager, was transferred in 1909. As a boy, he explored the countryside and collected beetles and other insects.

Burnet studied biology at Geelong College in Victoria, then obtained a scholarship to the University of Melbourne. He took his M.D. degree from the university in 1923 and later that year joined the Charles and Eliza Hall Institute, a research organization attached to the university and the Royal Melbourne Hospital. Except for two short stays in Britain, he remained with the institute for the rest of his working life, acting as its assistant director from 1934 to 1943 and its director from 1944 to 1965. Beginning in 1944, he was also a professor of experimental medicine at the University of Melbourne.

During the first of Burnet's British visits, from 1925 to 1927, he did research at the Lister Institute, part of the University of London. He earned a Ph.D. from that university in 1927. He also met his future wife, a fellow Australian named Linda Druse. They married in July 1928, soon after he went back to Australia. Burnet's second British sojourn took place in 1932 and 1933 at the National Institute of Medical Research in Hampstead.

Viruses were the center of Burnet's research for much of his career. At the beginning of that time, scientists knew very little about these microorganisms, partly because viruses were too small to see until electron microscopes were invented in the 1930s and partly because they could be grown only inside living cells. Burnet first studied bacteriophages, a group of viruses that infect bacteria. Then, while working at Hampstead, he invented a way to grow viruses in chicken embryos and the membranes surrounding them. This technique was much easier than previous methods of growing viruses and became a standard laboratory procedure for more than 20 years.

Burnet investigated numerous viruses that cause human and animal diseases, but his most extensive work was on the virus that causes influenza. During World War II, he tried unsuccessfully to develop a vaccine against the disease, and in the early 1950s he studied the virus's genetics. He learned that influenza virus genes mutate more often than those of most other viruses. Such changes make both natural immunity, acquired after suffering the disease, and immunity from vaccines useless within a few years because the alterations prevent the immune system from recognizing the virus, just as people might no longer recognize a criminal from a "wanted" poster if the criminal cut or dyed his hair. Burnet also showed that different strains of influenza virus can combine within cells, giving rise to new forms of the virus.

Burnet's studies of viruses led him to investigate the immune system, which defends the body against viruses and other invaders. He began studying immunity in the late 1940s, and in 1957 he stopped working on viruses and focused on the immune system full time. Scientists knew that cells in the system identify certain molecules on the surface of invaders, called antigens, and then form antibodies, other molecules that attach themselves to the antigens and signal different cells in the immune system to destroy whatever the antibodies cling to. Each antigen and its matching antibody are different from all others; the two fit together like two pieces of a puzzle.

Body cells also carry antigens, so Burnet wondered why the immune system does not form antibodies against these as well. He concluded that the immune system begins functioning fairly late in the process of development before birth, and all antigens present in the body at this time are identified as "self" and do not trigger the system to produce antibodies. Antigens that enter the body after that time, on the other hand, are considered "foreign" and will be attacked, whether they are on microbes or on tissues from other living things.

Burnet predicted that if tissue from another animal was grafted onto an embryo before its immune system became active, the embryo's immune system would accept, or tolerate, the antigens on this tissue as if it were part of the embryo's own body. When an embryo that had received such a transplant developed into an adult, it should be able to receive a second transplant from the same animal without attacking it. This phenomenon came to be known as acquired immunological tolerance.

In 1957, Burnet also proposed a theory to explain how the immune system could form many kinds of antibodies and how it knew which ones to make in large quantities to repel invaders. The genes of certain immune system cells, he said, are constantly changing, causing the cells and their offspring to produce millions of different kinds of antibodies. Normally, only one or a few cells make antibodies that fit any particular antigen. If an antibody locks onto its matching antigen, however—when a microbe carrying that antigen invades the body, for example—the cell that produced that antibody is stimulated to divide rapidly, making many copies, or clones, of itself, all able to generate the same kind of antibody. Burnet described his ideas about antibody production in *The Clonal Selection Theory of Acquired Immunity* (1959). He believed that this theory was his greatest contribution to science.

Burnet's research on immunity was the chief reason for his being awarded the 1960 Nobel Prize. He shared the prize with PETER BRIAN MEDAWAR, whose experiments provided proof for Burnet's two theories about the immune system. Burnet's ideas helped to explain how and why the system attacks transplanted organs and tissues, and later researchers built on these ideas to find ways to blunt the attacks, making organ transplants possible.

In the last years of his research career, Burnet studied conditions in which the immune system's normal tolerance fails and it attacks healthy body cells, causing serious illnesses such as arthritis. He retired from the Hall Institute in 1965, but for 12 years afterward he remained at the University of Melbourne as an emeritus professor. During this time he wrote 13 books, including his autobiography, *Changing Patterns* (1968), and discussions of such topics as aging and cancer.

Burnet won many awards in addition to the Nobel Prize, including the American Public Health Association's Albert Lasker Medical Research Award (1952), the EMIL VON BEHRING Prize (Germany's highest scientific award, 1954), and the Royal and Copley Medals of Britain's Royal Society (1947 and 1959). He was knighted in 1951, given a higher title (Knight Commander of the British Empire) in 1969, and awarded Australia's highest award, the title of Knight of Australia, in 1978.

Macfarlane Burnet died of cancer on August 31, 1985. He was survived by the son and two daughters he had had by his first wife, who had died in 1973; by his second wife, the former Hazel Jenkin, whom he had married in 1976; and by eight grandchildren. In a biographical sketch, coworker Frank Fenner called Burnet "Australia's greatest biologist."

Further Reading

Burnet, Frank Macfarlane. *Changing Patterns*. Melbourne: Heinemann, 1968.

Fenner, Frank. "Frank Macfarlane Burnet, 1899–1985." *Historical Records of Australian Science*, 7:1, 1987. Available online. URL: www.asap.unimelb.edu.au/bsparcs/aasmemoirs/burnet.htm. Updated 1998.

Sexton, Christopher. *The Seeds of Time: The Life of Sir Macfarlane Burnet*. New York: Oxford University Press, 1991.

⊠ Calvin, Melvin
(1911–1997)
American
Biochemist

Melvin Calvin worked out the steps in photosynthesis, the process by which green plants use energy from sunlight to change carbon dioxide and water into carbohydrates and oxygen. Almost all living things depend directly or indirectly on photosynthesis as a source of food (fuel) and oxygen.

Calvin was born on April 8, 1911, in St. Paul, Minnesota, the son of Russian immigrants Elias and Rose (Hervitz) Calvin. He attended the Michigan College of Mining and Technology, graduating in 1931 with a B.S. in chemistry. He earned a Ph.D. from the University of Minnesota in 1935. After two years of postdoctoral work at the University of Manchester in England, he returned to the United States and joined the faculty of the University of California at Berkeley in 1937. He remained there until his retirement in 1980, becoming an assistant professor in 1941, an associate professor in 1945, and a full professor in 1947.

Calvin was one of the first chemists to join Berkeley's Lawrence Radiation Laboratory, where at first he worked under the laboratory's founder, famed physicist Ernest O. Lawrence. Calvin became the director of the laboratory's bio-organic chemistry group in 1946, a post he held for 20 years. This group became the Laboratory of Chemical Biodynamics in 1960 and, later, the Structural Biology Division. On Calvin's retirement, the building that houses this division was named the Melvin Calvin Laboratory.

Calvin's early research did not focus on biology. During World War II, for instance, he carried out experiments connected to the top-secret Manhattan Project, which developed the atomic bomb. This work acquainted him with radioactive isotopes, forms of certain atoms that give off measurable radiation. Carbon, a key element in chemicals that make up living things, has a radioactive form called carbon 14. One story claims that, after the world war ended in 1945, Lawrence told Calvin, "Now is the time to do something useful with radioactive carbon."

Calvin had a proposal for what that might be. While studying in England, he had become interested in photosynthesis, about which little was known because the reactions involved in it take place very quickly and occur only in living cells. He reasoned that if he made carbon dioxide, one of the process's starting materials, radioactive with carbon 14, he might be able to track the radioactive atoms through different molecules in a plant as it carried on photosynthesis.

Around 1950, Calvin and a team of researchers at the Lawrence laboratories began exposing colonies of a single-celled green alga (water plant) called *Chlorella* to radioactive carbon dioxide for a few seconds. They killed the plants quickly at different times after the exposure and used chemical tests to identify the compounds into which the radioactive carbon atoms had been incorporated. In this way, Calvin worked out most of the steps in photosynthesis by 1957. Part of the process, a repeating chain of reactions, was named the Calvin cycle in his honor. Calvin described this research in *The Path of Carbon in Photosynthesis*, published in 1957, and *The Photosynthesis of Carbon Compounds* (with J. A. Bassham), published in 1962. His work won the Nobel Prize in chemistry in 1961. During the later 1960s, he attempted to recreate photosynthesis artificially in his laboratory.

In the 1970s and 1980s, Calvin investigated photosynthesis and plants as possible sources of energy for human use. He was involved with programs to obtain alcohol from waste plant matter and use it as automobile fuel, with artificial processes using solar energy that were modeled on photosynthesis, and with extraction of oil from certain kinds of plants. The U.S. Department of Energy started a research program on solar energy at his urging.

In addition to the Nobel Prize, honors Calvin received for his work included the National Medal of Science (1989), the Davy Medal from Britain's Royal Society, and the Priestley Medal from the American Chemical Society. Calvin was married to Marie Genevieve Jemtegaard, and they had two daughters and a son. Calvin died on January 8, 1997, at the age of 85.

Further Reading

Calvin, Melvin. *Following the Trail of Light: A Scientific Odyssey*. Washington, D.C.: American Chemical Society, 1992.

"Calvin, Melvin." *Current Biography Yearbook 1962*. New York: H. W. Wilson, 1962.

Yarris, Lynn. "In Memoriam: Melvin Calvin, 1911–1997: A Man of Exceptional Curiosity." *U.C. Berkeley Research Review* 37 (1997). Available online. URL: www.lbl.gov/Science-Articles/Research-Review/Magazine/1997/story12.html. Accessed 2003.

Carrel, Alexis
(1873–1944)
French/American
Surgeon

The techniques that Alexis Carrel developed for surgically joining blood vessels together and for keeping organs and tissues alive in the laboratory laid the groundwork for organ transplants. He was born in Sainte-Foy-les-Lyon, France, on June 28, 1873. His father, Alexis Carrel-Billiard, a cloth manufacturer, died when Alexis was only five years old, and his mother, Anne-Marie, had to earn money to care for him and his two younger siblings by doing embroidery for hire.

Carrel studied medicine at the University of Lyons, obtaining his M.D. in 1900, and decided to become a surgeon. While still a medical student, perhaps remembering the delicate sewing his mother had done, he asked a famous embroiderer in the city to teach him her craft. He then used this skill in developing techniques to reconnect blood vessels in ways that would reduce the risk of blood clots, one of the chief problems that arose after surgery. For instance, he folded back the ends of the vessels like cuffs before sewing them so that blood passing through the vessels would not be exposed to the rough surface produced by the stitches and form clots there. He described some of his methods in a journal article in 1902.

Frustrated by the Lyons professors' lack of interest in his advances, Carrel moved to Canada in 1904, planning to "forget medicine and raise cattle." He nonetheless gave a speech at a medical meeting soon after his arrival, and the chairman of the physiology department of the University of Chicago was so impressed with this talk that he offered the young man a job. Carrel accepted and remained at the university until 1906, when he was invited to join the newly formed Rockefeller Institute for Medical Research (later Rockefeller University), a prestigious research institute in New York City. Carrel became a full member of the institute in 1912 and, except around the time of World War I, remained there until his retirement.

At the University of Chicago and during his early years at the Rockefeller Institute, Carrel continued to improve his techniques for sewing blood vessels. His methods allowed surgeons to perform delicate operations that had not been possible before and paved the way for later advances such as reattachment of severed limbs. Carrel received the Nobel Prize in physiology or medicine in 1912 primarily for this work. He was the first scientist working in the United States to receive this prize.

Carrel and a coworker at Rockefeller, Charles Guthrie, used Carrel's techniques to transplant kidneys and other organs in laboratory dogs, proving that such transplants were possible from a surgical point of view. No matter how carefully he operated, however, Carrel found that organs transplanted from one animal to another usually withered and died after a short time. Later scientists such as PETER BRIAN MEDAWAR learned that this happened because the immune system attacks the transplanted tissue.

Carrel also developed ways to preserve tissues outside the body for study or transplantation. In 1910, he showed that blood vessels could be kept in cold storage for long periods before being used in transplants. His tissue culture methods kept part of the heart of a chicken embryo, which he first extracted in 1912, alive and growing in a laboratory flask for more than 30 years. Every January 17, the anniversary of the extraction, Carrel and some of the institute staff sang "Happy Birthday" to their chicken heart.

On a visit to France in 1913, Carrel married Anne-Marie de la Motte de la Meyrie, a widow with one son. World War I broke out the following year, and Carrel remained in France while it lasted, directing an army hospital while his wife worked as a surgical nurse. Carrel and Henry Dakin, an American chemist, invented a disinfectant solution for washing out deep battle wounds, for which the French government awarded them the Legion of Honor. Carrel returned to the United States in 1919.

In 1930, Charles Lindbergh, who had become world famous after making the first solo flight across the Atlantic in 1927, visited Carrel. Lindbergh's sister-in-law needed an operation to repair her heart, but he had been told that the operation was impossible because the heart could not be stopped long enough for the surgery to be performed. He hoped that Carrel, with his surgical skill and experience in transplantation and organ preservation, could change this gloomy picture. "Knowing nothing about the surgical problems involved," Lindbergh wrote later, "it seemed to me it would be quite simple to design a mechanical pump capable of circulating blood through a body during the short period required for an operation."

Carrel told Lindbergh that he had tried to build such a pump, but it had failed to work in living animals because it always became contaminated with dangerous microbes. He believed, however, that a version of it might be able to keep an organ alive outside the body. Lindbergh joined Carrel's laboratory and designed improved pumps, which Carrel tested on animals. In 1934, one of their pumps allowed circulation to be cut off from vital organs for up to two hours without producing permanent damage, and another kept a cat's thyroid gland alive for 18 days in 1935. Although these pumps were never used in humans, they were forerunners of the heart-lung machine, invented in the 1950s and used in surgery much as Lindbergh had imagined, and of artificial hearts, first used in the early 1980s. Carrel and Lindbergh described their pumps in a 1938 book, *The Culture of Organs*.

Although Carrel was hailed for his surgical advances, he had social ideas that were less popular. In a 1935 book called *Man the Unknown*, he proposed that an intellectual elite should rule society and that only the strongest and most intelligent people should be allowed to have children. This latter doctrine, called eugenics, was also popular in Nazi Germany. Criticism of Carrel increased when he returned to France after his retirement from the Rockefeller Institute in 1939 and established a research institution, the Foundation for the Study of Human Problems, with the help of the Vichy government, a puppet government that Germany set up after it conquered France in June 1940. Carrel died of a heart attack in Paris on November 5, 1944, just as the imposed government was losing power.

Although Alexis Carrel's personality and beliefs were controversial, he was widely honored for his work. He received the Nordhoff-Jung Cancer Prize in 1931 and the Newman Foundation Award of the University of Illinois in 1937, as well as the Nobel Prize and decorations from the governments of France and Belgium. Charles Lindbergh, who had become a close friend of Carrel's during their work together, said that Carrel had "one of the most brilliant, penetrating and versatile minds I have ever met."

Further Reading

Edwards, W. Sterling. *Alexis Carrel, Visionary Surgeon*. Springfield, Ill.: Charles C. Thomas, 1974.

"The Embroiderer Nobel Laureate." Who Named It? Available online. URL: www.charleslindbergh.com/heart/index3.asp. Accessed 2001.

Malinin, Theodore. *Surgery and Life: The Extraordinary Career of Alexis Carrel*. New York: Harcourt Brace Jovanovich, 1979.

Carson, Rachel Louise

(1907–1964)
American
Marine Biologist, Ecologist

Rachel Carson combined a professional knowledge of science with poetic language to write best-selling books about the sea and a famous warning that, unless human exploitation of the environment was curbed, much of nature might be destroyed. She was born in Springdale, western Pennsylvania, on May 27, 1907. Her father, Robert, sold insurance and real estate. Her mother, Maria, taught her to love nature.

When Carson entered Pennsylvania College for Women (later Chatham College), she planned to become a writer (her writing had first been published, in a children's magazine, when she was 10 years old). A biology class from an inspired teacher, however, made her change her major to zoology. After graduating in 1929, she did research at Johns Hopkins University in Baltimore, Maryland, and the Marine Biological Laboratory in Woods Hole, Massachusetts. She obtained a master's degree in zoology from Johns Hopkins in 1932.

Carson taught part time at Johns Hopkins and the University of Maryland for several years. Then, needing more money to support her mother and other relatives, she joined the U.S. Bureau of Fisheries as an aquatic biologist in 1936. Her supervisor assigned her to write radio scripts about marine biology and encouraged her to submit a version of one of them to the *Atlantic Monthly*. This article, "Undersea," appeared in the magazine's September 1937 issue. An editor at the publishing house of Simon & Schuster asked her to expand the piece into a book, and the result, *Under the Sea-Wind*, was published in 1941. Carson later said that this book was her favorite among all her work, but it sold poorly.

Carson became editor in chief of the publications division of the U.S. Fish and Wildlife Service in 1947. (This agency had been created in 1940 by a merger of the Bureau of Fisheries and the Biological Survey.) Soon afterward, she began a second book, *The Sea Around Us*, which described the physical nature of the oceans. Published in 1951, it became an immediate best-seller (readers wrote that it "overwhelmed [them] with a sense of the vastness of the sea" and "reduce[d] . . . man-made problems to their proper proportions") and received awards, including the National Book Award and the John Burroughs Medal. Now able to support herself by her writing, Carson left the Fish and Wildlife Service in 1952. Her third book, *The Edge of the Sea* (1955), described shore life and proved almost as successful as *The Sea Around Us*.

The book that gave Rachel Carson her place in history, however, was none of these. It grew out of an urgent letter that a friend, Olga Huckins, sent her in 1957 after a plane sprayed clouds of the pesticide DDT over the bird sanctuary that Huckins and her husband owned near

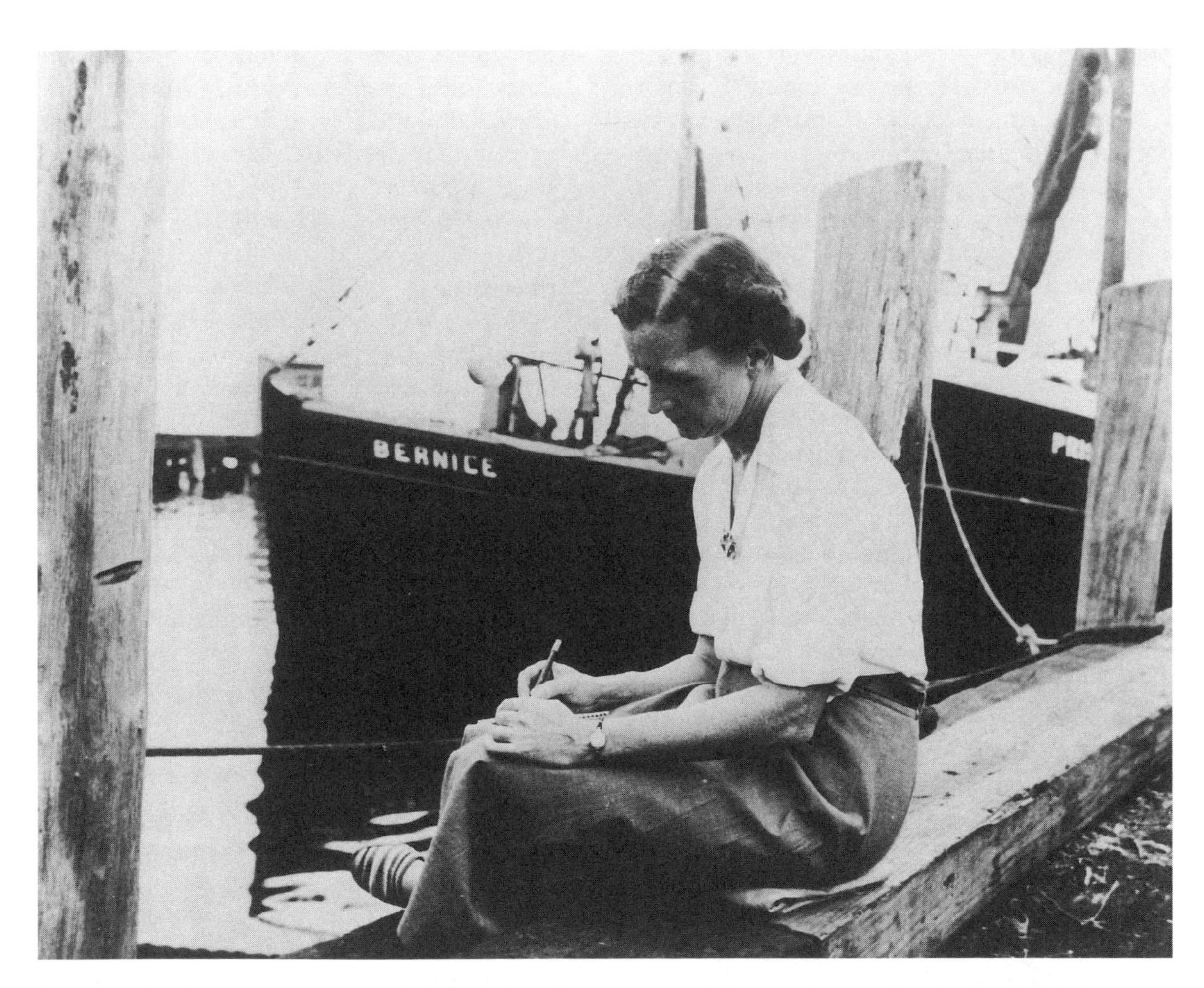

Rachel Louise Carson was a marine biologist before she became world famous for revealing the dangers that pesticides presented to the environment in *Silent Spring*, published in 1962. *(Yale Collection of American Literature, Beinecke Rare Books and Manuscripts Library, Yale University)*

Duxbury, Massachusetts. The morning after the plane passed over, Huckins found the bodies of seven songbirds that had "died horribly . . . their splayed claws . . . drawn up to their breasts in agony." Huckins asked Carson's help in alerting the public to the dangers of pesticides.

Carson had been concerned about these widely used chemicals for more than a decade, and she now began to research their effects in earnest. "The more I learned about pesticides, the more appalled I became," she wrote later. "Everything which meant most to me as a naturalist was being threatened." She came to believe that these compounds were doing terrible damage to wildlife and perhaps to humans as well, and she spent four years amassing scientific data to support her contentions.

Silent Spring, the book that resulted from Carson's research, appeared in 1962. It took its title from a "fable" at the book's beginning, which pictured a season that was silent because pesticides had destroyed singing birds and much other wildlife. The health of the human beings in this scenario was imperiled as well. The book stressed that pesticides were just one example of humans' abuse of nature. People failed to understand that all things in nature, including human beings, are interconnected, Carson wrote; damage to one therefore means damage to all. Carson used the word *ecology*, from a Greek word meaning "household," to describe this relatedness. She said that people needed to respect and work with nature rather than trying to conquer it.

Supreme Court Justice William O. Douglas called *Silent Spring* "the most revolutionary book since *Uncle Tom's Cabin*." Publicity sponsored by the powerful pesticide industry pictured Carson as an overemotional woman with no scientific background, ignoring her M.S. degree and years as a working biologist. The industry claimed that if Carson's supposed demand to ban all pesticides—a demand she never actually made—were followed, the country would plunge into a new Dark Age because pest insects would devour its food supplies and cause epidemics of disease. Some critics labeled her a Communist. Many scientists took Carson's side, however. For example, a 1963 report by a panel from President John F. Kennedy's Science Advisory Committee supported most of her conclusions.

Even while she finished the exhausting research and writing of *Silent Spring* and battled her critics after its publication, Rachel Carson was suffering from breast cancer. She died of the disease in Silver Spring, Maryland, on April 14, 1964. The trend she started, however, did not die. Chiefly because of it, a new federal department called the Environmental Protection Agency (EPA) was created in 1970, and DDT was banned in the United States in 1972. Most important, Carson's book reshaped the way the American public viewed nature. Today's environmental movement is her legacy.

Further Reading

Carson, Rachel. *Silent Spring*. Boston: Houghton Mifflin, 1962.

Glimm, Adele. *Rachel Carson: Protecting Our Earth*. New York: McGraw-Hill, 2000.

Lear, Linda J. *Rachel Carson: Witness for Nature*. New York: Holt, 1997.

Matthiessen, Peter. "Environmentalist: Rachel Carson." *Time:* 100 Scientists Who Changed the World. Available online. URL: www.time.com/time/time100/scientist/profile/carson.html. Accessed 2001.

Watson, Bruce. "Sounding the Alarm." *Smithsonian*, September 2002.

Chain, Ernst Boris

(1906–1979)
German/British
Biochemist

With HOWARD WALTER FLOREY and Norman Heatley, Ernst Chain turned ALEXANDER FLEMING's accidental discovery of a fungus (mold) that killed bacteria into the lifesaving antibiotic penicillin. Chain was born in Berlin on June 19, 1906. His father, Michael Chain, had emigrated to Germany from Russia. Michael Chain married a native German, Margarete Eisner, and became a well-to-do chemical manufacturer.

A talented pianist, Ernst Chain was still considering a career in music when he earned a Ph.D. in chemistry and physiology from Friedrich-Wilhelm University in Berlin in 1930. He began working for Berlin University's Charité Hospital, however, and decided to become a biochemist.

Chain left Germany soon after Adolf Hitler's Nazi Party seized control of the country's government in 1933. He knew that, as a Jew with Russian ancestry and leftist political views, he had little hope of career advancement under the Nazis' anti-Semitic and anti-Communist regime. He chose more wisely than he knew: His mother later died in a Nazi concentration camp, and his sister vanished and probably was also killed.

The displaced biochemist settled in England, becoming a British citizen in 1939. After a brief stay at the University of London, Chain joined Cambridge University. In 1935, his mentor at Cambridge, Frederick Hopkins, put him in touch with Florey, an Australian bacteriologist who had just moved to Oxford University and was looking for a biochemist to add to his research team. Florey was studying antibiosis, the process by which living things produce substances that kill others, such as bacteria. Chain agreed to become part of Florey's group.

For the next three years, Florey's team studied lysozyme, a bacteria-killing substance that Alexander Fleming had found in tears, mucus, and egg white. Chain purified this chemical and showed how it dissolved the outer walls of bacterial cells. While doing this research, he came across a 1928 article in which Fleming reported that a blue bread mold named *Penicillium notatum* made a substance that appeared able to stop the growth of bacteria. Chain told Florey about the article, and in 1938 the two decided to study the mold compound, which Fleming had named penicillin. With war looming on the horizon, they realized that a chemical that kept bacteria from causing infections in wounds could be immensely valuable.

The researchers' first task was to find a way to grow large amounts of *Penicillium* in their laboratory. Norman Heatley, another biochemist on Florey's research team, accomplished this. Chain then set about extracting and purifying the mold's germ-killing substance. Using freeze-drying, he finally succeeded around 1939. The group tested their tiny supply of the drug on mice and showed that it could stop certain bacterial infections and was fairly nontoxic. Then, in early 1941, they gave penicillin to 10 patients dying of infections caused by a common bacterium called staphylococcus. The compound saved eight of them.

World War II had begun by this time, and the British chemical industry was too involved in production of war-related materials to consider manufacturing a new product. Florey and Heatley, therefore, took their discovery to the United States. They persuaded the American government and pharmaceutical industry to mass-produce penicillin, and the drug began to be given to Allied soldiers in 1943. Chain, meanwhile, remained at Oxford and worked on determining the structure of the penicillin molecule. He proposed a structure in 1943, but the form he suggested was so unusual that many scientists refused to accept it until DOROTHY CROWFOOT HODGKIN confirmed it by X-ray crystallography in 1946. Chain, Florey, and Fleming shared the Nobel Prize in physiology or medicine in 1945 for their development of this lifesaving drug.

In 1949, the Italian State Institute of Public Health offered Chain an opportunity to organize and head its new International Research Center for Chemical Microbiology. Chain, bitter about what he felt was a lack of proper recognition at Oxford, accepted and moved to Rome along with his wife, fellow biochemist Anne Beloff, whom he had married late in 1948. (The couple later had three children.) In Italy, Chain produced and tested many types, or strains, of *Penicillium* in the hope of finding new versions of penicillin that would destroy bacteria resistant to the standard kind. He discovered one that, unlike the original drug, was not destroyed by acid in the stomach and therefore could be taken as a pill.

Chain returned to Britain in 1961 to direct the Wolfson Laboratories, part of the Imperial College of Science and Technology at the University of London, and to head the college's biochemistry department. He continued in these posts until his retirement in 1973, after which he did research as an emeritus professor until 1976.

In addition to penicillin, Chain studied subjects ranging from snake venoms to cancer biology during his long career. Awards he received, besides the Nobel Prize, included the Berzelius Medal of the Swedish Society of Physicians (1946), the Pasteur Medal from the Pasteur Institute in Paris, the Paul Ehrlich Centenary Prize (1954), and the Marotta Medal of the Italian Chemical Society (1962). He was also knighted in 1969. Chain died of heart failure in Ireland on August 12, 1979.

Further Reading

"Chain, Ernst Boris." *Current Biography Yearbook 1965*. New York: H. W. Wilson, 1965.

Clark, Ronald William. *The Life of Ernst Chain: Penicillin and Beyond*. London: Weidenfeld & Nicolson, 1985.

Cohen, Stanley

(1922–)
American
Biochemist

Stanley Cohen helped to work out the chemical identity of nerve growth factor, a substance discovered by RITA LEVI-MONTALCINI. He himself discovered epidermal growth factor, which affects skin cells.

Cohen was born on November 17, 1922, in Brooklyn, New York, to Russian Jewish immigrants Louis Cohen, a tailor, and his wife, Fanny. Brooklyn College's lack of tuition fees for residents made a college education possible for Stanley. After his graduation in 1943, he won a fellowship to Oberlin College in Ohio, from which he earned a master's degree in zoology in 1945. He took a Ph.D. in biochemistry from the University of Michigan in 1948.

Cohen did postdoctoral work at the University of Colorado Medical School in Denver until 1952, when he went to Washington University in St. Louis, Missouri. In 1954, Cohen joined Levi-Montalcini's laboratory at the university to help the Italian-born scientist determine the chemical nature of a mysterious substance she had found in mouse cancers. Because this substance made nerve fibers in chick embryos grow vigorously, Levi-Montalcini called it nerve growth factor, or NGF. The two scientists found that their skills and personalities complemented each other perfectly. "You and I [separately] are good," Cohen once told her, "but together we are wonderful."

Cohen thought that NGF was probably either a nucleic acid or a protein. To distinguish between these two possibilities, he treated a sample of NGF with a type of snake venom that destroys nucleic acids but does not affect proteins. The snake venom did not change NGF's effect on nerve growth, so Cohen concluded that the substance was probably a protein. At the same time, he was surprised to notice that the venom by itself stimulated nerve growth just as NGF did. He knew that poisonous snakes' venom comes from glands similar to those that make saliva in mammals, so, on a hunch, he tested the salivary glands of mice for substances that made nerves grow. He found that the glands of adult male mice were a rich source of NGF. Studying the compound became much easier once a good supply of it was available.

After failing to obtain a professorship as a biochemist in the zoology department, Cohen left Washington University in 1959 and joined the biochemistry department at Vanderbilt University School of Medicine in Nashville, Tennessee. He has remained there ever since. An assistant professor at first, he became a full professor in 1976 and a distinguished professor in 1986.

Cohen believed that NGF was just one of a family of growth factors that stimulate development of tissues and organs before birth and, under some conditions, in later life as well. In his work at Washington University, he had noticed that when salivary gland extract containing NGF was injected into newborn mice, the eyelids of the babies, which are closed at birth, opened sooner than they usually would. The infants' teeth also grew in ahead of schedule. Purified NGF, however, did not produce these effects. Cohen therefore suspected that the extract contained a second growth factor. After he moved to Vanderbilt University, he searched for this new growth factor, which he finished purifying in 1962. He named it epidermal growth factor, or EGF, because its chief effects were on cells in the outer layer of skin, or epidermis. He worked out EGF's chemical composition in the early 1970s.

Cohen has also made important discoveries about how growth factors work. He and others have shown that these chemicals must attach to molecules on the surface of cells, called receptors, before they can affect the cells. Different types of cells have receptors for different growth factors. Once a molecule of growth factor attaches to a receptor molecule, the pair is carried inside the cell and activates certain chemicals there, resulting in cell growth.

Cohen and Levi-Montalcini shared the 1986 Nobel Prize in physiology or medicine for their work on NGF. Cohen has also received other awards, some of which were likewise shared. These awards include the Louisa Gross Horwitz Prize of Columbia University (1983), the National Medal of Science (1986), and the Albert Lasker Medical Research Award (1986). Cohen retired in 2000 and is now a distinguished emeritus professor.

Further Reading

Cohen, Stanley. "Autobiography." *Les Prix Nobel 1986*. Available online. URL: www.nobel.se/medicine/laureates/1986/cohen-autobio.html. Last updated 2002.

Marx, Jean L. "The 1986 Nobel Prize for Physiology or Medicine." *Science*, October 31, 1986.

Cohen, Stanley N.

(1935–)
American
Geneticist

With HERBERT WAYNE BOYER, Stanley Norman Cohen performed the first transfer of genes from one kind of living thing to another and proved that the genes could function in their new location. Their discoveries showed that techniques of "gene splicing," or genetic engineering, could be used to change genes in living organisms.

Stanley Cohen was born in Perth Amboy, New Jersey, on February 17, 1935. He attended Rutgers University, graduating in 1956, and earned an M.D. degree from the University of Pennsylvania School of Medicine in 1960. He joined the faculty of the school of medicine at Stanford University, in northern California, in 1968. He is currently the Kwoh-Ting Li Professor of Genetics and professor of medicine at the university. His hobbies include sailing, skiing, and playing the five-string banjo.

At Stanford, Cohen began studying the genetics of bacteria. In addition to their main genome, some bacteria contain small, ring-shaped pieces of genetic material called plasmids. Each plasmid carries just one or a few genes. Bacteria in nature sometimes exchange plasmids, and Cohen developed a way to imitate this process, removing

Stanley N. Cohen of Stanford University, shown here, and Herbert Wayne Boyer of the University of California, San Francisco, made genetic engineering practical when they transferred genetic material from one kind of living thing into the genome of another in 1973 and showed that the material could produce a protein in its new location. *(Stanford University)*

plasmids from bacteria and making other bacteria take them up on demand.

When Cohen attended a scientific meeting in Honolulu, Hawaii, in November 1972, he heard Herbert Boyer, a researcher at the University of California at San Francisco, give a talk about his own research on bacteria, which focused on substances called restriction enzymes. Some bacteria use these chemicals to protect themselves against attacking viruses. Restriction enzymes act like molecular scissors, snipping the viruses' genetic material into pieces so that the viruses cannot reproduce. The enzymes cut DNA at any point where they find a certain sequence of bases within the DNA molecule.

During an evening walk that ended in a chat over corned beef sandwiches at a nearby delicatessen, Cohen and Boyer discussed their research. They found that their scientific interests fitted together as neatly as the pairs of bases in a DNA molecule. Each had expertise that the other could use.

Boyer told Cohen that he had learned that his molecular scissors did not snip cleanly. When they sliced through a double-stranded DNA molecule, they left a short single strand of bases at each end of the cut piece. The bases on these single strands attracted and attached to bases that made up the other half of their natural pairs, just as they do when the strands of DNA split apart and duplicate their missing halves before a cell divides. Any two pieces of DNA cut with the same restriction enzyme had single-stranded ends with the same sequences. In theory, therefore, two such pieces could be joined together, even if they came from different kinds of living things. Indeed, PAUL BERG, in a laboratory two floors above Cohen's at Stanford, was already using a different technique to join DNA from two kinds of microbes by means of single-stranded pieces.

Back in California in the spring of 1973, Boyer and Cohen began experiments that combined the techniques they had developed. They used a restriction enzyme from Boyer's laboratory to cut open some of Cohen's plasmids. The "sticky ends" produced by the process allowed them to join the two plasmids together to make a single larger one, even though the plasmids had come from two different strains of bacteria. They then used Cohen's technique to make other bacteria take up the new plasmids. They showed that the bacteria copied the plasmids along with their own genes each time they reproduced. Furthermore, they proved that the genes in the combined plasmid, which made bacteria resistant to two types of antibiotic functioned in their new location. Their research demonstrated that DNA fragments from other sources could be copied, or cloned, by linking them to a plasmid.

Cohen and Boyer went on to transfer plasmids from one species of bacteria to another and then, more daringly, to insert genes from a toad into bacteria. Their experiments provided the first proof that "gene splicing" could be used dependably in living things. After hearing an account of their work in June 1973, one researcher exclaimed, "Now we can put together any DNA we want to."

Boyer went on to pioneer commercial uses of genetic engineering, while Cohen remained at Stanford and concentrated on the new technology's application to research. Both men took advantage of the fact that bacteria multiply very rapidly, creating millions of identical individuals in a single day. Boyer used this fact to produce vats of bacteria that made substances specified by inserted genes, and Cohen applied the same technique to make multiple copies of particular genes that he or other scientists wanted to study.

Boyer and Cohen shared many awards for their work, including the Albert Lasker Medical Research Award (1980), the Swiss Helmut Horten Research Award (1993), and the Lemelson-MIT Prize for inventors (1996). Cohen also received the National Medal of Science, the National Medal of Technology, and the Wolf Prize in Medicine. He is a member of the U.S. National Academy of Sciences and was inducted into the National Inventors Hall of Fame in 2001. When awarding Boyer and Cohen the Lemelson-MIT Prize, Charles M. Vest, the president of MIT (Massachusetts Institute of Technology), said, "Boyer and Cohen's ingenuity has revolutionized the way all of us live our lives."

Further Reading

Yount, Lisa. *Genetics and Genetic Engineering.* New York: Facts On File, 1997.

Colborn, Theo E.

(1927–)
American
Zoologist, Ecologist

Like RACHEL LOUISE CARSON, Theodora E. Colborn has sounded a warning about poisons in our environment. Born on March 28, 1927, she studied pharmacy at Rutgers University in New Jersey, graduating in 1947. She worked as a pharmacist in New Jersey and then, starting in 1964, as a sheep rancher in Colorado. She also raised four children. Her love of nature, especially

In the 1990s, Theo Colborn warned that pollutants that affect hormones threaten animal and human health. *(World Wildlife Fund)*

birds, led her to become involved with the environmental movement. At age 51, she went back to college. She earned a Ph.D. in zoology from the University of Wisconsin, Madison, in 1985.

In 1987, Colborn went to work for the Conservation Foundation in Washington, D.C. While coauthoring a book about the condition of the Great Lakes, she reviewed numerous scientific papers on the health of wildlife and people in the region and discovered that 16 kinds of animals that ate fish from the lakes were having problems reproducing. Often, the adult animals appeared healthy, but they either bore deformed or sickly young that did not live long or else had no young at all. Some animals had reproductive organs that were part male and part female.

Colborn became convinced that substances in the lake water, including pesticides such as dieldrin and DDT and other pollutants such as dioxin and PCBs, were somehow derailing the development of young animals. After further research, she suggested that these and other pollutants might cause the problems by imitating or modifying the action of hormones. Hormones control many body processes, including reproduction and development before birth.

In 1991, Colborn set up a meeting of scientists to discuss the possible dangers of pollutants that affect hormones. The scientists discovered that hormone-related abnormalities in both animals and humans were being reported all over the world. Since then, researchers have found more than 500 types of chemicals in pesticides, plastics, cosmetics, toys, food and drink containers, and other common products that act like or interfere with hormones.

Colborn, now a senior researcher with the World Wildlife Fund and director of the organization's wildlife and contaminants program, continues to push for investigation and limitation of these pollutants. She emphasizes that exposure to even tiny doses of hormone-altering pollutants before or shortly after birth damages the developing brain and immune system as well as the reproductive system in humans, producing loss of intelligence, abnormal behavior, physical abnormalities, and lifelong harm to health. Some scientists question whether low doses of hormone-mimicking substances are as dangerous as Colborn and her supporters say they are, but her research has been supported by many scientific studies and has won several awards, including the Norwegian International Rachel Carson Prize (1999) and the Japanese International Blue Planet Prize (2000).

Further Reading

Colborn, Theo, Dianne Dumanoski, and John Peterson Myers. *Our Stolen Future: Are We Threatening Our Fertility, Intelligence, and Survival?* New York: Dutton Signet/Penguin, 1996.

Snell, Marilyn Berlin. "Theo Colborn." *Mother Jones*, March–April 1998.

Van Gelder, Sarah. "Toxic Legacy." Futurenet. Available online. URL: www.futurenet.org/6RxforEarth/ColbornInterview.htm. Accessed 2001.

Cori, Carl Ferdinand
(1896–1984)
Cori, Gerty Theresa Radnitz
(1896–1957)
Austro-Hungarian/American
Biochemists

The husband-and-wife research team of Carl and Gerty Cori worked out steps in the processes by which the bodies of living things store, use, and recycle the energy that they get from food. The Coris shared in the Nobel Prize in physiology or medicine in 1947.

It was no surprise that Carl Cori developed an interest in biology, since he was the son of a zoology professor. He was born in Prague, then part of the empire of Austria-Hungary and later the capital of Czechoslovakia and the Czech Republic, on December 5, 1896. He and his sisters grew up in Trieste, now part of Italy, where his father, Carl Isidore Cori, directed the Marine Biological Station.

Gerty Theresa Radnitz was also born in Prague, on August 15, 1896, just a few months before her future husband. Her father, Otto Radnitz, owned several beet sugar refineries. Cori and Radnitz met at the medical school of the German University in Prague, where they both enrolled in 1914. Joint work on a research project convinced them that they were ideal partners, and they married on August 5, 1920, two months after they earned their M.D. degrees. They had a son, Tom Carl, in 1936.

After a year of working separately in Vienna, the Coris moved to the United States in 1922. They did research at the New York State Institute for the Study of Malignant Diseases (later the Roswell Park Memorial Institute) in Buffalo for the next nine years, becoming American citizens in 1928. A project involving a study of the way cancers use carbohydrates (sugars and starches) interested the Coris in how the healthy body uses these groups of substances, which are the chief foods that living things break down to obtain energy.

Through years of painstaking experiments, the Coris worked out the basic cycle of carbohydrate use in the bodies of mammals. They first described this cycle, which came to be called the Cori cycle, in 1929. The two forms of carbohydrate in the Cori cycle are glucose, a simple sugar, and glycogen, the "sugar maker," a complex carbohydrate made of hundreds of glucose molecules bonded together. Glycogen (which CLAUDE BERNARD had discovered in 1857) is the form in which carbohydrate energy is stored, and glucose is the form that muscles break down to get energy. The body uses these chemicals over and over again, re-forming glycogen out of substances left over after energy production and also replenishing supplies of glucose and glycogen from carbohydrates in food.

The Coris moved to the Washington University School of Medicine in St. Louis, Missouri, in 1931. Typically for the time, the university hired Carl as a full professor and head of the pharmacology department (in 1942, he became a professor in biochemistry and head of the biochemistry department as well) but classified Gerty as a mere research associate and offered her only a fifth of the pay it gave Carl. She became a full professor only in 1947, the year they won the Nobel Prize. In their own laboratory, however, the Coris were equals. A *New York Post* reporter wrote that the pair's collaboration was so close

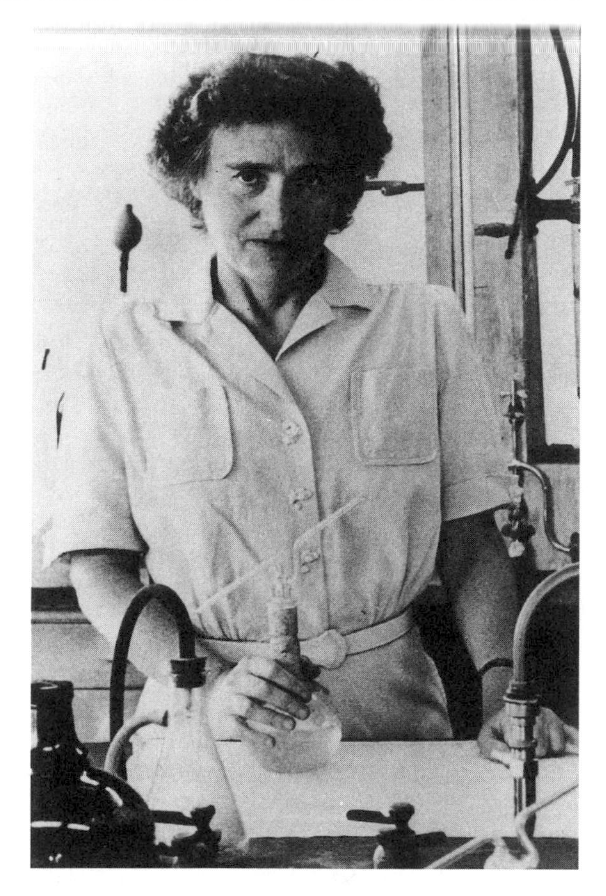

Gerty Theresa Radnitz Cori, shown here in her laboratory at the Washington University School of Medicine in St. Louis, Missouri, and her husband, Carl Ferdinand Cori, showed how the liver processes carbohydrates to release energy. *(Becker Medical Library, Washington University School of Medicine)*

that "it is hard to tell where the work of one leaves off and that of the other begins." William Daughaday, a coworker, said their abilities complemented each other: "Carl was the visionary, Gerty was the lab genius."

At Washington University, the Coris continued their work on the carbohydrate cycle, figuring out the details of the process they had described in broad outline before. They discovered several key compounds involved in the cycle, one of which (glucose-1-phosphate, or the Cori ester) was named after them. In 1939, Carl Cori also made glycogen in the laboratory for the first time. In other projects, the pair studied enzymes and hormones. For instance, they showed how certain hormones in the pituitary, a tiny gland in the brain, affect the glycogen cycle.

The Coris, along with Argentine scientist Bernardo A. Houssay, were awarded the 1947 Nobel Prize for their work on the carbohydrate cycle. They were the third married couple to receive the prize together. The Coris also shared awards such as the Squibb Award of the American Society of Endocrinology (1947). In addition, Carl Cori received several awards independently, including the Isaac Adler Prize (1943), the Albert Lasker Medical Research Award (1946), and the Sugar Foundation Prize (1947). Gerty Cori was awarded the Garvan Medal (1948) and the Sugar Research Prize (1950).

Just as the Coris shared the triumph of winning the Nobel Prize, tragedy struck: Gerty Cori learned that she had an incurable disease of the bone marrow, which makes all the cells in the blood. This illness slowly sapped her strength during the next 10 years, yet she continued to do research. She showed that several rare, inherited diseases were caused by the lack of certain enzymes involved in the carbohydrate cycle, the first time an inherited disease had been proved to be due to the lack of a particular enzyme.

Gerty Cori died on October 26, 1957. In 1967, Carl moved to Harvard Medical School and the Massachusetts General Hospital in Boston and married again, to Anne Fitzgerald Jones. He died on October 19, 1984.

Further Reading

"Carl Ferdinand Cori—Biography." *Nobel Lectures, Physiology or Medicine, 1942–1962*. Available online. URL: www.nobel.se/medicine/laureates/1947/cori-cf-bio.html. Last updated 2001.

"Cori, Carl F(erdinand); Cori, Gerty T(heresa Radnitz)." *Current Biography Yearbook 1947*. New York: H. W. Wilson, 1947.

McGrayne, Sharon Bertsch. *Nobel Prize Women in Science: Their Lives, Struggles, and Momentous Discoveries*. New York: Birch Lane Press, 1993.

⊠ Crick, Francis

(1916–)
British
Molecular Biologist

With JAMES WATSON, Francis Crick worked out the fact that molecules of DNA, the chemical of which genes are made, have the shape of a double helix. Fellow Nobel laureate PETER BRIAN MEDAWAR called this feat the greatest achievement of 20th-century science.

Francis Harry Compton Crick was born in Northampton, England, on June 8, 1916. His father, Harry Crick, ran a shoe factory, and his mother, the former Annie Wilkins, had been a teacher. As a child, Crick recalled in a 1989 interview, he "wanted to know what the world is made of."

Crick studied physics at University College, London, graduating in 1937, and then began working toward his Ph.D. His studies were interrupted when World War II began in 1939. He worked for the British admiralty, developing circuits for acoustic and magnetic mines, until 1947.

By this time, Crick was as interested in biology as in physics. He went to Cambridge, where in 1949 he joined the Medical Research Council unit in the Cavendish Laboratory. He met

Watson, a 23-year-old American who had come to the Cavendish to do postdoctoral work, in 1951. Crick wrote later, "Jim and I hit it off immediately, partly because our interests were astonishingly similar and partly, I suspect, because a certain youthful arrogance, a ruthlessness, and an impatience with sloppy thinking came naturally to both of us."

One of the pair's shared interests was in DNA. The work of OSWALD THEODORE AVERY and others had convinced them that this chemical carried inherited information, though many researchers still doubted that idea. They realized that if DNA did this job, it had to be able to reproduce itself so that each new cell could obtain a complete copy of a living thing's genetic information. To find out how DNA reproduced, they knew they would have to work out the shape of the DNA molecule.

At the time, scientists knew that DNA was a large, chainlike molecule called a polymer. It was made up of several kinds of smaller molecules: alternating molecules of sugar and phosphate, which formed a kind of "backbone," and four kinds of bases (adenine, thymine, cytosine, and guanine). No one was sure how these chemicals were arranged within the larger molecule, however. Several groups in addition to Watson and Crick were trying to find out. At King's College, London, New Zealand–born physicist MAURICE WILKINS and British chemist ROSALIND ELSIE FRANKLIN were examining the molecule with X-ray crystallography. In the United States, famed chemist LINUS CARL PAULING was trying out various structures by building three-dimensional models, a method that had helped him work out features of protein structure. Watson and Crick borrowed both approaches, building models based on data in X-ray photos taken by others.

Watson and Crick had a key stroke of luck in January 1953, when Wilkins, who had become a friend of Watson's, showed him an X-ray photograph of DNA that Franklin had made. Watson wrote later in *The Double Helix*, his memoir of the momentous discovery, that when he looked at this unusually clear photo, "my mouth fell open and my pulse began to race." He realized that the molecule must have the corkscrew shape of a double helix, with two sugar-phosphate backbones twining on the outside and the bases placed between them, like steps on a twisted ladder.

The question of how the bases were arranged remained. Experimenting with his models, Watson realized that an adenine-thymine pair would have the same overall shape as a cytosine-guanine pair. Biochemist Erwin Chargaff had shown that the amount of thymine in a molecule of DNA was always the same as the amount of adenine, and the same was true of cytosine and guanine. These, Watson therefore concluded, must be the pairs. Hydrogen bonds could hold each pair together.

Watson wrote in *The Double Helix* that at lunch on March 7, Crick, sure that he and Watson had discovered DNA's structure at last, "winged into the Eagle [a nearby bar] to tell everyone . . . that we had found the secret of life." The official announcement of their discovery, which appeared in the prestigious British science journal *Nature* on April 25, was more restrained. The brief paper concluded with what *Time* magazine later called "one of the most famous understatements in the history of science": "It has not escaped our notice that the specific pairing we have postulated immediately suggests a possible copying mechanism for the genetic material."

Crick and Watson elaborated on this statement in a second paper published about five weeks later. That paper explained that the hydrogen bonds holding the base pairs together are weak and break apart easily. Just before a cell divides, Watson and Crick theorized, each DNA molecule splits apart lengthwise like a zipper unzipping. Each half of the molecule then attracts the bases it needs to complete its pairs, along with pieces of sugar-phosphate "backbone," from free-floating molecules in the cell nucleus. The result is two DNA molecules identical to the first. Each of the two daughter cells formed by the division then can receive one complete set of DNA molecules. This theory of DNA reproduction was later confirmed by experiment.

Crick received his Ph.D. from Caius College, Cambridge, later in 1953. Remaining at Cambridge, he set about trying to determine how DNA encodes inherited information. GEORGE WELLS BEADLE and Edward Tatum had shown in the early 1940s that the basic job of genes is to tell cells how to make different kinds of proteins; each gene carries instructions for one protein. Crick believed that the instructions were coded in the order, or sequence, in which the bases were arranged within a DNA molecule. (Each molecule contains thousands of bases.) A problem, however, was that proteins are made of 20 kinds of smaller molecules called amino acids, whereas DNA has only four kinds of bases. How could so few bases specify all the amino acids?

In 1957, Crick and fellow Cambridge researcher Sydney Brenner proposed that the "genetic code" consists of sets of three bases, which they called codons. Such sets would provide 64 ($4 \times 4 \times 4$) possible combinations, more than enough to specify all 20 amino acids. Other scientists verified this idea and deciphered the code, determining which base combinations represented which amino acids, in the early 1960s.

Crick also studied how the information from DNA is translated into protein. Researchers were learning that DNA in the nucleus of the cell is first copied into a single strand of RNA, a second kind of nucleic acid. The RNA then moves into the cytoplasm, the jellylike substance that makes up the main body of the cell. Crick proposed that smaller pieces of RNA, following the instructions of this "messenger RNA," attach to single molecules of whichever kind of amino acid their code specifies. Towing their amino acid cargoes, they line up with the bases in the messenger RNA. In this way, the amino acids are assembled into a protein molecule in the order specified by the messenger RNA and, ultimately, the DNA that produced it. This theory, too, was later confirmed by experiment.

From the mid-1960s to the mid-1970s, Crick studied embryology. He then changed to another interest he had had for a long time, the workings of the brain. In 1977, he moved to the Salk Institute for Biological Studies in La Jolla, California, to do full-time research on this subject. Working mostly at a theoretical level, he has investigated the way mammals' brains interpret visual data and process information during dreaming. He has also written books on various subjects, including the possible origin of life and the nature of consciousness and the soul, as well as his autobiography, *What Mad Pursuit*.

Crick, Watson, and Wilkins received the Nobel Prize in physiology or medicine in 1962 for their work in discovering the structure of DNA. Crick's other awards, some of which were shared with Watson, include the Albert Lasker Medical Research Award (1960), the Prix Charles Leopold Meyer from the French Academy of Sciences (1961), and the Research Corporation Award (1962). Crick was married twice, to Ruth Dodd (they were married in 1940 and divorced in 1947), with whom he had a son, and to Odile Speed (in 1949), with whom he had two daughters.

Further Reading

Carolina Biological Supply Co. "A Visit with Dr. Francis Crick." 1989. Access Excellence. Available online. URL: www.accessexcellence.org/AE/AEC/CC/crick.html. Accessed 2003.

Crick, Francis. *What Mad Pursuit: A Personal View of Scientific Discovery*. New York: Basic Books, 1988.

"Crick, Francis (Harry Compton)." *Current Biography Yearbook 1983*. New York: H. W. Wilson, 1983.

Edelson, Edward. *Francis Crick and James Watson and the Building Blocks of Life*. New York: Oxford University Press, 1998.
Watson, James. *The Double Helix*. New York: New American Library, 1968.

Cushing, Harvey Williams
(1869–1939)
American
Surgeon

Harvey Cushing greatly reduced the death rate from operations on the brain and almost single-handedly created the specialty of neurosurgery. His interest in medicine came naturally, since men in his family had been physicians for three generations before him. He was born in Cleveland, Ohio, on April 8, 1869, the youngest of the 10 children of Dr. Henry Cushing and his wife, Betsey. He attended Yale University in New Haven, Connecticut, graduating in 1891, and then studied medicine at Harvard University. He received his M.D. degree in 1895.

Cushing made his first contribution to surgery while still a medical student. In 1895, upset because a patient to whom he had given anesthesia died on the operating table, he (with a fellow student) developed the "ether chart," on which the patient's pulse (heartbeat) and breathing rate were recorded continuously. This chart let the surgeon and anesthetist know the patient's condition at all times and helped them detect signs of trouble quickly.

Cushing was always eager to embrace others' new technology as well as to create his own. At Massachusetts General Hospital in 1896, for instance, he used X rays, discovered by physicist WILHELM CONRAD RÖNTGEN just the year before, to locate a bullet that had lodged in a woman's spinal cord. He later became the first surgeon to use X rays to locate brain tumors. In 1897, after he had transferred to the new Johns Hopkins University Medical School in Baltimore, Maryland, he began to carry out major operations using only a local anesthetic, injected cocaine, to numb nerves in the area to be operated on. This procedure was less risky than making the patient unconscious with ether. Cushing was one of the first surgeons to use cocaine in this way.

At Johns Hopkins, Cushing studied under famous surgeons and physicians, such as William S. Halsted and Sir William Osler. In 1900 and 1901, on the advice of these mentors, he took a year off to meet renowned surgeons and researchers in Europe as well. The year after his return to the United States, he married a childhood friend, Katherine (Kate) Crowell. They later had five children.

Cushing stayed at Johns Hopkins until 1912, when he became the Moseley Professor of Surgery at Harvard Medical School and chief surgeon at the new Peter Bent Brigham Hospital in Boston. He remained at the hospital until 1932, training many surgeons who later became famous as well as carrying out hundreds of operations himself.

Brain surgery had been Cushing's special interest since his final year at Yale. In those days, surgeons tried to avoid operating on the brain because patients who received such surgery almost always died either during the operation or soon afterward. When brain operations were performed, usually to remove tumors, most surgeons made little effort to identify different types of tumors, to note their patients' symptoms before the operations, or to perform autopsies on the patients who died and track what happened in subsequent years to those who survived.

Cushing did all these things. His studies of patients and tumors let him and other surgeons make better predictions about which patients would benefit from surgery. By adding measurements of blood pressure (a technique he had learned in Europe) to his ether charts, developing new methods of controlling bleeding during surgery, and using meticulous care and cleanliness while operating, he reduced the death rate for brain surgery from 90 percent to 8 percent by 1915.

Within brain surgery, Cushing made a special study of the pituitary, a tiny gland buried in the brain. He began his research in 1908. The pituitary produces several hormones, including one that controls growth. Cushing showed that when the pituitary makes too little growth hormone, the result is a dwarf, a person with a normal-sized head and trunk (torso) but very short arms and legs. If the gland produces too much of this hormone, on the other hand, it causes a condition called acromegaly, in which the bones of the hands, feet, and face grow abnormally and produce a giant. Cushing identified several other conditions produced by abnormalities in the pituitary, one of which is named after him, and developed a way to reach and remove pituitary tumors, which had not been possible before.

During World War I, Cushing went to France with a Harvard surgical team. On some days, he worked for 16 hours at a time with "black earth thrown up like a geyser" by shells exploding all around him. He developed new techniques, such as the use of a magnet to remove shell fragments from head wounds. He also worked out a system for classifying head injuries that is still used.

Cushing's staff held a party for him on April 15, 1931, after he removed his 2,000th brain tumor. He retired from active surgery 15 months later. In 1933, he became Sterling Professor of Neurology at Yale, a post he kept until 1937. During these years he gave lectures, served on advisory committees, and expanded an already active writing career. (For example, he wrote a biography of Sir William Osler that won a Pulitzer Prize in 1926.) Cushing died on October 7, 1939, after a heart attack.

Further Reading

Denzel, Justin S. *Genius with a Scalpel: Harvey Cushing*. New York: Messner, 1971.

Kutz, Scott, and Patrick O'Leary. "Harvey Cushing: A Historical Vignette." Louisiana State University Health Sciences Center. Available online. URL: www.medschool.lsuhsc.edu/Nsurgery/cushing.html. Accessed 2001.

Thomson, Elizabeth Harriet. *Harvey Cushing, Surgeon, Author, Artist*. New York: Henry Schuman, 1950.

Cuvier, Georges, Baron

(1769–1832)
Wurttemberger/French
Naturalist, Paleontologist

Like GEORGES-LOUIS BUFFON, Georges-Léopold-Chrétien-Frédéric-Dagobert Cuvier did not believe in evolution, yet he provided much evidence that later thinkers, such as CHARLES ROBERT DARWIN, would use to support evolutionary theory. He founded the sciences of paleontology and comparative anatomy. French admirers called him the ARISTOTLE of biology.

Cuvier was born on August 23, 1769, in Montbéliard, a French-speaking village near what is now the Swiss city of Basel. It was then part of an area controlled by Württemberg, a German state. Cuvier's father was a Swiss soldier who had served in the French army.

Cuvier's unusual intelligence attracted the attention of the Duke of Württemberg, who sent him to the university in the German city of Stuttgart when he was only 14 years old. He remained in Stuttgart from 1784 to 1788, when he obtained a degree. He then became a tutor to the children of a noble family in Caen, Normandy, a country post that kept him safe during the French Revolution. A traveling priest gave him a letter of introduction to Étienne Geoffroy Saint-Hilaire, an influential biologist, and in 1795 Geoffroy (as he was called) hired the young man as an assistant anatomy professor at the Musée National d'Histoire Naturelle, the great natural history museum in Paris that was then the world's largest research facility.

In addition to advancing at the Natural History Museum, Cuvier was professor of natural history at the Collège de France from 1799 on. He became permanent secretary of the French Academy of Sciences in 1803 and later was chancellor of the University of Paris as well. He also managed to hold powerful government posts during a series of very different political regimes. Napoleon put him in charge of redesigning the country's public school system in 1802, for example. The emperor also made him a councillor of state, a position he kept through the reigns of three subsequent kings. Cuvier was knighted in 1811, awarded the Legion of Honor in 1817, and made a baron in 1831. Along with his public duties and scientific studies, he somehow found time to marry and have four children.

As a biologist, Cuvier held several beliefs that guided all others. One was that all the parts and functions of a living thing's body are closely related and interdependent. This correlation of parts, as he called it, was one reason why he rejected the idea that types of living things could have changed significantly over time. "None of these separate parts can change their forms without a corresponding change on the other parts of the same animal," he maintained. Any single change, he assumed, would have thrown organisms out of balance so badly that they could not have survived. Cuvier described his ideas about animal anatomy in *Leçons d'anatomie comparée* (*Lessons on Comparative Anatomy*), a book based on lectures he had given between 1800 and 1805.

Cuvier's belief in the correlation of parts also led him to conclude that "each of these parts, taken separately, indicates all the other parts to which it has belonged." An animal with sharp teeth and claws, for instance, is likely to be a carnivore, or meat eater. Therefore, Cuvier said, it can be predicted to have certain other characteristics needed for a carnivore's lifestyle, such as good vision and the ability to move quickly.

Cuvier drew on this power of prediction to reconstruct ancient animals from fossil fragments in the rocks around Paris with almost magical precision. One of his contemporaries, famed writer Honoré de Balzac, exclaimed, "Is Cuvier not the greatest poet of our century? Our immortal naturalist has reconstructed worlds from blanched [bleached] bones. He picks up a piece of gypsum [a kind of rock] and says to us, 'See!' Suddenly stone turns into animals, the dead come to life, and another world unrolls before our eyes."

Cuvier's *Les Ossements fossiles des quadrupèdes* (*Fossil Bones of Quadrupeds* [four-footed animals]), published in 1812, was the first book to describe and classify fossil vertebrates in a systematic way. It essentially established paleontology, the study of fossils, as a separate branch of biology. Cuvier described 168 new kinds of fossil vertebrates, including strange lizardlike creatures now known as dinosaurs.

Fossils had been known since ancient times, but people had never been sure how to interpret them. This was partly because most people thought that Earth and all the kinds of living things on it had been created—exactly as described in the Bible—only about 6,000 years before. Geologists, however, were beginning to offer evidence that the planet was far older and had changed with time. Certain types of rocks, they pointed out, existed in layers, or strata, that apparently had been laid down one on top of another over eons.

Cuvier showed that each layer in the Parisian rocks contained types of animals not found in other layers. Some fossil bones in the top strata were similar to the bones of types of animals that still existed, but they were not exactly like the bones of any known living animals. Cuvier believed that the world had been well enough explored to reveal all existing kinds of large animals. Therefore, he said, the fossil animals must have been types that had died out completely, or become extinct. Cuvier was the first to accept extinction as a scientific fact.

As modern evolutionary biologist STEPHEN JAY GOULD has pointed out, the recognition that types of animals could die out was an important step in biological thinking. If some living things had existed at certain times in the Earth's past but not at other times, fossils and fossil-containing rocks could be used as a kind of clock to reveal the past history of the planet. Cuvier accepted this idea, noting that the fossil animals in the upper rock layers were more like types of living animals than those in deeper layers, which he recognized as being older. This fact did not make him conclude, as later scientists did, that living things had changed over time, but he believed it did show that the Earth was far older than religious leaders claimed.

Some geologists held that changes in rocks had occurred gradually, but Cuvier disagreed. He thought that Earth had undergone repeated natural disasters, such as floods and giant volcanic eruptions. Each "great and terrible event" had wiped out all living things existing at the time, at least in particular places, and new types of plants and animals had later been created or had moved in from other parts of the world to replace them. Cuvier first proposed this theory, which came to be called catastrophism, in his book on fossil bones. He described it in more detail in an 1825 book, *Discours sur les révolutions de la surface du globe (Discourse on the Revolutions of the Surface of the Globe)*.

Cuvier classified animals differently from other biologists of his time. Most arranged living things in a single "Great Chain of Being," with the simplest organisms at the bottom or beginning of the chain and the most complex kind of organism—human beings—at the top. In *Le Règne animal distribué d'après son organisation* (*The Animal Kingdom, Distributed According to Its Organization*) (1817), however, Cuvier divided animals into four large groups—vertebrates, jointed animals, molluscs, and radiates (animals with a circular body plan, such as sea stars). He did not see any of these groups as "higher" or "lower" than the others, or, indeed, as related to the others in any way. He called the groups branches, but they were later expanded in number and termed phyla.

Cuvier argued intensely with other prominent biologists of his day, such as his old patron, Geoffroy, and JEAN-BAPTISTE LAMARCK. His conservative views continued to be widely accepted long after his death from cholera in Paris on May 13, 1832. In time, however, the ideas of the evolutionists supplanted them. "As far as Cuvier . . . was concerned, he won every battle with his evolution-minded opponents," biologist and historian ERNST MAYR has written. "He did not live long enough to realize that he had lost the war."

Further Reading

Gould, Stephen Jay. "The Stinkstones of Oeningen." *Natural History*, June 1982.

Outram, Dorinda. *Georges Cuvier: Vocation, Science, and Authority in Post-Revolutionary France*. Manchester, England: Manchester University Press, 1984.

Rudwick, Martin J. S. *Georges Cuvier, Fossil Bones, and Geological Catastrophes: New Translations and Interpretations of the Primary Texts*. Chicago: University of Chicago Press, 1997.

D

Dale, Henry Hallett

(1875–1968)
British
Biochemist, Pharmacologist

Henry Hallett Dale, along with German pharmacologist Otto Loewi, proved that nerves transmit messages by means of chemicals and isolated one such chemical, acetylcholine. For this and other work, Dale and Loewi shared the Nobel Prize in physiology or medicine in 1936.

Dale was born in London on June 9, 1875. His father was a businessman. Dale studied biology at Cambridge, graduating in 1898, and earned his M.D. degree from St. Bartholomew's Hospital in London around 1907. He did postdoctoral research under ERNEST HENRY STARLING in England and PAUL EHRLICH in Germany.

In 1904, Dale joined the Wellcome Physiological Research Laboratories, sponsored by the large drug company Burroughs Wellcome. He became the laboratory's director two years later and held this post until 1914. Much of Dale's early research at Wellcome was on ergot, a fungus (mold) that infects rye and other grasses. In 1910, Dale and a coworker, George Barger, purified a substance from ergot that they named histamine. All plants and animal cells can make this substance, which has powerful effects on the body. In humans, it is most often produced during allergic reactions, when the immune system overreacts in attempting to defend the body against harmless substances such as plant pollen. Dale and Barger's identification of histamine helped scientists understand and control these reactions, which can be fatal.

Dale isolated a second chemical from ergot, which he named acetylcholine, in 1914. He showed that this chemical counteracted the effects of adrenaline, a hormone. Scientists knew that the human body contains two systems of nerves, called sympathetic and parasympathetic nerves, that have effects opposite to one another. The effects of sympathetic nerves were associated with their release of adrenaline. In the early 1920s, Dale and Loewi proved that parasympathetic nerves released acetylcholine and that this chemical was responsible for the nerves' effects on the heart and other tissues. Their work provided the first clear proof that nerves transmit signals to tissues by means of chemicals.

Dale became an eminent administrator as well as a researcher. He was secretary of the Royal Society, Britain's premier institution of scientists, from 1925 to 1935 and president of the society from 1940 to 1945. He headed the department of biochemistry and pharmacology at the National Institute for Medical Research from 1914 to 1928. He then became director of the entire institute, a post he held until he retired in 1942.

After that, he was professor of chemistry and a director of the Davy-Faraday Laboratory at the Royal Institution, London, until 1946. He was also chairman of the board of the Wellcome Trust, which supports medical research and scholarships, from 1938 to 1960.

Dale received many awards in addition to the Nobel Prize, including the Copley Medal of the Royal Society (1937) and the Haly Medal of the Royal College of Physicians, London. The British government knighted him in 1932 and gave him an Order of Merit in 1944. The governments of Belgium, West Germany, and the United States also awarded him honors. Dale married Ellen Harriet Hallett, a first cousin, in 1904. He died on July 23, 1968, in Cambridge.

Further Reading

"Sir Henry Hallett Dale—Biography." *Nobel Lectures, Physiology or Medicine, 1922–1941*. Available online. URL: www.nobel.se/medicine/laureates/1936/dale-bio.html. Last updated 2001.

Darwin, Charles Robert

(1809–1882)
British
Naturalist, Evolutionary Biologist

Probably no other individual has changed biology as much as Charles Darwin. Although other biologists had concluded that species, or types of living things, had changed, or evolved, during Earth's history, Darwin was the first to describe a convincing mechanism that could drive such changes. His work fundamentally altered the way not only biologists but everyone in Western society saw themselves and their world. In *Great Thinkers of the Western World*, Mark T. Riley writes, "No area of human thought, scientific or popular, has remained unaffected by [Darwin's] theory of evolution."

Darwin was born on February 9, 1809, in Shrewsbury, England. His father, Robert Waring Darwin, and grandfather, Erasmus Darwin, were respected physicians. Erasmus Darwin had also written papers supporting the idea of evolution. Charles Darwin's mother, Susannah, was the daughter of the famous industrial potter Josiah Wedgwood and brought some of his money into the Darwin family. Charles was the fifth of Robert and Susannah's eight children.

As a young man, Darwin was an unimpressive student. He began studying medicine at the University of Edinburgh, Scotland, in 1825 but quit because he found some parts of the subject boring and others disgusting. His father then sent him to Cambridge University to train as a minister, but, as Darwin described in his autobiography, he went to more hunting and drinking parties than church services during his college years. He barely passed his final examinations in 1831. Nonetheless, he did obtain a good background in botany and geology. His geology professor, Adam Sedgwick, introduced him to the writings of CHARLES LYELL, who maintained that the earth was very old and had changed gradually over time. John Henslow, his botany professor, found Darwin his first job.

At Henslow's recommendation, the captain of the HMS *Beagle* took Darwin along as an unpaid naturalist (general scientist) on a voyage commissioned by the British government to survey the coasts of South America. The *Beagle*'s journey lasted from 1831 to 1836 and included parts of New Zealand and Australia as well as South America. Because of its impact on Darwin, Mark T. Riley calls this sea voyage "the single most important event . . . in the history of biology."

Darwin often left the *Beagle* for weeks or even months at a time, exploring various areas and making endless notes about the rocks, plants, animals, and people he saw. He made some of his most important observations on the Galápagos Islands, off the coast of what is now Ecuador. There, he saw groups of animals that were basically similar, yet had slightly different features on different islands. Birds called finches, for instance, had different sizes and bill shapes, which allowed them to eat slightly different diets. The finches appeared to be separate species, but they were so much alike overall—and so similar to other finches Darwin saw on the South American mainland—that he began to think that they must have descended from a single ancestor. He suspected that, living apart from each other, they had become more and more different as centuries passed until, finally, they could no longer interbreed.

The idea that species might have changed over time "haunted" Darwin, as he wrote later, but he could not imagine what might make these changes come about until 1838, when he read economist THOMAS ROBERT MALTHUS's *Essay on the Principle of Population*. Malthus said that animals reproduce until the number of individuals outstrips the available food supply. Individuals then compete for the limited food, and only those that are strongest, healthiest, or otherwise best able to compete will survive. Darwin realized that this "struggle for existence" put pressure on species to vary and thus could provide the driving force for change that he was looking for. As he later wrote, "The slightest advantage in certain individuals . . . over those with which they come into competition . . . will, in the long run, turn the balance [between survival and extinction]."

Darwin came to believe that chance variations within a species arise constantly. A deer might be born, for example, with legs slightly longer than those of other members of its species, allowing it to run faster. If the deer's extra speed helps it outrun predators to which other deer fall victim, it will be more likely to live long enough to have offspring. Darwin knew that offspring usually are similar to their parents. Over generations, therefore, if pressure from predators continues, more and more deer will be born with longer legs. The long-legged deer may eventually replace the others, creating a new species.

Darwin concluded that nature was acting just like human breeders of plants and animals. Breeders deliberately mate individuals with characteristics that the breeders consider desirable, eventually creating varieties in which these characteristics predominate. Darwin believed that competition, combined with factors in particular environments, has the same effect in the wild. He therefore called his theory evolution by natural selection. It differed from JEAN-BAPTISTE LAMARCK's theories about evolution in that it credited chance variation and inheritance of characteristics, not the use or disuse of body parts during an individual's lifetime, with determining how species change.

In *On the Origin of Species by Means of Natural Selection,* published in 1859, Charles Robert Darwin explained how the combination of variations within species and competition for resources caused types of living things to change, or evolve, over time. *(Edgar Fahs Smith Collection, Annenberg Rare Book and Manuscript Library, University of Pennsylvania)*

In 1839, Darwin married his first cousin, Emma Wedgwood, thereby acquiring enough additional Wedgwood money to support his household permanently. He moved to a country home called Down House, in the village of Downe in Kent, where he lived for the rest of his life. He also began to suffer from ill health, including headaches and stomach problems, and this, too, became permanent. Some historians believe that his illness was mostly psychological, but others think that Darwin may have had Chagas's disease, a parasitic infection that he could have acquired in South America. He and Emma nonetheless had 10 children, seven of whom survived to adulthood.

Darwin spent his first years at Down House writing up the notes he had taken during the *Beagle* voyage, which were published in 1839 as *Journal of Researches into the Geology and Natural History of the Various Countries Visited by the H.M.S.* Beagle. He then wrote a book on the formation of coral reefs (1842) and a series of papers on small sea creatures called barnacles (1851–54). Only after these projects were completed did he devote his full attention to his theory of natural selection.

Aware that his ideas would most likely arouse harsh criticism, Darwin spent years amassing examples from his and others' experience to illustrate and indirectly prove them. He might have gone on doing so indefinitely if he had not received a short manuscript from an obscure fellow naturalist named ALFRED RUSSEL WALLACE in June 1858. The two had never met, but to Darwin's amazement, Wallace's article described exactly the same theory that he himself had been propounding. Darwin arranged for papers describing their theory to be presented at a meeting of the Linnaean Society, a respected scientific group, on July 1, 1858. He also quickly finished what he saw as a mere summary of the "big book" he had hoped to write about evolution. It was published in November 1859 as *On the Origin of Species by Means of Natural Selection, or the Preservation of Favoured Races in the Struggle for Life*.

The book caused a sensation. Many religious leaders opposed it, not only because it proposed that changes had occurred and were continuing to occur in a supposedly "perfect" creation, but because it insisted that chance and the environment, not God, were responsible for those changes. Some scientists also criticized the book, pointing out that Darwin had left many questions unanswered. He did not explain, for example, how slight variations could produce a complex organ such as an eye.

Darwin's illness, or his personality, made him unable to face large crowds, so he did not publicly defend his book. Less shy supporters such as Thomas Henry Huxley (nicknamed "Darwin's Bulldog") did so with vigor, however. In fact, despite the stir they created at first, Darwin's basic ideas came to be accepted within less than a decade, partly because they fitted with the Victorian era's belief in progress. Thinkers such as Herbert Spencer translated Darwinism into social terms, maintaining that the wealthy deserved to be so because their material success showed that they were "fitter" than others, and these ideas also proved popular.

Darwin stirred up trouble again with an 1871 book, *The Descent of Man*, in which he insisted that "the . . . difference in mind between men and the higher animals, great as it is, is certainly one of degree and not of kind." This conflicted with the commonly held belief that humans were completely separate from other living things. Worse still in the eyes of many, Darwin proposed that humans and apes, such as chimpanzees and gorillas, had descended from a common ancestor. This notion was often popularly misinterpreted as a claim that humans had descended from apes.

In his last years, Darwin wrote only about uncontroversial topics such as plants and earthworms. By the time he died of a heart attack in his home on April 19, 1882, most people had forgotten that his work had once been so shocking. Indeed, he was so highly regarded that he was buried in Westminster Abbey, next to Lyell and eminent physicist Sir Isaac Newton.

Further Reading

Browne, Janet. *Charles Darwin: The Power of Place*. New York: Alfred A. Knopf, 2002.

Darwin, Charles. *The Autobiography of Charles Darwin 1809–1882*. Edited by Nora Barlow. Reprint, New York: W. W. Norton, 1993.

———. *Origin of Species*. 1859. Edited by Greg Suriano. Reprint, New York: Gramercy Books, 1998.

———. *The Voyage of the Beagle*. 1839. Reprint, New York: Penguin, 1989.

Patent, Dorothy Hinshaw. *Charles Darwin: The Life of a Revolutionary Thinker*. New York: Holiday House, 2001.

Dawkins, Richard

(1941–)
Kenyan/British
Evolutionary Biologist, Philosopher of Science

Clinton Richard Dawkins is best known for his defense and popularization of a modern version of CHARLES ROBERT DARWIN's theory of evolution. He has described evolution and behavior, including human behavior, in terms of genes, ideas, and computers.

Dawkins was born in Nairobi, Kenya, on March 26, 1941. His father, Clinton, worked for the British colonial government as an agricultural educator. When Richard was eight, his father inherited a farm in England and returned there to raise dairy cattle. Richard first learned about Darwin's ideas when he was 16, an experience he describes as "mind-blowing."

Dawkins went to college at Oxford University, beginning an association with Oxford that, except for two years (1967–69) at the University of California at Berkeley, has lasted all his adult life. He earned his bachelor's degree in zoology in 1962 and his Ph.D. in 1966. Since 1995, Dawkins has been the Charles Simonyi Professor of Public Understanding of Science, a position that frees him from teaching and research duties so that he can concentrate on explaining evolutionary ideas to the public through writing and speaking.

Dawkins's first book, *The Selfish Gene* (1976), made him famous. It described evolution in terms of genes, picturing them as if they were independent living things with behavior that could be studied much as ethologists like NIKO TINBERGEN, Dawkins's mentor at Oxford, studied the behavior of animals. In this book and later writings and interviews, Dawkins stated that living things, including human beings, are nothing but "temporary survival machines, robot vehicles blindly programmed for . . . [the] benefit" of "selfish" genes intent only on reproducing.

Dawkins has never claimed that genes control human behavior completely, however. On the contrary, he maintains that humans are the only living things that can choose to defy their genes, for instance by using contraception. Just as important as genes in shaping human behavior and evolution, he says, are what he calls memes—ideas or beliefs that "reproduce" by spreading from person to person. Memes can mutate and evolve just as genes can, Dawkins believes. Indeed, he sees memes as the cultural equivalent of genes. Both are "replicators," bending the behavior of their carriers to achieve their goal of reproduction.

Dawkins's ideas, like Darwin's before him, aroused considerable debate. Some critics disliked his open hostility to religion, which he has labeled "an enemy of truth." Other people were disturbed by being called "robots" with no purpose except to be manipulated by genes or memes. On the other hand, many scientists hailed Dawkins for describing evolution in a clear and unusually vivid way.

Dawkins has written numerous other books, most of which have expanded on ideas in his first one. *The Extended Phenotype* (1982) states that genes can affect more than the bodies in which they reside. For instance, parasites, which take their nourishment from other living creatures without immediately killing them, can change the behavior of those living things in the process of ensuring their own replication. This book also claims that the artifacts animals and people build (such as a bird's nest or a computer), their social organizations, and the environments they create are expressions of their genes just as much as their bodies are and should be considered part of their evolution.

The Blind Watchmaker (1986), Dawkins's third book, shows how even the most complex features can develop from simpler structures by chance mutations. It describes a computer program that Dawkins designed to mimic evolution, creating "biomorphs" that bear an uncanny resemblance to insects and other living things. Dawkins's other books include *River Out of Eden* (1995), *Climbing Mount Improbable* (1996), and *Unweaving the Rainbow* (1998).

Dawkins has been married three times and has one daughter. His current wife, actress and artist Lalla Ward, illustrated two of his books. His awards include the Royal Society of Literature Award (1987), the Royal Society Michael Faraday Award (1990), and the Nakayama Prize for Human Science (1994). He was named Humanist of the Year by the American Humanist Association in 1996 and elected a fellow of the Royal Society, Britain's premier scientific organization, in 2001.

Richard Dawkins has said he would like to "explain [science] so that the reader feels it in the marrow of his bones." Most of his readers would agree that, for better or worse, he has succeeded.

Further Reading

Bass, Thomas A. *Reinventing the Future*. Reading, Mass.: Addison-Wesley, 1994.

Dawkins, Richard. *The Selfish Gene*. 1976. Reprint, New York: Oxford University Press, 1990.

"Dawkins, Richard." *Current Biography Yearbook 1997*. New York: H. W. Wilson, 1997.

"Revolutionary Evolutionist." *Wired*, July 1995. Available online. URL: www.world-of-dawkins.com/Dawkins/Biography/bio.htm.

Delbrück, Max

(1906–1981)
German/American
Molecular Biologist

Although he trained as a physicist, Max Delbrück's discoveries about viruses that infect bacteria and his influence on other scientists make him a founder of molecular biology. Delbrück was born on September 4, 1906, in Berlin, Germany, the youngest of seven children. His father, Hans, was a professor of history at the University of Berlin, and his mother, the former Lina Thiersch, was a granddaughter of Justus von Liebig, a famous German chemist.

Delbrück studied physics at the Universities of Tübingen, Bonn, and Göttingen, obtaining his Ph.D. from Göttingen in 1930. In the early 1930s, he did research with famous physicists in several countries, including Niels Bohr in Denmark and Lise Meitner in Germany. A 1932 lecture of Bohr's interested him in biology, which he and Bohr believed could be studied by the same methods that were applied to physics and chemistry.

Delbrück came to the California Institute of Technology (Caltech) in Pasadena in 1937 to study fruit fly genetics under THOMAS HUNT MORGAN. After he arrived, however, another Caltech professor, Emory Ellis, interested him in viruses called bacteriophages, which infect bacteria. Delbrück liked to choose a simple system that showed phenomena he wanted to examine and then form a team to study that system intensely, and he believed that viruses offered the simplest possible system for studying genes. He once called bacteriophages "the hydrogen atoms of biology."

When funding for his studies at Caltech ran out in 1940, Delbrück became a physics instructor at Vanderbilt University in Nashville, Tennessee, where he remained until 1947. In 1941 he married Mary Adeline Bruce; the couple later had four children. He became a United States citizen in 1945.

During his years at Vanderbilt, Delbrück persuaded scientists at several other universities, including SALVADOR LURIA and ALFRED DAY HERSHEY, to study bacteriophage genetics. Delbrück and Luria formed what they called the Phage Group in 1943. In that same year, they discovered that bacteria exposed to bacteriophages can undergo spontaneous genetic mutations that make them able to resist infections by the viruses. This was the first study of bacterial genetics. Delbrück and Hershey, working independently, found in 1946 that bacteriophages could exchange or combine genes, producing new viruses different from both of the original viruses. This was the first time that viruses had been shown to engage in a process somewhat like sexual reproduction, which also allows living things to reshuffle their genes.

Delbrück returned to Caltech in 1947 and remained there as a professor of biology until his death. Beginning around 1950, he became interested in sensory physiology and attempted to study vision by examining a mold (fungus) called *Phycomyces*, which was attracted to light. In this case, however, Delbrück failed to choose a good system. *Phycomyces* proved too different from more complex living things to reveal anything useful about their visual sense.

Delbrück shared the Nobel Prize in physiology or medicine with Luria and Hershey in 1969 for their work on bacteriophages. Commentators have said that Delbrück won the award at least as much for his inspiration of other scientists as for his own research. In addition to founding the Phage Group, which established much of the basis of molecular biology, Delbrück later interested fellow scientists in the study of viruses that cause cancer in animals, which proved important to understanding this disease. Delbrück died of cancer in Pasadena on March 10, 1981.

In the 1940s, Max Delbrück of Vanderbilt University and the California Institute of Technology helped to found molecular biology by studying the genetics of bacteriophages, viruses that infect bacteria. *(National Library of Medicine)*

Further Reading

Fischer, Ernst Peter, and Carol Lipson. *Thinking About Science: Max Delbrück and the Origins of Molecular Biology*. New York: W. W. Norton, 1988.

De Vries, Hugo

(1848–1935)
Dutch
Botanist, Geneticist

Hugo Marie De Vries expanded on CHARLES ROBERT DARWIN's theory of evolution by natural selection by proposing genetic change, or mutation, as the source of the variations on which natural selection acts to produce new species. He was also one of three scientists who rediscovered GREGOR MENDEL's laws of heredity and thus helped to found genetics.

De Vries was born in Haarlem, the Netherlands, on February 16, 1848. His father, Gerrit De Vries, held several high government posts, including that of prime minister. Maria Reuvens, his mother, came from a scholarly family. De Vries studied medicine at the Universities of Heidelberg (Germany) and Leiden (the Netherlands), obtaining a Ph.D. from the latter in 1870. He then moved to the University of Würzburg in Germany, where he did research on plant physiology, chiefly the ways plants use water and react to loss of water.

Dutch botanist Hugo De Vries was one of three scientists who independently rediscovered Gregor Mendel's paper on the laws of heredity in 1900; De Vries also described mutations, or sudden changes in inherited traits. *(National Library of Medicine)*

After teaching at a high school in Amsterdam and at the University of Halle and writing papers on agriculture for the Prussian government in Würzburg, De Vries became a lecturer in plant physiology—the first teacher in this field in the Netherlands—at the University of Amsterdam in 1877. He stayed at the university for the rest of his career, becoming a full professor in 1881 and a senior professor in 1896. He retired in 1918 and died in Lunteren, near Amsterdam, on May 21, 1935.

De Vries continued to study plant physiology in his first years at the University of Amsterdam, but he was also interested in Darwin's ideas, especially in how variations within species arose and were passed on to offspring. Darwin had believed that changes in species and formation of new species occurred very gradually, but many other scientists held that life had not existed on Earth long enough for such slow changes to produce modern species. This disagreement cast doubt on Darwin's theory.

Looking for a type of plant in which he could study variation effectively, De Vries considered more than 100 species, but he found nothing that suited him until one evening in 1886, when he passed an abandoned potato field full of yellow wildflowers called evening primroses. He noticed that the plants varied considerably in height and other features and decided that they might be just what he had been seeking. Between 1892 and 1900, he raised thousands of primrose plants in his experimental garden.

Each type of primrose dependably passed its special characteristics on to its offspring, De Vries found. Some characteristics, however, appeared more often than others. After many breedings, he determined that when plants with two forms of the same characteristic, such as height, were mated, some forms appeared on average in three out of every four of the plants' offspring, whereas other forms appeared in only one out of every four offspring. De Vries eventually demonstrated that a number of other plant species also showed this pattern.

In addition, De Vries noticed that, occasionally, individual plants with characteristics unlike those in any of their ancestors suddenly appeared. These characteristics could be dependably passed on to the plants' own offspring. Plant and animal breeders took advantage of such variations, which they called sports, to create new varieties, such as flowers with unusual color combinations. De Vries eventually found about 20 examples of these apparently spontaneous variations, which he considered to be new species.

De Vries gave the name *mutations* to these sudden changes in traits, which he came to believe were the source of the variations on which natural selection acts. He maintained that, if a mutation gave its possessor a survival advantage, it could create a new species in a single generation. "They came into existence at once, fully equipped, without preparation or intermediate steps," he wrote of his apparent new species of primroses. "No series of generations, no selection, no struggle for existence was needed." De Vries described his ideas about mutations at greatest length in *Die Mutationstheorie* (*The Mutation Theory*), published from 1901 to 1903. Supporters of Darwin's theory rejected his proposals at first but came to accept them when they realized that, rather than contradicting that theory, they repaired a major weakness in it.

Before scientists publish accounts of their experiments and conclusions, they check to see whether any earlier researchers have discovered something similar. In the course of doing this, De Vries ran across a paper about breeding experiments in peas that Gregor Mendel, a monk in what is now the Czech Republic, had published in an obscure journal in 1866. To the Dutch scientist's amazement, he found that the Czech monk had discovered the same 3:1 ratio of inheritance in his peas that he himself had found in his primroses. When he wrote a description of his experiments in early 1900, he gave credit to Mendel's earlier research.

Two German scientists, Karl Correns and Erich Tschermak von Saysenegg, separately published reports at almost exactly the same time that also mentioned Mendel's paper and reported confirmation of his results. Evolutionary biologist and science historian ERNST MAYR calls this three-way coincidence "one of the most extraordinary events in the history of biology." The three papers brought Mendel's long-forgotten work to the attention of the scientific community, and it became the basis for the new science of genetics.

Geneticists today have a somewhat different understanding of mutation than De Vries did. The term is now used to refer to a change in a gene. (*Mutation* can be either a verb, meaning the process of genetic change, or a noun, referring to the result of such a change.) Science historians believe that most of the "mutations" De Vries saw in his primroses resulted from changes in the number of chromosomes in the plant cells rather than from true changes in genes. Nonetheless, De Vries's work was useful in focusing attention on new characteristics that were inheritable, and the basic idea of his mutation theory is still accepted. Ernst Mayr says that De Vries "synthesized Darwin and Mendel."

Further Reading

Mayr, Ernst. *The Growth of Biological Thought*. Cambridge, Mass.: Belknap Press of the Harvard University Press, 1982.

Stanhuis, Ida H., Inno G. Meijer, and Erik J. H. Zavenhuizen. "Hugo de Vries on Heredity, 1889–1903: Statistics, Mendelian Laws, Pangenes, Mutations." *Isis*, June 1999.

Doll, Richard
(1912–)
British
Epidemiologist

William Richard Shaboe Doll played a key role in linking cigarette smoking with lung cancer. Doll, born in 1912, first wanted to be a mathematician, but, when told that there were few jobs for people trained in mathematics, he decided to study medicine instead. He later combined his two interests by specializing in epidemiology, the study of the way disease spreads through populations, which depends heavily on statistics.

Doll took his medical training at St. Thomas Hospital Medical School, part of the University of London, and obtained his M.D. degree in 1937. He served in the Royal Army Medical Corps during World War II. From 1946 to 1969, Doll worked for the Statistical Research Unit of the British government's Medical Research Council. He became the unit's director after Sir Austin Hill, the former director and Doll's mentor, retired. In 1969, Doll became Regius Professor of Medicine at Oxford University, a title he held until he became the first warden of Oxford's Green College in 1979.

Doll became famous for a study that he and Hill published in 1950, which showed that smokers were 14 times more likely to develop lung cancer than nonsmokers. A few earlier studies had linked other forms of tobacco use with other kinds of cancer; pipe smoking seemed to increase the chances of developing lip cancer, for example. Doll and Hill's study and another issued at about the same time by two United States scientists, Ernst Wynder and Evart Graham, were among the first to focus on cigarette smoking, however.

Doll and his protégé, Richard Peto, began a lengthy research project in 1951 that confirmed the earlier study's results. They asked 35,000 British physicians whether they smoked and then checked up on them at intervals for 40 years to find out the causes of their deaths. Early results from this study showed that smoking was associated with increased risk of death from heart disease as well as lung cancer. This and other research eventually resulted in campaigns to discourage smoking. (Doll himself said in 1997 that "the whole promotion of tobacco ought to be banned.")

Doll went on to evaluate the risk of developing cancer associated with exposure to environmental factors, including asbestos (a fire-retardant mineral formerly used in some building materials), fluoride in drinking water, low-dose radiation from atomic bomb tests and nuclear plants, natural radiation from such sources as radon gas, and electromagnetic radiation from electric power lines. These studies contributed to changes in national and international law, such as the banning of asbestos and creation of treaties to stop bomb tests that released radioactive debris into the atmosphere.

Although Doll's early research tended to blame environmental factors, including pollutants, for most cases of cancer, many of his later reports claimed that such factors as living near a nuclear power plant do not significantly increase cancer risk. Some environmental groups have questioned these conclusions. They claim that many of the research organizations and charities that Doll has worked for receive money from sources such as the automobile and pharmaceutical industries, which manufacture or benefit from the pollutants being examined. They therefore believe that Doll's results might be biased. In an article for *The Ecologist*, Martin Walker called Doll "one of the most powerful and influential promoters of entrenched industrial and political interests." Doll, in turn, has referred to the environmental movement as "the anti-science Mafia."

Despite these criticisms, mainstream European science and government have honored Doll highly. For example, he has received the United Nations Award for Cancer Research

(1962), the Gold Medal of the British Medical Association (1983), the Royal Medal of the Royal Society (1986), and the Gold Medal of the European Society of Cardiology (2000). The British government knighted Doll in 1971 and gave him a higher title, Companion of Honor, in 1996.

Even after his retirement from Green College in 1983, Doll has continued to do research as an honorary member of several research units at Oxford. Doll celebrated his 90th birthday in December 2002.

Further Reading

Bower, Hilary. "No Sign of Slowing Down." *British Medical Journal*, March 8, 1997.

Walker, Martin. "Sir Richard Doll: A Questionable Pillar of the Cancer Establishment." *The Ecologist*, March–April 1998.

Domagk, Gerhard
(1895–1964)
German
Pharmacologist, Pathologist

Gerhard Johannes Paul Domagk discovered the compound that gave rise to the sulfa drugs, the first substances able to kill disease-causing bacteria inside a living body. He was born on October 30, 1895, in Lagow, then part of a state called Brandenburg in Germany but now in Poland. His father, Paul, was a teacher, and his mother, the former Martha Reimer, came from a farming family.

Domagk studied medicine at the University of Kiel. His studies were interrupted by World War I, in which he was wounded. He returned to Kiel after the war and obtained his M.D. degree in 1921. After several years of postdoctoral research, he became a teacher of pathological anatomy at the University of Greifswald in 1924 and then at the University of Münster in 1925. He also did research on cancer at the two universities. He married Gertrud Strübe in 1925, and they later had four children. In 1927, while still retaining his position at Münster, Domagk also went to work for a dye manufacturing plant, the Farbenfabriken Bayer in Wuppertal-Elberfeld, which belonged to the large German dye company I. G. Farbenindustrie. He became a professor of general pathology and pathological anatomy at Münster in 1928 and the director of a new laboratory for experimental pathology and bacteriology at the dye plant in 1929.

The executives of the dye company, and Domagk himself, hoped that their products could be used for more than coloring carpets. They knew that more than 20 years before, PAUL EHRLICH had started his search for "magic bullet" drugs that killed disease-causing microbes inside the body by examining coal-tar dyes like the ones the company made, and he had found a few that had such an effect. Ehrlich and others had gone on to develop a small number of drugs that killed certain kinds of microbes, such as the spirochetes that cause the serious sexually transmitted disease syphilis, but they had not found any that were effective against bacteria, one of the most common causes of disease. In the hope of discovering such a drug, Domagk began testing the factory's dyes on mice and rabbits infected with streptococcus, a common kind of bacteria that causes serious wound infections.

In 1932, Domagk found that a red dye named Prontosil cured some of his mice. Just as he was beginning to test the substance on humans, his daughter, who had been helping him set up the tests, accidentally pricked her finger and developed a deadly streptococcus infection of her own. After other treatments failed, Domagk gave her the dye, and she recovered. He gave Prontosil equally effectively to other patients with similar infections. Meanwhile, Britain's Medical Research Council had also heard about the drug and began conducting human tests.

Domagk published an account of his research in 1935, and the British group reported equal success a year later. At the same time, scientists at the Pasteur Institute in France analyzed Prontosil and showed that the part of it that actually killed the bacteria was a compound called sulfanilamide. This was good news (except, perhaps, for Domagk's employers) because sulfanilamide was cheaper to make than Prontosil, could not be patented, and did not dye patients' skins red, an annoying side effect of Prontosil.

Other scientists created variations of sulfanilamide, some much more effective than the original compound, and by 1938 a whole family of sulfonamide, or sulfa, drugs began to be used in hospitals against a variety of bacterial diseases. According to an article in the British medical journal *Lancet*, they "revolutionized treatment of bacterial infections." The sulfa drugs were mostly replaced by penicillin and other antibiotics after 1945, but they are still sometimes used against bacteria that have become resistant to antibiotics.

In 1939, Gerhard Domagk was awarded the Nobel Prize in physiology or medicine for his discovery of the germ-killing powers of Prontosil. However, Adolf Hitler, whose Nazi government then controlled Germany, had decreed that no German could accept a Nobel Prize. Domagk was therefore unable to claim the prize until 1947, after the Nazis had fallen from power. Because so much time had passed by then, he was given only a scroll and a gold medal; following Nobel rules, his $35,000 prize money had been reabsorbed into the organization's general fund.

In addition to his work on Prontosil and streptococcus, Domagk searched for drugs to treat other diseases, most notably tuberculosis and cancer. He found some drugs that were used as treatments for tuberculosis in the late 1940s and 1950s. He also won other awards, including the Emil Fischer Memorial Plaque of the German Chemical Society (1937), the Cameron Prize of the University of Edinburgh, Scotland (1938), and the Paul Ehrlich Prize of the University of Frankfurt (1956). Several countries, including Spain and Japan, gave him honorary titles. He died in Burgberg, West Germany, on April 24, 1964.

Further Reading

"Domagk, Gerhard (Johannes Paul)." *Current Biography Yearbook 1958*. New York: H. W. Wilson, 1958.

"Gerhard Domagk: Biography." *Nobel Lectures: Physiology or Medicine 1922–1941*. Available online. URL: www.nobel.se/medicine/laureates/1939/domagk-bio.html. Last updated 2001.

Earle, Sylvia Alice

(1935–)
American
Botanist, Marine Biologist

Sylvia Earle has spent more than 6,000 hours underwater, including living in an undersea "habitat" for two weeks, and has dived more than two and a half miles down into the sea. Admirers have nicknamed Earle "Her Royal Deepness."

Earle was born on August 30, 1935, in Gibbstown, New Jersey, and spent her childhood on a farm near Camden. Her mother, Alice, a former nurse, taught her to love nature. Earle has told interviewers that she first fell in love with the sea after a wave knocked her into the ocean when she was three years old. "I found it exhilarating rather than frightening," she says. "I found my feet and plunged back in." She had more opportunities to indulge her fascination with the water after her father, Lewis, an electrical engineer, moved the family to Dunedin, Florida, when she was 12 years old. Earle made her first ocean dive, in the Gulf of Mexico, when she was 17 and, she has said, "practically had to be pried out of the water."

Earle earned a B.S. from Florida State University in 1955 and an M.S. in botany from Duke University in 1956. She married a zoologist named John Taylor around 1957 (they divorced in 1966) and had two children. She began full-time undersea research in 1964. She collected algae (seaweeds and related plants) in the Gulf of Mexico for her Ph.D. project at Duke, which she finished in 1966. Unlike most marine biologists of the time, she dived to study undersea life in its own habitat rather than dragging it up to the deck of a ship in nets.

In 1970, Earle lived underwater for two weeks as part of a project called Tektite, sponsored by NASA (the National Aeronautics and Space Administration). The project's name came from a type of glassy meteoric rock often found on the seafloor. Earle headed a crew of four other women scientists. They lived in a two-pod "habitat" 50 feet under the Caribbean Sea and spent up to 10 hours a day in the water, studying ocean life. Although the group did nothing that male Tektite crews had not also done, they attracted far more publicity. Earle saw this as reverse discrimination, but it also made her aware that being a woman scientist gave her a unique opportunity to reach and educate the public.

Ironically, a greater achievement of Earle's won much less attention than her Tektite stint. In September 1979, she donned a heavy plastic and metal "Jim suit" (named after a diver who tested an early version of it), a sort of underwater space suit, and dived 1,250 feet into the water near Hawaii. A submarine lowered and then released her. No other diver had gone this deep without being attached to a cable. Earle remained submerged for two and a half hours under water pressure of 600 pounds per square inch, observing such creatures as "a lantern fish . . . with lights along its sides, looking like a miniature passenger liner." *Current Biography Yearbook 1992* called this feat "possibly the most daring dive ever made."

While preparing for the Jim suit dive, Earle met British engineer Graham Hawkes, who had designed the suit. The two formed two companies, Deep Ocean Technology and Deep Ocean Engineering, in 1981. One of their products was a one-person submersible called *Deep Rover*, which Earle piloted down to about 3,000 feet in 1985, the deepest any solo diver had gone. The couple also married in 1986, but they have since divorced. (This was Earle's third marriage; between 1966 and 1975 she had been married to Giles Mead and had had a third child.)

In addition to working with Hawkes, Earle took part in many research projects during the 1970s and 1980s, including diving with humpback whales. She was also curator of phycology (the study of algae) at the California Academy of Sciences in San Francisco from 1979 to 1986.

In 1990, President George Bush chose Earle to be the chief scientist of the National Oceanic and Atmospheric Administration (NOAA). She was the first woman to hold this post. She hoped to use the position to encourage ocean conservation projects, but she found most of her time taken up by assessing damage that the Gulf War caused to sea life in the Persian Gulf. She resigned from NOAA in 1992, saying, "I think I can be more effective [in preserving the oceans] if I am on the loose." Among her activities during the 1990s was the founding in 1992 of Deep Ocean Exploration and Research (DOER), a company that consults on, operates, and designs manned and robotic underwater systems.

During her long career, Sylvia Earle has led more than 50 oceanic expeditions and won many awards for her work. They include the Society of Women Geographers Gold Medal (1990), the Director's Award of the National Resources Council (1992), the Kilby Award (1997), and induction into the National Women's Hall of Fame (2000). *Time* magazine named her a "hero for the planet" in 1998.

Beginning in 1998, Earle's major project has been directing the Sustainable Seas Expeditions, a five-year study of the 12 National Marine Sanctuaries (the underwater equivalent of national parks) in the United States. The study is sponsored by NOAA and the National Geographic Society, for which Earle is an Explorer in Residence. Earle also spends a great deal of time speaking and writing about overfishing, pollution, and other human activities that harm ocean life. She points out that threats to the ocean are threats to humanity as well. The time for halting harm to the seas before it becomes irreparable, she says, is passing quickly: "If we don't wake up soon to the damage we are doing, it may be too late."

Further Reading

Baker, Beth. *Sylvia Earle, Guardian of the Sea*. Minneapolis, Minn.: Lerner Publications, 2000.

"Earle, Sylvia A." *Current Biography Yearbook 1992*. New York: H. W. Wilson, 1992.

Earle, Sylvia A. *Dive: My Adventures in the Deep Frontier*. Washington, D.C.: National Geographic Society, 1999.

———. *Sea Change: A Message of the Oceans*. New York: Putnam, 1995.

Rosenblatt, Roger. "Call of the Sea." *Time*, October 5, 1998.

Edwards, Robert

(1925–)
British
Embryologist

Robert Geoffrey Edwards, with surgeon Patrick Steptoe, developed the technique of in vitro fertilization (fertilization in the laboratory, outside of a living body), which has allowed the birth of "test-tube babies." Edwards was born in 1925 in Yorkshire, England, and grew up in the city of Manchester. He served in the British army during World War II. He was educated at the Universities of North Wales and Edinburgh (Scotland), obtaining a Ph.D. in animal genetics from the latter in 1957. After short postdoctoral periods at several universities, he joined Cambridge University in 1963. He remained a professor of physiology there until 1985, when he became professor of human reproduction. He retired, becoming an emeritus professor, in 1989. He and Steptoe established the Bourne Hallam Clinics for treatment of infertility, and he was the scientific director of the clinics from 1988 to 1991.

From the beginning of his career, Edwards was interested in embryology and especially in fertilization, the process in which an egg (female sex cell) and a sperm (male sex cell) join together to begin a new living thing. While he was still at Edinburgh, he and his future wife, Ruth Fowler, worked out a method for treating female mice with hormones to make the mice release mature eggs on demand. This made the laboratory study of fertilization and reproduction much easier. He went on to fertilize the eggs, implant them into the uteruses of the female mice, and show that they developed into normal offspring. He did similar procedures on a variety of other mammals in the early 1960s.

Edwards made his first attempts to fertilize human eggs outside the body in 1965. He hoped that being able to do this would allow doctors to treat some forms of infertility. For example, some women cannot become pregnant because their fallopian tubes, which normally carry eggs from the ovaries (where eggs mature) to the uterus (where the developing child will grow), are blocked. Edwards reasoned that if he could remove mature eggs from such a woman's ovaries surgically, fertilize them in the laboratory, and then implant one in her uterus, the child might develop normally.

Such a procedure proved to be more easily imagined than accomplished, however. Edwards and his coworkers had to perform many technical feats, such as developing a nutrient solution in which human eggs could live after they had been removed from the body. Even when these things had been done, he was not sure how eggs would be obtained in the first place. His experiments had used eggs from ovary tissue that surgeons had removed for other reasons, but he knew that few women would be willing to undergo major abdominal surgery for this purpose alone.

Edwards glimpsed a possible solution to this problem in 1967, when he read a paper by Patrick Steptoe, a surgeon who specialized in the female reproductive system. Steptoe described a new device called a laparoscope, invented in France and Germany, which allowed abdominal organs to be viewed through a tiny, keyhole-shaped incision near the navel. A slender tube containing optical fibers was inserted through the hole and used as a kind of telescope to see the organs. Steptoe was the first English-speaking surgeon to use this device.

Edwards realized that the laparoscope might be adapted to remove eggs in a relatively minor operation. He contacted Steptoe, and in 1968 the two began working together to create such a procedure. Somewhat as Edwards had done with his mice, they treated volunteer women with hormones to make their eggs mature at a certain time. They developed a device that Steptoe could use with his laparoscope to remove the eggs. Edwards then fertilized them in his laboratory, using sperm provided by the women's husbands. By 1971, Edwards and Steptoe had obtained fertilized eggs and showed that they appeared to begin normal development.

The researchers were now ready to try the last step, placing a fertilized egg in a woman's uterus to complete its growth. This proved to be very difficult. They first succeeded in 1975, but they had to terminate the pregnancy because the egg implanted itself in the woman's fallopian tube, where its further development would have threatened the mother's life. Only in 1977, after 102 failed attempts, did Edwards and Steptoe work out the best time for implanting the egg and succeed in beginning a normal pregnancy. Steptoe delivered Louise Joy Brown, the world's first test-tube baby, by cesarean section at his clinic in Oldham on July 25, 1978.

Some religious leaders objected to in vitro fertilization because it separated sex from procreation. Some scientists also questioned whether it was safe. Infertile couples, however, hailed it as a great advance, and controversy over it died down as more and more test-tube babies were born without incident. Almost a million such babies were born in North America and Europe during the next 20 years. The procedure also helped researchers develop new techniques such as tests to identify unborn children who will suffer from inherited diseases.

In his later years, Edwards continued his research into fertility and ways to help infertile couples, including work on male infertility and on preserving embryos by freezing. Meanwhile, he was honored as a pioneer in the field. For example, he received the prestigious Albert Lasker Clinical Medical Research Award in 2001. (Steptoe was not eligible because he had died in 1988.) When Britain issued a set of postage stamps in 2000, honoring the four scientists who had made what the government felt were the greatest British contributions to clinical medicine during the previous 1,000 years, Edwards was one of those pictured.

Further Reading

Edwards, Robert G. "The Rocky Road to Human In Vitro Fertilization." *Nature Medicine*, October 2001.

———, and Patrick Steptoe. *A Matter of Life: The Story of a Medical Breakthrough*. New York: 1980.

Goldstein, Joseph L. "Comments at the Awards Ceremony." Lasker Foundation. Available online. URL: www.laskerfoundation.org/awards/library/2001remarkjg.shtml. Posted 2001.

Ehrlich, Paul

(1854–1915)
Prussian/German
Immunologist, Bacteriologist, Pharmacologist

Paul Ehrlich founded or helped to found three modern biomedical research specialties: immunology, the study of the body's defense system; hematology, the study of the blood; and chemotherapy, the prevention or treatment of disease with drugs. He developed the first drug designed to kill a particular kind of disease-causing microorganism inside the body.

Ehrlich was born in Strehlen, then a part of Silesia, in the German state of Prussia (now Strzelin, Poland), on March 14, 1854, to Ismar and Rosa Weigert Ehrlich. His father was an innkeeper. Ehrlich was a restless and uninspired student, studying at the Universities of Breslau, Strassburg, Freiburg, and Leipzig in the 1870s. He earned a medical degree from Leipzig in 1878. His final student project, a study of how certain animal tissues could be stained, established two important themes in his life: It involved him with dyes, especially the aniline dyes that Europeans had begun making from coal tar in 1853, and introduced him to the idea that chemicals could affect different kinds of cells in different ways.

After obtaining his M.D., Ehrlich became an assistant in the medical clinic of the University of Berlin and continued his research on dyes. He first used them to identify different kinds of cells in the blood, thus creating the speciality of hematology. Then, in 1882, he developed a new way to stain the bacteria that bacteriologist ROBERT KOCH had just identified as the cause of tuberculosis. Later scientists applied a modified form of his technique to a wide range of bacteria, and it is still used today. Ehrlich married Hedwig Pinkus in 1883, and they later had two daughters. In 1884, he became head physician of the medical clinic at the Charité Hospital in Berlin and a professor at the university.

During his research on tuberculosis bacteria, Ehrlich himself caught the disease. There were no drugs to treat it in those days, so doctors usually told people who had it to spend time in a warm, dry climate. Ehrlich therefore went to Egypt for two years (1886–88), during which he recovered. In 1890, soon after his return, he joined Koch's new Institute for Infectious Diseases, where he became one of the first scientists to study the immune system. He showed that the system kills microbes by means of chemical reactions.

At this time, scientists were beginning to harness the immune system to treat certain diseases. They injected disease-causing bacteria into horses, which made the horses' immune systems form chemicals that counteracted poisons made by the bacteria. The chemicals were released into the serum, the liquid part of the horses' blood. The serum could then be harvested and given as a treatment—a so-called antitoxin or antiserum—to people who had the disease.

The German government built a new Institute for Serum Research and Investigation in Steglitz, a suburb of Berlin, and put Ehrlich in charge of it in 1896. EMIL VON BEHRING, another German scientist trained by Koch, had developed an antiserum for diphtheria, a bacteria-caused disease often fatal to children, and Ehrlich worked out a way to measure the strength of the antitoxin (which varied in different batches of serum) so that it could be standardized, making the treatment more dependable. His method was later applied to other kinds of antisera as well.

Trying to determine more precisely how antisera worked, Ehrlich continued his research on the basic actions of the immune system. He showed that certain substances on the surface of microbes or their toxins attach to other chemicals on the surface of particular immune system cells. Borrowing a term from his studies of dye chemistry, he called the cell surface substances side chains. After this attachment occurs, he said, the cells make multiple copies of the side chains and release them into the blood. He named these free-floating side chains *antibodies*.

At the beginning of the 20th century, German scientist Paul Ehrlich revealed new facts about the immune system and developed the "magic bullet" Salvarsan, the first drug that destroyed a particular type of disease-causing microorganism when taken internally. *(National Library of Medicine)*

Each kind of antibody matches only one kind of microbe or other invader, just as a key fits only a particular lock. If that type of microbe invades the body a second time, the antibodies attach to the microbes and mark them for destruction by immune system cells. Ehrlich concluded that antisera work because they contain antibodies formed by the horse's immune system, which attach to the bacteria and their toxins and spur the human immune system into destructive action. For this "side-chain theory of immunity," as well as his work in standardizing antisera, Ehrlich shared the 1908 Nobel Prize in physiology or medicine with Russian-French researcher ILYA ILYICH MECHNIKOV, another founder of immunology.

Ehrlich became the public health officer for the city of Frankfurt in 1897, and two years later he became director of the Royal Institute of Experimental Therapy, which the German government built in Frankfurt just for him. He also headed the Georg Speyer House, a private laboratory built next door to the institute.

With these extensive facilities and staff at his command, Ehrlich turned his research in a new direction. Combining his discoveries about antibodies with his student observation that dyes—that is, chemicals—affected different kinds of cells in different ways, he began a search for drugs that would be drawn to particular kinds of microbes, just as antibodies and stains were. He coined the term *chemotherapy* for the use of such specific drugs and said that the drugs would be "magic bullets," attacking microbes without harming body cells.

Ehrlich searched for his magic bullets by having his staff test hundreds of different compounds on animals that had been infected with particular kinds of microbes. These compounds included not only known substances, such as dyes, but new ones that he had his chemists make by analyzing the composition of existing chemicals and then varying that composition in systematic ways. He first tried this technique against microscopic parasites called trypanosomes, which cause illnesses such as sleeping sickness, and then against a related group of microbes named spirochetes because of their corkscrew shape. In 1906, other scientists determined that a spirochete caused syphilis, a widespread human disease that produced disfigurement, insanity, and eventual death. Ehrlich thereafter focused on this disease, for which no effective treatment existed.

The substances that Ehrlich and his assistants tested against spirochetes were mostly variants of a drug called atoxyl, which other scientists had shown to kill some kinds of trypanosomes. Atoxyl was a compound of arsenic and, like its parent element, could be very poisonous. Ehrlich hoped to find, or create, a form of atoxyl that would be both less toxic and more effective than the original drug.

Patiently, stubbornly, Ehrlich and his staff worked their way through 605 chemicals related to atoxyl with no luck. Then, on August 31, 1909, the 606th compound cured a rabbit that had been infected with syphilis microbes. The following year, after hundreds of experiments on animals and tests on human patients had confirmed his discovery, Ehrlich publicly announced that this compound, which he had named Salvarsan, could cure syphilis. Salvarsan was widely hailed and quickly became very popular, remaining the best treatment for syphilis until penicillin became available in the early 1940s. Perhaps even more important, Ehrlich's idea of "magic bullets" became the new philosophy of drug development.

Ehrlich was greatly honored in his later years. In addition to the Nobel Prize, he was awarded the Prize of Honor at the 15th International Congress of Medicine at Lisbon in 1906, the Liebig Medal of the German Chemical Society in 1911, and the Cameron Prize of the University of Edinburgh in 1914. He was elected to the Privy Medical Council of Prussia (part of Germany) in 1897 and raised to the highest rank of the council in 1911.

In the final years of his life, Ehrlich did research on cancer. He predicted that scientists would eventually produce vaccines against cancer, much like those used against diseases caused by microbes; researchers are still pursuing this idea. Stress from controversies about Salvarsan's safety and the outbreak of World War I in 1914 took its toll on the aging scientist, however. He had a slight stroke in December 1914 and then, while on vacation in Bad Homburg, suffered a second stroke, which killed him on August 20, 1915.

Further Reading

Baumler, Ernest. *Paul Ehrlich: Scientist for Life*. New York: Holmes & Meier, 1984.

de Kruif, Paul. *Microbe Hunters*. New York: Harcourt Brace, 1926.

"Paul Ehrlich—Biography." *Nobel Lectures, Physiology or Medicine, 1901–1921*. Nobel Foundation. Available online. URL: www.nobel.se/medicine/laureates/1908/ehrlich-bio.html. Last updated 2001.

Silverstein, Arthur M. *Paul Ehrlich's Receptor Immunology: The Magnificent Obsession*. San Diego, Calif.: Academic Press, 2002.

Elion, Gertrude Belle

(1918–1999)
American
Chemist, Pharmacologist

Although Nobel science prizes are usually awarded for basic research, the 1988 prize in physiology or medicine went to three people in applied science—also known as drug developers. "Rarely has scientific experimentation been so intimately linked to the reduction of human suffering," the 1988 *Nobel Prize Annual* said of their work. One of the honored researchers was Gertrude Belle Elion.

Elion, whom everyone called Trudy, was born on January 23, 1918, in New York City to parents who had immigrated from Lithuania and Poland. Her father, Robert, was a dentist. The family moved to the Bronx, then a suburb, in 1924. Trudy spent much of what she calls a happy childhood reading, especially about "people who discovered things."

In 1933, the same year Elion graduated from high school at age 15, her beloved grandfather died painfully of stomach cancer. She determined to find a cure for this terrible disease. There was no money to send her to college, however, because her father had lost his savings in the 1929 stock market crash. Elion therefore enrolled at New York City's Hunter College, which offered free tuition to qualified women. She graduated with a B.A. in chemistry and the highest honors in 1937.

Elion failed to win a scholarship to graduate school, so she worked at several short-term jobs, meanwhile taking courses at New York University for her master's degree. She then did her degree research on evenings and weekends while teaching high school and finally completed the degree in 1941.

World War II removed many men from workplaces, making employers more willing to hire women, and in 1944 the doors of a research laboratory finally opened to Gertrude Elion. Burroughs Wellcome (now GlaxoSmithKline), a New York drug company, hired her as an assistant to GEORGE HERBERT HITCHINGS. Researchers at the time normally developed drugs more or less by trial and error, but Hitchings thought that an approach based on cell chemistry could produce better results. He wanted to look for or create compounds that interfered with essential processes taking place only (or at least more often) in cells that cause disease, such as cancer cells or bacteria.

Scientists had not yet learned that DNA carries the inherited information on which all cells depend, but they did know that DNA molecules must somehow reproduce themselves each time a cell divides. They also knew that the large DNA molecule includes several kinds of smaller molecules, some of which belong to families of compounds called purines and pyrimidines. DNA must take up these compounds from the cell in order to reproduce. Hitchings reasoned that if DNA in, say, cancer cells could be made to take up substances that were similar to purines and pyrimidines, yet also slightly different, these chemicals might block DNA reproduction and ultimately kill the cell, much as an ill-fitting part can jam and even destroy a machine.

Around 1947, Hitchings and his team began to search for such compounds and send them to the Sloan-Kettering Institute for testing as possible anticancer drugs. Hitchings set Elion to work synthesizing "almost"-purines, while other laboratory workers did the same for pyrimidines.

The laboratory's first major success was 6-mercaptopurine (6-MP), which Elion created in 1950. It was one of the first drugs to fight cancer by interfering with cancer cells' DNA. It worked especially well against childhood leukemia, a blood cell cancer that had formerly killed its victims within a few months. When combined with other anticancer drugs, 6-MP now cures about 80 percent of children with some forms of leukemia.

6-MP's effects on leukemia proved to be just the beginning of its powers. Scientists in other laboratories discovered that, by acting on the same cells that are overproduced in leukemia, the drug halts some actions of the immune system. The immune system's attacks on "foreign" substances protect the body against invaders such as bacteria, but they also destroy transplanted organs, except in identical twins. At Elion and Hitchings's recommendation, researchers even-

Gertrude Belle Elion, shown here in her laboratory at Burroughs Wellcome (now GlaxoSmithKline), created drugs that destroyed cancer cells and made organ transplants possible. *(GlaxoSmithKline and Jonathan L. Elion)*

tually tested not only 6-MP but azathioprine, a related compound that proved able to suppress the immune system even more effectively than 6-MP did, as possible drugs to prevent the rejection of organ transplants. Boston surgeon Joseph Murray used azathioprine in the first successful kidney transplant between unrelated humans in the early 1960s. It was the breakthrough drug that made organ transplants practical.

Another compound Elion developed in her cancer research that proved to have other uses was called allopurinol. Because it can prevent the formation of uric acid, allopurinol has become the standard treatment for a painful disease called gout, in which crystals of uric acid are deposited in a person's joints. In the early 1970s, Elion also helped to develop acyclovir, the first drug to successfully combat infections by a dangerous group of viruses called herpesviruses. Indeed, Michael Colvin in *Science* magazine calls it "the first truly effective antiviral compound" of any kind.

As Elion and Hitchings developed drug after drug, they advanced together within Burroughs Wellcome. Finally, in 1967, Elion was made head of her own laboratory, the newly created Department of Experimental Therapy. Although she had always enjoyed working with Hitchings, she was glad to have more independence. When Burroughs Wellcome moved to Research Triangle Park, North Carolina, in 1970, Elion moved with it. Her laboratory became a "mini-institute" with many sections.

Gertrude Elion officially retired in 1983, but she remained as busy as ever. She traveled often and spoke widely, especially to encourage young people to enter science. "We've got to tell them how much fun it is," she said. She worked with young scientists in programs at Duke University, where she was a research professor of medicine and pharmacology, and the University of North Carolina at Chapel Hill, where she was an adjunct professor of pharmacology. Elion's scientific legacy continued as well. For instance, workers from her team, using approaches she had developed, discovered AZT, the first drug approved for the treatment of AIDS.

Elion's greatest honors came after her retirement. In 1988, she shared the Nobel Prize in physiology or medicine with Hitchings and British drug researcher JAMES WHYTE BLACK for their creation of "rational drug design." She was inducted into the Inventors' Hall of Fame, the first woman to be so honored, in 1991 and also received the National Medal of Science that year. She won the Lemelson/MIT Lifetime Achievement Award in 1997. She is included in the National Women's Hall of Fame and the Engineering and Science Hall of Fame as well. Although she was glad to have such prizes, Elion said, "My rewards had already come in seeing children with leukemia survive, meeting patients with long-term kidney transplants, and watching acyclovir save lives and reduce suffering." Elion died in Chapel Hill, North Carolina, on February 21, 1999.

Further Reading

Bouton, Katherine. "The Nobel Pair." *New York Times Magazine*, January 29, 1989.

"Elion, Gertrude B." *Current Biography Yearbook 1995*. New York: H. W. Wilson, 1995.

"Gertrude B. Elion—Autobiography." *Les Prix Nobel 1988*. Available online. URL: www.nobel.se/medicine/laureates/1988/elion-autobio.html. Last updated 2002.

McGrayne, Sharon Bertsch. *Nobel Prize Women in Science: Their Lives, Struggles, and Momentous Discoveries*. New York: Birch Lane Press, 1993.

St. Pierre, Stephanie. *Gertrude Elion: Master Chemist*. Vero Beach, Fla.: Rourke Enterprises, 1993.

Enders, John Franklin

(1897–1985)
American
Bacteriologist, Virologist

John Enders developed techniques for growing viruses in the laboratory that aided research on these extremely tiny microorganisms and made possible the development of lifesaving vaccines such as those against measles and polio. Enders was born on February 10, 1897, in West Hartford, Connecticut. His father, John Ostrom Enders, was a well-to-do banker. His mother was the former Harriet Whitmore.

In his student days, Enders had trouble deciding what career he wanted to follow. He entered Yale in 1915, left to serve as a pilot and flight instructor during World War I, briefly sold real estate, and then went back to Yale, graduating in 1920. He began graduate school at Harvard in English literature and philology, the study of languages, and earned an M.A. in 1922, but then he changed his focus and earned a Ph.D. in bacteriology and immunology in 1930. He married Sarah Bennett in 1927, a marriage that produced two children and lasted until her death in 1943. He married Carolyn Keane in 1951.

Enders joined the Harvard Medical School faculty as an assistant in 1929, even before he received his doctoral degree, and remained there throughout his career. His first research was on bacteria, but in 1938 he began studying viruses that infect mammals, including humans. Viruses can reproduce only inside living cells, and at the time Enders started his research, scientists believed that they had to be grown inside whole animals. This made viruses very difficult to study and also held back development and mass production of vaccines against virus-caused diseases.

In 1946, university officials asked Enders to establish and head a research division on infectious diseases at Children's Hospital Medical Center in Boston, which is affiliated with Harvard. There, in the late 1940s, he and his coworkers showed for the first time that viruses could be grown in cells kept in laboratory tissue culture. They used the newly discovered antibiotic penicillin to prevent bacteria from destroying the cells. Enders used this technique first on the viruses that cause mumps and measles and then, working with Frederick Robbins and Thomas Weller, extended it to chicken pox virus in 1948. They grew the chicken pox virus in skin and muscle tissue taken from naturally aborted human embryos, choosing those tissues because they are the ones that the virus normally infects.

Then, as often happens in science, a piece of luck occurred—luck not only for Enders, but for all the families who dreaded summer because the season brought epidemics of poliomyelitis (polio, for short), a virus-caused disease that had killed or crippled thousands of children and some adults, including President Franklin Roosevelt. Researchers had long wanted to develop a vaccine against polio but had been unable to do so largely because, as far as anyone knew, the virus could grow only in the brain and nerve tissue of living monkeys. The virus could be extracted from the tissue and stored, but the amount produced was very tiny.

A vial of poliovirus happened to be in a cabinet next to the area where Enders and the others were working on the chicken pox virus. They had more dishes of embryo tissue than they needed for those experiments, so they decided to try the poliovirus on it. After some time had passed, they injected the cell preparation into monkeys. The monkeys developed polio, showing that the virus had reproduced in the mixture. The success of this and subsequent experiments not only proved that poliovirus could grow outside a living body and outside nerve tissue but created a way to produce large amounts of it fairly easily.

Enders's discovery transformed polio research. JONAS SALK and ALBERT BRUCE SABIN used his technique (adapted to work with monkey rather than human tissue) to develop vaccines against the disease in the early 1950s and to produce the tremendous amounts of virus used in the vaccines. By demonstrating that poliovirus could live in tissues besides nerves, Enders's research also helped other scientists learn how the disease was spread. They discovered that the virus entered the body through the mouth and multiplied in the intestines before entering the bloodstream and traveling to the brain.

Enders shared the 1954 Nobel Prize in physiology or medicine with Robbins and Weller for their development of ways to grow viruses, especially poliovirus, in the laboratory. He also received the Kyle Award of the U.S. Public Health Service (1955), the Presidential Medal of Freedom (1963), and the Science Achievement Award of the American Medical Association (1963). Meanwhile, he studied other viruses, for instance developing a vaccine against measles in 1957. He became a full professor of bacteriology and immunology at Harvard in 1956 and a university professor, the highest faculty level, in 1962.

Enders retired from the Harvard faculty in 1968 and from his post at Children's Hospital in 1972, after which he became chief of the hospital's virus research unit. He died of heart failure in Waterford, Connecticut, on September 8, 1985. "That uncountable children were saved from disability because of the work John Enders and his colleagues did is probably his greatest testimony," medical historian Walt Schalick told *Harvard Gazette* writer Alvin Powell in 1998.

Further Reading

"John Franklin Enders—Biography." *Nobel Lectures, Physiology or Medicine 1942–1962*. Available online. URL: www.nobel.se/medicine/laureates/1954/enders-bio.html. Last updated 2001.

Powell, Alvin. "John Enders' Breakthrough Led to Polio Vaccine." *Harvard Gazette*, Oct. 8, 1998. Available online. URL: www.news.harvard.edu/gazette/1998/10.08/JohnEndersBreak.html. Accessed 2003.

Fabre, Jean-Henri

(1823–1915)
French
Entomologist

Jean-Henri Fabre made some of the first detailed studies of insect behavior in the wild and wrote books that inspired a wide audience to share his fascination with the natural world. Famed French writer Victor Hugo, referring to the ancient Greek writer of epic poems, called Fabre the "Homer of the insects."

Fabre was born in the village of Saint-Léons, in the south of France, on December 23, 1823, to a poor family. He spent his first years in the countryside with his grandmother. As a young man, he had to work odd jobs, such as selling lemons at a fair, to earn money to continue his education. Teaching himself for the most part, he earned degrees in mathematics and physical sciences in the 1840s and then a doctorate of science from the Sorbonne in Paris in 1855. He also married in that year, the first of two marriages. He had five children by his first wife.

Fabre spent the first half of his adult life as an elementary and high school teacher, chiefly (for about 20 years, beginning in 1852) at a high school in Avignon, where he taught physics and chemistry. In 1862, he started writing popular books about nature as a way of adding to his meager salary, and he eventually composed 95 of them.

Fabre's books sold well, and the provincial teacher began to earn a reputation. In 1867, for instance, the French emperor Napoleon III awarded him the Legion of Honor, and in the following year the education minister asked him to tutor the emperor's son. (Disliking court life, Fabre declined.) Established scientists ignored him, however, both because he had no university position and because he observed and experimented on living things in their natural surroundings rather than in a laboratory. Fabre, in turn, criticized them as theorists who understood little of nature.

Fabre's eccentric ways brought apparent disaster in 1870. Unlike most people of his time, he believed that girls as well as boys deserved an education, and he taught local girls in the evenings. Catholic clergy criticized him for mentioning sexual reproduction in these classes, even though his discussions involved only plants. This criticism made his religious-minded landladies evict him and his family. The disaster, however, proved to be a blessing in disguise. It pushed Fabre to borrow enough money from a friend, British philosopher John Stuart Mill, to be able to give up teaching and devote his full time to observing and writing.

He lived first in the village of Orange, and then, after 1879, in a house next to a *harmas*, or abandoned field, near Sérignan, in the district of Provence in southeastern France. The field became his outdoor laboratory.

Fabre's chief interest was insects, a supposedly lowly form of life that few other scientists had bothered to study. He showed that these and similar creatures could carry out extremely complex behavior, yet had little ability to deal with anything that changed or interrupted their routine. If he destroyed part of a spider's web, for instance, the spider could not repair the damage but instead started a whole new web. These observations proved that the instinct guiding such behaviors was very different from human intelligence. Fabre's studies helped to lay the groundwork for ethology, the study of animal behavior.

Some of Fabre's ideas were ahead of his time, but others were just the opposite. In some ways as pious as his unforgiving landladies, he could never accept CHARLES ROBERT DARWIN's theory of evolution by natural selection, although he and Darwin exchanged letters and respected each other. Fabre preferred to believe that a divine being had placed the directions for all their behavior into insects' brains. Scientists today would say that instinctive behavior is programmed by genes.

Fabre's most famous works were a set of 10 books called *Souvenirs Entomologiques* (*Entomological Memories*). These books, published between 1879 and 1907, make the lives of insects and spiders as thrilling as a novel and as moving as poetry. They also present a vivid picture of Fabre's own enthusiastic, opinionated personality. They are still popular. His exquisite paintings of mushrooms are also much admired.

Although Fabre was quite well known by the end of his life—leading scientists from all over Europe came to honor him at a celebration in 1910—he received few awards, probably because he was not part of orthodox academic circles. He was nominated for the Nobel Prize in literature but never received it. Perhaps the happy years in his field and the knowledge that generations of scientists and nonscientists alike had been inspired by his writings were rewards enough. Fabre died on October 11, 1915, at the ripe age of 92.

Further Reading

Anderson, Margaret J. *Children of Summer: Henri Fabre's Insects*. New York: Farrar, Straus & Giroux, 1998.

"Biography of Jean-Henri Fabre." Translated by Sue Asscher. Available online. URL: www.e-fabre.net/. Accessed 2001.

Fabre, Jean-Henri. *Fabre's Book of Insects*. Translated by Alexander Teixiera de Mattos. Reprint, Mineola, N.Y.: Dover Publications, 1998.

Pasteur, Georges. "Jean Henri Fabre." *Scientific American*, July 1994.

Fisher, Ronald Aylmer

(1890–1962)
British
Statistician, Geneticist, Evolutionary Biologist

By reconciling GREGOR MENDEL's laws of genetic inheritance with CHARLES ROBERT DARWIN's theory of evolution by natural selection, statistician Ronald Fisher created the mathematical basis for modern genetics. Evolutionary biologist and science writer STEPHEN JAY GOULD wrote, "No scientist is more important [than Fisher] as a founder of modern evolutionary theory."

Fisher was born to George and Katie (Heath) Fisher in East Finchley, London, on February 17, 1890. His father was an art auctioneer. Fisher attended Gonville and Caius College at Cambridge University, graduating in 1912. Although his undergraduate major was astronomy and he also studied physics, his chief interest was mathematics, especially the relatively new science of statistics. He married Ruth E.

Guiness in 1917 and, like his own parents, had eight children.

Fisher had strong opinions and did not take criticism easily. (*The Cambridge Dictionary of Scientists* describes him as "small, forceful, eloquent, and eccentric.") He refused a faculty position at Cambridge in 1919 because it would have required him to work under a statistician who had made negative comments about his work. Instead, he joined the Rothamsted Experimental Station, an agricultural research facility in the countryside, and set about analyzing more than 60 years' worth of data about farming experiments.

Fisher's work at Rothamsted led him to invent concepts such as the analysis of variance and maximum likelihood, which are considered to be at the heart of modern statistics. His methods helped scientists deal with the variation that is unavoidable in biological research and showed them how to make precise estimations and reach valid conclusions on the basis of relatively small samples. He earned a doctorate in science for some of this work in 1926. The techniques described in two of his books, *Statistical Methods for Research Workers* (1925) and *Design of Experiments and Statistical Methods* (1934), are still used constantly in scientific research.

Fisher's most important direct contribution to biology is *Genetical Theory of Natural Selection*, published in 1930. This book uses statistics to reconcile Mendel's and Darwin's theories, which many geneticists had seen as being in conflict. According to Gould, Fisher's work is "the keystone for the architecture of modern Darwinism." Fisher is considered to have founded the field of population genetics, or genetics applied to large groups.

In 1933, Fisher became Galton Professor of Eugenics at University College, London. He was a strong believer in eugenics, the idea that humanity could be improved by selective breeding—encouraging people with supposedly desirable characteristics to reproduce (positive eugenics) and discouraging or preventing those with undesirable traits from reproducing (negative eugenics). Fisher concentrated on positive eugenics. Eugenics has since been discredited, but at the time it was a popular belief, even among scientists.

At University College, Fisher studied the inheritance of human blood groups, especially the Rh groups. These differences in substances on the surface of blood cells, discovered by KARL LANDSTEINER, can cause immune reactions in people given transfusions and in some pregnant women. Fisher returned to Cambridge as a professor of genetics in 1943, a post he kept until he retired in 1957. He then served for two years as president of Gonville and Caius College, after which he moved to Adelaide, Australia, and did research for the Commonwealth Scientific and Industrial Research Organization (CSIRO).

Some of Fisher's personal beliefs drew criticism, such as his support of eugenics and his attempts in the 1950s to disprove the idea, then being proposed for the first time by RICHARD DOLL and others, that cigarette smoking might cause lung cancer. Nonetheless, his work in statistics won many awards, including three medals from the Royal Society, Britain's top science organization: the Royal Medal (1938), the Darwin Medal (1948), and the Copley Medal (1956). Fisher was knighted in 1952 and died in Adelaide on July 29, 1962.

Further Reading

Box, Joan Fisher. *R. A. Fisher, The Life of a Scientist*. New York: Wiley, 1978.

Fisher, Ronald Aylmer. *The Genetical Theory of Natural Selection*. Edited by J. H. and Henry Bennett. 1930. Reprint, New York: Oxford University Press, 2000.

"Sir Ronald Aylmer Fisher." University of Minnesota. Available online. URL: www.mrs.umn.edu/~sungurea/introstat/history/w98/RAFisher.html. Accessed 2001.

Fleming, Alexander
(1881–1955)
British
Bacteriologist

In perhaps the best-known story of accidental discovery in the history of science, Alexander Fleming's recognition of something "funny" about a spoiled culture dish in his laboratory led to the development of the lifesaving antibiotic penicillin. Fleming and the two scientists who put his discovery into practical form, HOWARD WALTER FLOREY and ERNST BORIS CHAIN, shared the Nobel Prize in physiology or medicine in 1945.

Fleming was born into a poor sheep-farming family on August 6, 1881, at Lochfield in Ayrshire, Scotland. He was one of the four children of Hugh and Grace (Morton) Fleming. He moved to London in 1894 and worked in a shipping office for four years to earn money for an advanced education. A small inheritance from a relative, followed by a scholarship, finally let him study at St. Mary's Medical School, part of London University. He graduated in 1906 and obtained M.B. and B.S. degrees, along with a gold medal, in 1908. He married Sarah McElroy, an Irish nurse, in 1915, and they had one son, Robert. Sarah died in 1949, and in 1953 Fleming married Amalia Koutsouris-Voureka, a Greek-born coworker in his laboratory.

Fleming worked as a lecturer at St. Mary's Hospital until World War I, when he served in the British Army Medical Corps in France. He was frustrated to find that treating soldiers' wounds with disinfectants, which killed bacteria on the body's surface, often failed to prevent fatal infections from developing inside the tissues. When he returned to St. Mary's after the war, Fleming began a search for bacteria-destroying compounds that, unlike disinfectants, could safely be taken internally. He became director of the hospital's department of systematic bacteriology in 1920.

British bacteriologist Alexander Fleming discovered penicillin in 1928 when he noticed something "funny" about a speck of mold growing in a dish of bacteria. *(National Library of Medicine)*

For a while, beginning in 1922, Fleming thought his search might be successful. One day, he accidentally sneezed on a laboratory culture plate covered with bacterial colonies and noticed that clear spots, showing a lack of bacterial growth, developed wherever droplets of mucus had hit the plate. He therefore tested mucus and other secretions, such as tears and saliva, for antibacterial activity. He found that these and many body tissues contained a substance that could kill bacteria by dissolving their cell walls. He named this chemical lysozyme. Lysozyme probably acts as a mild natural antiseptic, but the only bacteria it proved able to kill when given as a drug were harmless, so Fleming had to conclude that it was useless for this purpose.

Famed French scientist LOUIS PASTEUR was fond of saying that "chance favors the prepared mind." Fleming's discovery of lysozyme may have prepared his mind to recognize the importance of a second laboratory accident that occurred in 1928, the year he attained the title of professor. Although he was a meticulous experimenter, he was less careful about cleaning up, especially after his assistant, Merlin Pryce, left him to work elsewhere. When Fleming went on a two-week vacation in late August, therefore, he left behind some used culture plates that he had not gotten around to washing out.

Pryce stopped by for a visit soon after Fleming's return, and Fleming was showing him the stack of plates when he suddenly stopped and said, "That's funny." Fleming had noticed that in addition to a yellow carpet of staphylococcus, a common kind of bacteria that can cause serious wound infections, one dish in the stack contained a patch of yellow-green mold—and around the patch was a clear ring, something like the ones that had led Fleming to lysozyme. He guessed that spores of the mold had drifted in through the open window from another laboratory on the floor below, where molds were being studied.

At first Fleming thought his mold might be making lysozyme, but staphylococcus normally is not harmed by lysozyme, so he concluded that the germ-killer must be something else. After identifying the mold as an unusual strain of a common fungus called *Penicillium notatum*, which sometimes grows on stale bread, he named its microbe-killing "mold juice" penicillin.

Fleming injected the mold juice into healthy mice and rabbits and eventually even into his new lab assistant, Stuart Craddock, and showed that it was not poisonous. Nonetheless, he found many reasons to conclude that penicillin, like lysozyme, would be worthless as a drug. He could extract only a tiny amount of the substance from each batch of mold, and he could not purify it. It broke down quickly after being extracted, and when he combined it with blood in test tubes, it became inactive. It showed only weak antiseptic action when he applied it to an infected wound on a woman's leg. Fleming therefore did not bother to try it on mice with bacterial infections. Instead, he published a brief paper about it in 1929, gave some samples of the mold to other laboratories in case anyone else might want to study it, and then more or less forgot about it.

During the 1930s, Fleming continued to investigate disinfectants and to devise ways to identify types of bacteria that infect wounds. He was amazed to read in the British medical journal *The Lancet* in 1940 that Florey, Chain, and their coworkers at Oxford University were in the process of turning penicillin into a useful drug. He visited them to "see what you are doing with my old penicillin," he said, and was greeted by the startled Chain's exclamation, "We thought you were dead!"

When penicillin became famous in the mid-1940s, so did Fleming. People loved the dramatic story of the mold on the culture plate, and the white-haired Scottish doctor in his neat suit and bow tie equally seemed to love chatting with reporters and posing for photographs. He was knighted in 1944 and received many awards in addition to the Nobel Prize, including the Cameron Prize of the University of Edinburgh (1945), the Gold Medal of the Royal College of Surgeons (1946), and the Gold Medal of the Royal Society of Medicine (1947). Partly because Florey preferred to avoid publicity, Fleming sometimes received more credit for the wonder drug than he deserved, although he usually tried to put what he called "the Fleming myth" into perspective. He once said, "My only merit [in the discovery of penicillin] is that I did not neglect the observation [of the mold in the culture dish] and that I pursued the subject as a bacteriologist."

Fleming remained at St. Mary's until his retirement in 1948. He served as Rector of Edinburgh University from 1951 to 1954 and died of a heart attack at his home in London on March

11, 1955. As befitted a national hero, he was buried in St. Paul's Cathedral.

Further Reading

Gottfried, Ted. *Alexander Fleming: Discoverer of Penicillin*. New York: Franklin Watts, 1997.

Ho, David. "Bacteriologist Alexander Fleming." *Time*, March 29, 1999.

Macfarlane, Gwyn. *Alexander Fleming: The Man and the Myth*. Cambridge, Mass.: Harvard University Press, 1984.

⊠ Florey, Howard Walter
(1898–1968)
Australian/British
Pathologist, Pharmacologist

Howard Florey, with the help of ERNST BORIS CHAIN, Norman Heatley, and others on his team at Oxford University, turned ALEXANDER FLEMING's accidental discovery of a mold that killed bacteria into the lifesaving antibiotic penicillin. Florey was born into a well-to-do family in Malvern, a suburb of Adelaide, South Australia, on September 24, 1898. His father, Joseph Florey, was a shoe manufacturer. His mother was Bertha Florey, Joseph Florey's second wife.

Florey became interested in scientific research at around age 12. He attended St. Peter's College in Adelaide, where he excelled in sports as well as academics, and graduated with honors in 1916. He then studied medicine at Adelaide University. While there he met Ethel Reed, a fellow medical student, and they became friends. They were separated, however, when Florey won a Rhodes scholarship to attend Oxford University in Britain after his medical school graduation in 1921.

Florey obtained a bachelor of science degree from Oxford in 1924, then did postgraduate studies at Cambridge, Britain's other most famous university, and at several universities in the United States. He earned his Ph.D. from Cambridge in 1927. While in England he had continued to write to Reed, and he finally asked her to join him there. She did so, and they were married on October 19, 1927. They later had a daughter and a son.

Florey researched and taught briefly at Cambridge (1927–31) and Sheffield University (1932–35). In 1935, he once again joined Oxford, as head of the university's Sir William Dunn School of Pathology, and he remained there for the rest of his career. At Oxford, he began to look for substances that, unlike antiseptics, could be taken internally and would kill disease-causing bacteria without harming patients. Most researchers at the time believed that such compounds would most likely come from the chemist's laboratory, but Florey looked instead to the natural world. Certain molds and other microorganisms were known to make bacteria-killing substances as part of the competition involved in evolution, a phenomenon called antibiosis ("against life"). He thought one of these chemicals might prove to have potential as a drug.

Florey was also unusual for his time in that he decided not to work alone but instead gathered a team of experts in different fields to research this question. Chain and Heatley, for instance, were biochemists. The group first investigated lysozyme, an antibacterial substance in tears, mucus, saliva, and body tissues that Fleming had discovered in the early 1920s. Like Fleming, they eventually concluded that lysozyme was not very useful for treating infections, so they looked through scientific literature to see what other examples of antibiosis had been discovered. During this search, Chain found a paper that Fleming had written in 1929, in which he described the bacteria-killing effects of liquid made by a mold called *Penicillium notatum*. Fleming had named his "mold juice" penicillin, but he had not done extensive research on it.

Florey learned that Fleming had given samples of his mold to several research institutions, including the Dunn School, and that the school still had its sample. In 1939, his team began studying the mold. After much effort, Chain succeeded in extracting small amounts of the active compound from the mold's fluid, a task at which Fleming had failed. The group tested it first on healthy mice and rabbits and found that it did them no harm. Then, on May 25, 1940, they injected eight mice with streptococcus, a common kind of bacteria that can cause serious infections in wounds, and, an hour later, gave four of the animals shots of the group's tiny supply of penicillin. By the following day, the four untreated mice were dead, but the ones that had received penicillin were healthy. Even Florey, usually known for understatement, said, "It looks like a miracle."

The miracle seemed to be arriving just in time. Britain had gone to war with Nazi Germany in September 1939, and, Norman Heatley said later, Florey now "began to think of the implications [of penicillin] for war wounds." It seemed almost impossible to make enough penicillin to even test on humans, however, let alone enough to treat thousands of soldiers. Heatley tried growing mold in containers ranging from laboratory dishes to metal cookie boxes and discovered that the best *Penicillium* farms were hospital bedpans. He extracted penicillin from the mold in jury-rigged devices whose parts included a butter churn, a bookcase, a bathtub, and aquarium pumps.

At last, the team thought they had enough drug to try. In case it proved to have unexpected side effects, they wanted to use it first on someone who was already expected to die shortly. They chose Albert Alexander, a policeman who had developed an infection after being pricked by a rose thorn. Alexander received his first shot of penicillin on February 12, 1941, and a day later his fever dropped. The group stretched their supply of the drug as far as they could by reextracting it from Alexander's urine, but the supply still ran out before he was completely healed. The infection then returned and killed him. After that Florey tested only children, because their smaller size meant that they needed less of the drug. Five out of the next six patients survived.

Florey's group now knew they had an important discovery, but British drug companies were too involved in war work to try something new. The constant German bombing raids also would have threatened any penicillin factory. Indeed, fearing that his team might have to flee if the Germans invaded Britain, Florey had them smear the linings of their coats with *Penicillium* mold so that they could start their research again elsewhere.

Because of these difficulties in Britain, Florey and Heatley went to the United States in June 1941 to seek help in developing technology to mass-produce penicillin. They found a warm response, especially after the United States entered World War II in December. Commercial production of penicillin began in 1943, and Florey flew to North Africa to give the drug its first test on soldiers. Britain also began making the drug at about this time. By the time the Allies invaded France on D day (June 6, 1944), there was enough penicillin to treat all wounded Allied soldiers. Rocko Fasanella, a Yale professor who has studied penicillin's history extensively, says that "American penicillin won the war" just as much as the atomic bomb. When the antibiotic became available for civilian use shortly afterward, it had almost as much effect on the war against bacteria.

Florey and his team were hailed as heroes. Florey was knighted in 1944 and shared the Nobel Prize in physiology or medicine with Fleming and Chain in 1945, becoming the first Australian to receive this prize. He avoided as much media attention as he could, however, and continued his scientific work at Oxford. He searched for new antibiotics to attack bacteria that resisted penicillin and helped to develop one called cephalosporin. He also did research on the immune system.

Although Florey never returned permanently to Australia, he kept in close contact with scientific organizations there, and in the late 1940s he helped to establish the John

Curtin School of Medical Research at the new Australian National University in Canberra. He became chancellor of the university in 1965. He was also elected provost of Queen's College, Oxford, in 1962, the year he retired from active research. He served as president of the Royal Society, Britain's top science organization, from 1960 to 1965 and was made Baron of Adelaide and Marston and a member of the Order of Merit in 1965. In 1967, a year after Ethel Florey's death, Florey married Margaret Jennings, a member of his research team and longtime friend. He died of heart disease at his home in Oxford on February 21, 1968. Former Australian Prime Minister Sir Robert Menzies said that Florey had had more effect on world welfare than any other Australian.

Further Reading

Bickel, Lennard. *Howard Florey, the Man Who Made Penicillin.* Melbourne, Australia: Melbourne University Press, 1972.

Carleton, Sharon. "Howard Florey." Australian Broadcasting Corp., Radio National, "Norman Swan Health Report," September 14 and 21, 1998.

Florey Centenary Committee. "Florey Home Page." Available online. URL: www.tallpoppies.net.au/florey/. Accessed 2003.

Kiester, Edwin, Jr. "A Curiosity Turned into the First Silver Bullet Against Death." *Smithsonian,* November 1990.

Macfarlane, Gwyn. *Howard Florey, the Making of a Great Scientist.* New York: Oxford University Press, 1979.

Folkman, Moses Judah

(1933–)
American
Surgeon, Pharmacologist

Judah Folkman needed almost 30 years to convince fellow researchers that his idea of a new way to treat cancer had merit, but in the late 1990s it generated more excitement than any other proposed cancer treatment had done in years. Folkman was born in Cleveland, Ohio, on February 24, 1933. He was one of three children of Jerome Folkman, a rabbi, and his wife, Bessie. When seven-year-old Judah had been good, his father took him along on ministerial visits to hospitals as a reward. There, Judah observed that physicians as well as religious leaders could help people, and he decided to become a doctor.

Folkman attended Ohio State University, graduating in 1953. He then attended Harvard Medical School and graduated in 1957. While still a medical student, he helped to invent the pacemaker, an electrical device implanted in the heart to keep the heartbeat steady. This work won several prizes, including the Boylston Medical Prize and the Borden Undergraduate Award in Medicine.

Planning on a career in surgery, Folkman began his internship at Boston's Massachusetts General Hospital. In 1960, he married Paula Prial; they later had two daughters. In that year, he also began two years of research at the National Naval Medical Center in Bethesda, Maryland, during which he had the key insight about cancer that shaped the rest of his career.

While testing blood substitutes on rabbit thyroid glands maintained in glass chambers, Folkman decided to see what happened if he transplanted cancer cells into the glands. He noticed that the cancers never grew larger than the size of a pinhead. Over the next several years, he thought about how this tissue differed from the same tissue inside the body and realized that it lacked a blood supply. Perhaps, he reasoned, cancers could not grow beyond a very small size unless they had their own blood supply. He continued to think about this idea as he advanced in Harvard Medical School, Boston City Hospital, and the Children's Hospital Medical Center in Boston after his navy stint was over. In 1967, he became a full professor of surgery at Harvard and head of surgery at Children's Hospital.

Folkman described his ideas about tumors and blood vessels in an article published in the *New England Journal of Medicine* in 1971—after being rejected by many other medical journals because, the editors said, he had too little evidence to back up his theories. He proposed that all but the smallest cancers need blood vessels to bring oxygen and nutrients into the tumors. He claimed that the cancers persuade new vessels to grow, a process called angiogenesis, by secreting a substance that he named tumor angiogenesis factor, or TAF, even though he had not yet actually isolated it. He suggested that if this factor could be blocked, tumors might be prevented from growing, just as had happened in his rabbit thyroids. They might also be prevented from spreading, since blood vessels inside tumors give cancer cells a way to enter the bloodstream, travel through the body, and set up secondary tumors (metastases) at distant sites.

No one believed Folkman's theory at first because he was unable to prove the existence of either a substance that made blood vessels grow or one that made such growth stop. "For 10 years there was almost nothing but criticism every time I gave a paper," he recalled to *New York Times* reporter Nicholas Wade in late 1997. Nonetheless, he persisted. "In research, there's a very fine line between persistence and obstinacy," Folkman told Nancy Linde, producer of a *Nova* television program about him, in a 2001 interview. "You do not know whether if you're persistent a little while longer you'll make it, or whether you're just being obstinate, [and it] doesn't exist. And, of course, you can . . . stay with an idea too long—[that's] called pigheadedness." During the 1970s, many of Judah Folkman's fellow researchers placed him firmly in the "pigheaded" category.

Folkman believed in his research enough to step down from his surgical position at Children's Hospital in 1981 in order to devote more time to it—and finally, in the decade that followed, his efforts began to pay off. In 1983, Michael Klagsbrun and Yuen Shing isolated a stimulator of blood vessel growth from cancerous tissue in Folkman's laboratory. Then, in an accidental discovery much like the one that led British bacteriologist ALEXANDER FLEMING to penicillin in 1928, Don Ingber found one of the first angiogenesis inhibitors in a mold that had contaminated a culture dish of blood vessel cells. Robert D'Amato, another scientist in Folkman's laboratory, identified a disused drug called thalidomide as an angiogenesis inhibitor in 1992. Thalidomide had become infamous in the early 1960s for causing terrible birth defects—probably, D'Amato now realized, because it had interfered with blood vessel growth in the unborn babies.

Michael O'Reilly, a third Folkman researcher, found angiostatin and endostatin, the two angiogenesis inhibitors that were to make the laboratory famous, in the mid-1990s. Used together, these compounds stopped the growth of more than 95 percent of human tumors transplanted into mice. Just as important, the drugs seemed free of the side effects that make treatment with most anticancer drugs such an ordeal for patients. Blood vessel tissue also did not develop resistance to them, as cancers often did to standard drug treatments.

Only scientists noticed when a paper describing the mouse tests on angiostatin and endostatin appeared in a November 1997 issue of the prestigious British science journal *Nature*, but virtually everyone paid attention when the same test results were proclaimed in a front-page story in the *New York Times* on May 3, 1998. Suddenly Folkman found himself hailed as having discovered a cure for cancer. He and other cancer specialists tried to tone down the excitement, pointing out that, as Folkman had told Nicholas Wade, "wonderful things in the laboratory [often do] not make it to the clinic." In any case, they warned, the drugs would have to undergo years of animal and human testing before the federal Food and Drug Administration, which must approve all medications, would permit them to be sold.

Folkman presently directs the surgical research laboratories at Children's Hospital and is the Andrus Professor of Pediatric Surgery at Harvard Medical School, an appointment he has held since 1968. He is also a professor of anatomy and cellular biology at the medical school. He and his coworkers continue to investigate angiogenesis inhibitors, as do many other researchers and biotechnology companies. Most believe that such drugs, even if they pass all their tests, will be used along with the standard treatments for cancer—surgery, radiation, and other drugs—rather than as a single "magic bullet" cure. They might have to be given throughout a cancer patient's life to keep new tumors from growing.

By 2001, some 300 angiogenesis inhibitors were being studied, and at least 20 were being tested on humans. Angiostatin and endostatin have passed the first stage of such testing, which focuses on safety, but it is too soon to tell whether they will be as effective in humans as they are in mice. Even if they are not, as Folkman says, "the principle of angiogenesis is well established" and is sure to prove a fruitful new subject for medical research—thanks to Judah Folkman's "pigheadedness."

Further Reading

Begley, Sharon, and Claudia Kalb. "One Man's Quest to Cure Cancer." *Newsweek*, May 18, 1998.

Cooke, Robert. *Dr. Folkman's War: Angiogenesis and the Struggle to Defeat Cancer*. New York: Random House, 2001.

"Folkman, Judah." *Current Biography Yearbook 1998*. New York: H. W. Wilson, 1998.

Linde, Nancy. "Cancer Warrior: Dr. Folkman Speaks." NOVA Online. Available online. URL: www.pbs.org/wgbh/nova/cancer/folkman.html. Updated 2001.

Fossey, Dian
(1932–1985)
American/Rwandan
Primatologist

Living like a hermit in a mountain rain forest in Africa, Dian Fossey learned more about the endangered mountain gorilla than had ever been known. As more and more gorillas were killed by poachers, she turned from scientist to fierce conservationist. She finally gave her life for the animals she loved.

Fossey was born in San Francisco, California, in 1932 and grew up there. Her father, George, taught her to love nature, but her parents divorced when she was six, and her mother, Kitty, and stepfather, Richard Price, did not let her have pets. She loved animals anyway, and in 1950 she entered the University of California at Davis with plans to become a veterinarian. After two years, however, she transferred to San Jose State College, where she trained as an occupational therapist. She obtained her B.A. in 1954. In 1956, she moved to Louisville, Kentucky, and became head of the occupational therapy department at Kosair Crippled Children's Hospital.

"I had this great urge, this *need* to go to Africa," Fossey once told a *Chicago Tribune* interviewer. In 1963, she borrowed money to finance a seven-week safari to the continent. She met British anthropologist LOUIS S. B. LEAKEY and saw her first mountain gorillas during this trip. When she saw Leakey again in Louisville in 1966 and he said he was looking for a woman to do a long-term study of mountain gorillas like the one his protégée JANE GOODALL was doing with chimpanzees, Fossey eagerly volunteered.

Mountain gorillas are much rarer than lowland gorillas. They live only in the Virunga Mountains, a group of volcanoes in east central Africa shared among the countries of Rwanda, Uganda, and the Democratic Republic of Congo (formerly Zaire). Fossey began her research in Zaire in 1966, but the country was involved in a civil war, and soldiers drove her out of her camp and imprisoned her after she had been there only six months. She escaped and fled to Rwanda, where she was able to settle down to her studies at last. Local people soon called her *Nyirmachabelli*, "the woman who lives alone in the forest."

Fossey found that, contrary to gorillas' fearsome "King Kong" image, mountain gorillas were very shy. She finally learned to soothe their fears by imitating the sounds they made while eating. She observed nine groups, each consisting of five to 19 members, and made close contact with four. Her studies uncovered details of the gorillas' family life, diet, and communication that had never been observed before. She used them to obtain a doctorate in zoology from Britain's Cambridge University in 1974.

Fossey estimated in 1970 that there were only 375 gorillas left in the Virunga Mountains, and she became increasingly determined to protect these few remaining animals. It was no easy task: Seeking more land, Rwandan farmers often invaded the gorillas' supposedly protected habitat. Worse still, poachers sometimes slew the animals themselves.

When poachers killed Digit, Fossey's favorite gorilla, in 1977, she felt as if a beloved family member had been murdered. The fol-

Dian Fossey studied and attempted to protect endangered mountain gorillas in the volcanic mountains of Rwanda in the late 1960s and 1970s. *(Dian Fossey Gorilla Fund International, www.gorillafund.org)*

lowing year she established a fund in Digit's name to pay Rwandan guards to track and drive off poachers. She also began a personal war against the intruders, reputedly using tactics that ranged from scaring them with a Halloween mask to kidnapping their children. Her approach angered local people, Rwandan government officials, and even some wildlife protection groups.

Desperate for funds to continue her work, Fossey went to the United States in 1980. For three years, she taught and lectured at Cornell University, meanwhile writing a popular book about her experiences called *Gorillas in the Mist*. The book, published in 1983, became a best-seller (it was later made into a 1988 movie of the same name starring Sigourney Weaver as Fossey) and earned enough money to let her return to Rwanda. Fossey now suffered from emphysema (a lung disease) and other health problems, however, and had to turn her gorilla research over to assistants. Her growing moodiness and obsession with fighting the poachers also isolated her from those around her.

On December 27, 1985, one of Fossey's Rwandan guards found her in her hut, slashed to death by a machete. The murderer was never identified. Fossey's friends buried her in the graveyard she had set up for the slain gorillas, under a tombstone that reads, in part, "No one loved gorillas more." The Digit Fund, renamed the Dian Fossey Gorilla Fund International in 1992, continues her research and protection work.

Further Reading

Fossey, Dian. *Gorillas in the Mist*. Boston: Houghton Mifflin, 1983.

Hayes, Harold T. P. *The Dark Romance of Dian Fossey*. New York: Simon & Schuster, 1990.

Krystek, Lee. "Dian Fossey and the Gorillas of the Virunga Volcanoes." Virtual Exploration Society. Available online. URL: www.unmuseum.org/fossey.htm. Posted 1999.

Matthews, Tom L. *Light Shining Through the Mist: A Photobiography of Dian Fossey*. Washington, D.C.: National Geographic Society, 1998.

Mowat, Farley. *Woman in the Mists*. New York: Warner, 1987.

⊠ Franklin, Rosalind Elsie
(1920–1958)
British
Chemist

Deoxyribonucleic acid, or DNA, carries the inherited information in the genes of most living things. In the early 1950s, scientists realized that the key to finding out how this information was stored and reproduced lay in the structure of DNA's complex molecules. Rosalind Franklin took X-ray photographs that gave two rival scientists, JAMES WATSON and FRANCIS CRICK, the clues they needed to work out the structure of DNA.

Franklin was born on July 25, 1920, in London. Her father, Ellis, was a well-to-do banker, and her mother, Muriel, did volunteer social work as well as raise five children. Rosalind decided at age 15 that she wanted to be a scientist. Her father objected, believing that higher education and careers made women unhappy, but she finally overcame his resistance. She studied chemistry at Newnham, a women's college at Cambridge University, and graduated in 1941.

During World War II, Franklin did research on the structure of carbon molecules for the Coal Utilization Research Association (CURA). She earned a Ph.D. from Cambridge on the basis of this work in 1945. While at the French government's central chemical research laboratory between 1947 and 1950, she learned the technique of X-ray crystallography, in which a beam of X rays passes through a crystal and strikes photographic film. To the trained eye, the pattern of black dots on the resulting photograph reveals the three-dimensional arrangement of the atoms in the crystal.

Chemists had developed ways to use X-ray crystallography on amorphous compounds, which do not form obvious crystals. Franklin became expert at photographing such substances, which include most biological chemicals. In 1950, she joined a group of researchers at King's College, part of the University of London, who were trying to work out the structure of one such compound, DNA.

Scientists at the time knew that the complex DNA molecule included a long chain, or "backbone," made of alternating molecules of sugar and phosphate (a phosphorus-containing compound). Four kinds of other molecules called bases were attached to the backbone. No one could tell, however, whether the chain was straight or twisted, how the bases were arranged on it, or how many chains were in each molecule. The King's College group hoped Franklin could take X-ray photographs that would reveal this information.

By mid-1952, Franklin's research had led her to conclude that there was more than one chain in each DNA molecule and that each chain had the twisted shape of a helix, like the threads of a screw. She also believed that the phosphate backbone was on the outside of the chain and the bases on the inside. These ideas were correct, but they did not answer all the questions about the molecule. Some of Franklin's admirers think she might eventually have done so if she had had a scientist of her own caliber with whom to talk over her ideas. Unfortunately, she and MAURICE WILKINS, the leader of the King's College group and the obvious choice for such a role, disliked each other intensely. Isolated, she was unable to follow up on her conclusions.

Watson and Crick, a young American and an older British scientist working together at Cambridge University, were also trying to determine the structure of DNA. Watson became friends with Wilkins, and in January 1953, Wilkins showed him an exceptionally fine X-ray photograph that Franklin had made in 1952—without asking her permission. When Watson saw the photo, he wrote later, "my mouth fell open and my pulse began to race." He hurried back to Cambridge to describe it to Crick, and they then became able to work out the remaining aspects of the molecule's shape. Although the Cambridge scientists had insights that went beyond the evidence in Franklin's photograph, science historian Horace Freeland Judson has written that they "absolutely had to have the information" in the photograph in order to solve the puzzle of DNA.

Watson and Crick published a paper describing the structure of DNA in Britain's chief science journal, *Nature*, on April 25, 1953. Neither then nor later did they fully credit Franklin for the important part her work had played in their discovery, and Franklin herself probably never realized her role. By the time the Cambridge scientists' paper appeared, she had stopped working on DNA. She had transferred to Birkbeck, another college at the University of London, and was beginning an X-ray study of a common plant virus called tobacco mosaic virus.

In 1956, Rosalind Franklin discovered that she had ovarian cancer. She died of this disease on April 16, 1958. Four years later, Watson, Crick, and Wilkins shared the 1962 Nobel Prize in physiology or medicine for their work on DNA. Nobel Prizes are never awarded after a person's death, but debate still rages about whether Franklin would have received a share of the prize if she had lived. As it is, she is remembered in the high praise of some of her colleagues. J. D. Bernal, under whom Franklin worked at Birkbeck, wrote of her, "As a scientist Miss Franklin was distinguished by extreme clarity and perfection in everything she undertook. Her photographs are among the most beautiful X-ray photographs . . . ever taken."

Further Reading

Judson, Horace Freeland. "The Legend of Rosalind Franklin." *Science Digest*, January 1986.

Maddox, Brenda. *Rosalind Franklin: The Dark Lady of DNA*. New York: Harper Collins, 2002.

McGrayne, Sharon Bertsch. *Nobel Prize Women in Science: Their Lives, Struggles, and Momentous Discoveries*. New York: Birch Lane Press, 1993.

Sayre, Anne. *Rosalind Franklin and DNA*. New York: Norton, 1975.

Watson, James D. *The Double Helix*. New York: New American Library, 1959.

Frisch, Karl von
(1886–1982)
Austrian/German
Zoologist, Ethologist

Karl von Frisch showed how bees sense, navigate to, and tell their hive-mates about food sources. With KONRAD LORENZ and NIKO TINBERGEN, with whom he shared the 1973 Nobel Prize in physiology or medicine, he helped to found ethology, the study of animal behavior.

Frisch was born on November 20, 1886, in Vienna, Austria, to Anton von Frisch, head of surgery at the Vienna General Polyclinic, and the former Marie Exner, who came from a scholarly family and helped to interest her son in research. Frisch liked animals from childhood on, forming and carefully classifying a "little zoo" when he was young, but his father pressured him to study medicine because it was more likely than zoology to offer a steady, high-paying career.

Frisch began studies at the University of Vienna's medical school, but after two years he transferred to the zoology department. He obtained his Ph.D. in 1910 for experiments on the way fish perceive light and color. He then moved to the Zoological Institute of the University of Munich in Germany. One of his assignments there was to study bees, and he developed a lifelong fascination for these insects.

As a teaching and research assistant at the Zoological Institute, Frisch continued his research on fish, proving that they were not color-blind as had been thought, and began studying bees color sense. For instance, he placed food sources on squares of blue paper until the bees grew used to flying to paper with that color. He then showed that, even after he removed the food and changed the position of the papers, the bees flew to squares that were blue rather than any other color. He presented his work at a scientific meeting in 1914.

Frisch's poor eyesight kept him out of the army during World War I, but he helped his brother treat patients at a hospital in a Vienna suburb and also taught bacteriology to nurses there. One of the nurses was Margarethe Mohr, whom Frisch married in 1917. They later had four children.

After the war, Frisch returned to the Zoological Institute as an assistant professor. He continued his research on bees, showing that they could distinguish among smells and tastes as well as colors. He also found that they could perceive, and indeed are strongly drawn to, ultraviolet light, which humans cannot see. During these studies he noticed that when a new source of food was put out, the first "scout" bee that detected it flew back to the hive, and moments later a swarm of other workers followed the scout to the bonanza. "It was clear to me that the bee community possessed an excellent intelligence service," he wrote later, and he determined to find out how it worked.

Frisch had a breakthrough insight about this communication in the spring of 1919, when he watched a scout bee perform a circling "dance" on the vertical side of the hive. After intensive experiments during the next three years, he concluded that bees perform two kinds of dances, which he called the round dance and the waggle dance. The round dance indicates a nearby food source, while the waggle dance is done for food sources farther away. The waggle dance shows the direction of the food source relative to the angle of the sun's rays and also how far away it is. Frisch described the bees' dance "language" in a scientific report in 1924 and a book published in 1927, issued later in English as *The Dancing Bees*.

While doing his bee research, Frisch taught at zoological institutes at the University of Rostock (1921–23) and the University of Breslau (1923–25). He returned to Munich as the director of the Zoological Institute in 1925 and oversaw construction of elaborate research facilities there. In addition to continuing work on bees, he investigated the hearing of fish, for instance training a catfish to obey his whistled signals.

Frisch turned his bee research in a more practical direction during the first part of World War II, trying to find out about an epidemic that was killing the insects all over Europe. Later in the war, he left Munich to avoid Allied bombing, which destroyed his beautiful new laboratory. He became the head of the zoology department of the University of Graz, Austria, in 1946, but he was happy to return to the rebuilt Munich facility in 1950. He retired in 1958 but continued his research as an emeritus professor. During this time, he observed, for instance, that different subspecies of bees give somewhat different "dances" and cannot interpret the dances of other subspecies. He concluded from this that the dance patterns are inherited.

Frisch received the Nobel Prize at the ripe age of 88. He was also awarded other prizes, such as the Kalinga Prize for the popularization of science and the American Philosophical Society's Magellan Prize. He died on June 12, 1982.

Further Reading

Frisch, Karl von. *A Biologist Remembers*. New York: Pergamon Press, 1967.

———. *The Dance Language and Orientation of Bees*. Reprint, Cambridge, Mass.: Harvard University Press, 1993.

"Frisch, Karl von." *Current Biography Yearbook 1974*. New York: H. W. Wilson, 1974.

Funk, Casimir
(1884–1967)
German/British
Biochemist

Casimir Funk showed that several diseases were due to lack of tiny amounts of nutrients in food, gave these nutrients the collective name of vitamins, and identified two of them. He was born in Warsaw, in what is now Poland, on February 23, 1884. He was interested in medical and biological research from an early age, and his father, Jacques, a dermatologist (physician who treats skin conditions), persuaded him to focus on the then-new field of biochemistry.

Funk entered Berne University in Switzerland when he was only 16 and earned his Ph.D. in 1904. He went on to do research at the Pasteur Institute in France and the University of Berlin in Germany. In 1910, after being promised a laboratory of his own, Funk joined the Lister Institute of Preventive Medicine in London.

One of Funk's assignments at the Lister Institute was to study beriberi, a nerve disease common among poor people in Asia. Several scientists had concluded that this disease was linked to the people's diet, which consisted mostly of polished rice (rice with the outer husks removed). After reviewing existing research and conducting some of his own on pigeons, Funk decided that rice husks contained some substance necessary for the health of nerves that the grains alone lacked.

Funk attempted to isolate and purify this substance, an extremely difficult task, and in 1911 he succeeded in making a crude extract of rice hulls that could cure a beriberi-like disease in chickens. Later scientists purified the active chemical in this extract and named it thiamine, or vitamin B_1. Funk also made a relatively pure extract of a second nutrient, niacin or vitamin B_6, the absence of which causes a disease called pellagra.

Funk named these nutrients, and similar ones yet to be identified, *vitamines*, short for "vital amines" (compounds containing carbon, hydrogen, and nitrogen). The word was changed to *vitamins* in 1920, when other scientists showed that some of the compounds did not contain nitrogen. Funk described his ideas about vitamins and their relationship to disease in a paper called "The Etiology [Cause] of the Deficiency Diseases," published in 1912 in the British *Journal of State Medicine*. He expanded his paper into a book, *The Vitamines*, published in 1913. In his writings, Funk showed that vitamins were different from one another and that the lack of each caused a specific disease. In an era when most diseases were blamed on microorganisms, the idea that lack of trace amounts of chemicals in the diet could cause sickness startled many researchers.

Funk was awarded a doctor of science degree for his work in 1913. He then headed the biochemistry department of London's Cancer Hospital Research Institute for two years. In 1914, he married a Belgian woman, Alix Schneidesch; they later had a son and a daughter. He moved to the United States in 1915 and, after a year of research at Cornell Medical College, joined the H. A. Metz Company, a drug company. He worked for this organization from 1917 to 1923, making, among other things, the first artificial version of adrenaline, a hormone made by small glands above the kidneys. He was an associate professor of biochemistry at Columbia University Medical School in New York City from 1921 to 1923 and became a naturalized American citizen in 1920.

In 1923, Funk accepted an offer to direct the biochemical department of the State Institute of Hygiene in his native Warsaw. There, he chiefly did research on hormones. He showed that the pituitary, a tiny gland inside the brain, produces at least two different hormones, one that affects the way the body uses water and one that acts on muscles. In 1927, feeling that Poland was a politically "unhealthy" place, Funk moved to France and established a laboratory called Casa Biochemica near Paris. He and his coworkers made a crude extract of the male sex hormone androsterone from human urine and showed that sex hormones could be used to treat certain diseases. He also continued to study thiamine, and in 1936 he determined its molecular structure and made it artificially.

When Germany invaded France in 1940, Funk returned to the United States. He worked as a research consultant for the United States Vitamin Corporation and also did studies on the relationship of cancer to diet. Beginning in 1940, he headed the Funk Foundation for Medical Research. Funk died on November 19, 1967.

Further Reading

"Funk, Casimir." *Current Biography Yearbook 1945*. New York: H. W. Wilson, 1945.

Galdikas, Birutė

(1946–)
Canadian/Indonesian
Primatologist

Birutė Marija Filomena Galdikas studies and protects one of humanity's closest cousins, the red-haired Asian ape called the orangutan. She has discovered much of what is known about this solitary animal. She believes she inherited her love of nature from her ancestors in Lithuania, the heavily forested central European country where her parents, Antanas and Filomena Galdikas, grew up. They fled the country separately during World War II and met in a refugee camp. They were married in 1945, and Birutė, their oldest child, was born on May 10, 1946.

After the war the Galdikases immigrated to Canada, where Antanas Galdikas worked as a machinist and, later, a contractor. Birutė grew up in Toronto with two brothers and a sister. In high school, she first read about orangutans, whose name means "people of the forest" in the language of their homeland, Indonesia.

Galdikas's family moved to the United States in 1964. She did undergraduate work, followed by graduate studies in anthropology, at the University of California at Los Angeles (UCLA). During those years she heard about two young women, JANE GOODALL and DIAN FOSSEY, who were doing groundbreaking studies of chimpanzees and gorillas by observing the animals in their natural habitat. Galdikas became "obsessed" with the idea of doing similar work with orangutans. She found a path to her dream in 1969, when she attended a lecture given by British anthropologist LOUIS S. B. LEAKEY. Leakey had sponsored Goodall and Fossey, and Galdikas persuaded him to do the same for her.

Galdikas reached Indonesia, a string of islands off the southeast Asian coast, in September 1971. Rod Brindamour, a fellow Canadian whom she had met at UCLA and married in 1969, went with her. Officials sent them to Tanjung Puting, a forest reserve on the southern coast of the country's largest island, Borneo, where many orangutans lived.

At first, the couple seldom saw the orangutans, which moved silently through the trees far above their heads. With time and patience, however, Galdikas learned to spot the orange apes and follow their progress. The orangutans, in turn, slowly came to ignore the intruders. Galdikas and Brindamour remained in the forest for four years, observing 58 orangutans. Among other things, they verified that, unlike gorillas and chimpanzees, adult orangutans usually lived alone. Galdikas eventually obtained a Ph.D. from UCLA on the basis of her research, which described details of orangutans' daily lives that had never been reported before.

Almost as soon as she began observing wild orangutans, Galdikas also started rehabilitating young orangutans seized from people who had captured them illegally. Forestry officials asked Galdikas and Brindamour to provide a haven for these repossessed babies and help them return to life in the wild. During her career, Galdikas and her coworkers have returned hundreds of formerly captive orangutans to the wilderness.

While struggling with these "unruly children in orange suits," Galdikas was also raising her own son, Binti Paul, born in 1976. Brindamour divorced Galdikas, remarried, and returned to Canada in 1979, and Galdikas took Binti back to Canada the following year to live with his father. Galdikas herself married again in 1980. Her second husband, Pak Bohap bin Jalan, is a Dayak, one of the aboriginal people of Borneo. He and Galdikas have two children, Frederick and Jane.

Galdikas, who became an Indonesian citizen in the mid-1990s, today divides her time between overseeing research and rehabilitation work in Indonesia, teaching at Simon Fraser University in British Columbia, Canada, and raising funds for Orangutan Foundation International (OFI), a Los Angeles-based group that she founded in 1986. OFI works to protect the world's 10,000 to 20,000 remaining orangutans, most of which live on Borneo or another Indonesian island, Sumatra. Their habitat is threatened by land clearing, illegal mining and logging, and drought. Galdikas tries to educate people about the need to preserve these gentle, intelligent animals and their forest home, "a world which is in grave danger of vanishing forever." In 1995, she was made an Officer of the Order of Canada, and in 1997 she won the Tyler Prize for Environmental Achievement. She also won Indonesia's most prestigious environmental prize, the Kalpataru, in 1997. Her research, now in its 31st year, is one of the longest continuous studies of wild animals in the history of science.

Biruté Galdikas studies and protects orangutans in Indonesia; she is shown here with an infant orangutan. *(Orangutan Foundation International, www.orangutan.org)*

Further Reading

Galdikas, Biruté M. F. "A Message from the President of OFI, Dr. Biruté Galdikas." 1999. Orangutan Foundation International. Available online. URL: www.orangutan.org/index.html. Updated 2003.

———. *Reflections of Eden*. Boston: Little, Brown, 1995.

———, and Nancy Briggs. *Orangutan Odyssey*. New York: Harry N. Abrams, 1999.

Gallardo, Evelyn. *Among the Orangutans: The Biruté Galdikas Story*. San Francisco: Chronicle Books, 1993.

Montgomery, Sy. *Walking with the Great Apes*. Boston: Houghton Mifflin, 1991.

Galen

(ca. 130–201)
Greek/Roman
Physician

More than anyone else except perhaps HIPPOCRATES, the Greco-Roman physician Galen laid the foundations of Western medicine. He was born in Pergamum, or Pergamon (now Bergama, Turkey), around 130 A.D. His father, Nikon, a wealthy Greek architect, made sure that the boy received the best possible education. When Galen was about 16, Nikon had a dream in which Asklepios, the Greek god of healing, told him that his son should become a physician. The god had a large temple in Pergamum, so Galen was sent there to start his training.

The Greco-Roman physician Galen's medical books recommended that physicians learn about the body with their own eyes, but for more than a millennium, other physicians relied on the sometimes-mistaken information in his works rather than taking his advice. *(National Library of Medicine)*

After Nikon died four years later, Galen left Pergamum and continued his studies in Smyrna (in present-day Turkey), Corinth (in Greece), and Alexandria (in Egypt). Returning to Pergamum in 158, he obtained the respected post of physician to the local gladiators. Treating these professional fighters' terrible wounds taught him much about the structure of the human body and the ways that damage to different body parts affected function.

In 162, apparently feeling that he had learned all he could from his work with the gladiators, Galen departed for Rome. There, he quickly acquired noble patients, including even the Roman emperor, Marcus Aurelius, and became a highly regarded public figure. Except for a return to Pergamum for about a year around 166, he remained in Rome for the rest of his life. His writings suggest that he was an opinionated and quarrelsome man, quick to praise himself and criticize others. ("It is I, and I alone, who have revealed the true path of medicine," he once bragged.) Nonetheless, he became court physician to Marcus Aurelius and also held high posts under Marcus's son Commodus and several later emperors. Galen died around 201.

Galen started writing while still a youth (according to one account, he composed three books before he was 13), and he was a prolific author all his life. He composed some 300 books about medicine, more than 130 of which still survive at least in part. His works summed up the medical knowledge of the Greeks and Romans up to his time and described his own discoveries.

Hippocrates had recommended careful observation of individual patients, but Galen was the first major thinker to emphasize that physicians should also study the basic structure of the body (anatomy) and the functions of its parts (physiology). He stressed that disease, in which the structure and functions of the body become distorted, can be understood and treated only by learning about the healthy body. This learning should be done by direct observation. Galen wrote, "If anyone wishes to observe the works of Nature, he should put his trust not in books . . . but in his own eyes."

In books such as *De Usu Partium* (*The Uses of the Parts of the Body*), Galen published detailed descriptions of human anatomy and physiology, based both on his observation of human patients and on dissections of and experiments on apes and other animals, both dead and living. (He did not dissect human corpses because doing so was illegal in Rome.) Especially in anatomy, he provided meticulous descriptions and corrected many errors of previous thinkers. For instance, earlier medical writers had maintained that the vessels called arteries were filled with air, but Galen showed that they carry blood. He also proved that urine is made in the kidneys rather than the bladder, that nerves coming from the brain and spinal cord control body movements, and that the voice originates in the throat rather than in the heart.

At the same time, Galen made numerous mistakes of his own. Some came from unrecognized differences between humans and the animals he dissected. Others grew out of his uncritical acceptance of incorrect theories proposed by ARISTOTLE, Hippocrates, and other early thinkers, such as the belief that most disease is caused by imbalances among four supposed body fluids called humors. Galen's errors included misunderstandings about the function of major organs such as the liver, which he thought created blood from food, and the heart, which he thought heated the blood.

Galen's advances were in understanding the body rather than in treating disease. Unlike Hippocrates, he leaned heavily on drugs, complex mixtures of plant parts and other substances. He diagnosed illness mostly by studying the pulse, or heartbeat, and recommended bleeding (believed to drain off an excess of blood, one of the four humors) as a treatment for many conditions.

Galen was not a Christian and, indeed, disapproved of all religions, at least to the extent that they required belief in the supernatural or acceptance of ideas on the basis of faith rather than observation. Like the Christians, however, he believed in a single Supreme Being who had designed everything in nature for a purpose. This belief made later Christian leaders more comfortable with his writings than with those of most other ancient thinkers, with the result that Galen's works were among the few preserved and taught to physicians throughout the Middle Ages.

Indeed, by the fourth century, European scholars considered Galen to be the chief authority on medicine, and he remained so for more than a thousand years. His ideas were almost as sacred to medieval physicians as those in the Bible. As a result, medicine was held back by Galen's mistakes, not to mention those of his many translators and interpreters. That began to change only after about 1500, when thinkers such as ANDREAS VESALIUS and WILLIAM HARVEY returned to the personal observation and experimentation that Galen himself had insisted upon.

Further Reading

Galen. *Selected Works*. Translated by P. N. Singer. New York: Oxford University Press, 1997.

Nuland, Sherwin B. *Doctors: The Biography of Medicine*. New York: Alfred A. Knopf, 1988.

Gallo, Robert

(1937–)
American
Virologist

Robert Charles Gallo discovered the first virus shown to cause cancer in humans and did important work in identifying the virus that causes AIDS and developing a test to detect the virus in

Robert C. Gallo discovered the first virus shown to be responsible for a human cancer and showed that the virus later known as HIV was the cause of AIDS. *(Robert Gallo)*

blood. He was born on March 23, 1937, in Waterbury, Connecticut. His parents were Francis Gallo, the owner of a welding company, and Louise (Ciancuilli) Gallo. The death of his younger sister, Judy, from leukemia, a blood cell cancer, during his early teens and a friendship with one of her doctors helped to steer him toward medical research.

Gallo obtained a B.A. in biology from Providence College in Rhode Island in 1959 and an M.D. from Jefferson Medical College in Philadelphia in 1963. He married Mary Jane Hayes in 1961, and they later had two sons. He did his internship and residency (postmedical training) at the University of Chicago. While still a medical student, he did research that impressed officials at the National Cancer Institute, part of the National Institutes of Health (NIH), and they hired him as a clinical associate when he finished his training in 1965. He was promoted to the rank of senior investigator in the laboratory of tumor cell biology in 1968 and became head of that laboratory in 1972.

No virus had been proved to cause cancer in humans, but certain viruses, especially those known as retroviruses, were known to produce the disease in animals, and some researchers, including Gallo, thought that retroviruses capable of infecting humans might also exist. After first developing a way to keep cancerous blood cells from leukemia patients alive and multiplying in the laboratory, Gallo's group finally extracted a virus from some of these cells in 1979. They named it HTLV-1, short for Human T cell (a type of white blood cell) Leukemia Virus 1. HTLV-1 was the first human retrovirus to be discovered and the first virus shown to cause cancer in humans. Shortly after the group published an account of its work in 1980, they found a variant of this virus in other cells and called it HTLV-2. Like other retroviruses, the HTLVs have the unusual power to insert their genes directly into the genetic material of the cells they infect.

After testing samples of leukemia cells from all over the world, Gallo (and other workers, independently, in Japan) found that leukemia caused by these viruses was relatively common in Japan, the Caribbean, and Africa. The viruses seemed to be spread by intimate contact, such as that between mother and baby during breast feeding or between sexual partners. They could also be transmitted through blood transfusions.

Gallo was still working with his leukemia viruses in 1981, when physicians in San Francisco and New York began reporting an unusual cluster of diseases that occurred mostly among homosexual men and people who injected illegal drugs. The diseases were caused by microorganisms that the immune system normally can destroy. This suggested that some underlying agent had damaged the patients' immune systems.

Scientists had many theories about what might cause this condition, but the most common suspect was an unknown virus. Gallo recognized some similarities between this proposed virus and his leukemia viruses, such as its geographical distribution (many cases of the new disease had appeared in Africa and the Caribbean, as well as in the United States), the kind of immune system cells it apparently infected, and its methods of spread, including sex and procedures that involved blood. He proposed that the new disease, which the Centers for Disease Control and Prevention designated as AIDS (Acquired Immunodeficiency Syndrome) in September 1982, was caused by a virus related to but distinct from his HTLVs—in other words, a new human retrovirus.

When NIH created a task force to investigate AIDS in 1982, it made Gallo the group's director. His laboratory began applying the techniques it had used on the leukemia viruses to samples of immune system cells from patients with the new disease. In May 1984, Gallo's team announced that they had isolated a new type of virus from the cells of numerous AIDS patients and gave convincing evidence that it caused AIDS. They called the virus HTLV-3 because it seemed to be related to the viruses that they had found earlier. The group also said that they had developed a test that could identify the virus in blood, an important advance because the disease was being spread partly through virus-contaminated blood transfusions. The U.S. government obtained a patent on the test a year later, and the test began being used in March 1985.

A year before the Gallo laboratory's announcement, LUC MONTAGNIER and other researchers at the Pasteur Institute in Paris, France, had written a paper describing their isolation of a similar virus, which they called LAV. They, too, worked on a blood test for the virus, and they applied for a patent on the test in the United States seven months before the American group did. (Gallo maintains that the French test was not as accurate and that the Pasteur researchers, unlike his own group, did not succeed in permanently growing the virus in the laboratory, "a prerequisite for a workable global blood test.") The Montagnier and Gallo viruses were soon shown to be of the same type and were given the new name HIV (human immunodeficiency virus).

The Pasteur group filed a lawsuit against NIH and the United States government in December 1984, demanding a share of the credit for discovering HIV and of the royalties for the blood test. The bitter conflict between the rival laboratories continued until April 1987, when the governments of the United States and France signed an agreement saying that the two groups would divide the credit and the royalties equally. Later investigations determined, however, that credit for first isolating the virus actually belonged to Montagnier's team. The French scientists had sent Gallo two samples of infected cells in late 1983, and Gallo had unknowingly used LAV from one of these samples. Nonetheless, Gallo points out, his laboratory proved that they had isolated HIV from many other samples as well.

Furthermore, even the French admitted that Gallo's research on human retroviruses, his suggestion that AIDS would prove to be caused by a retrovirus, and his laboratory's discovery of a method of growing human T cells in the laboratory with the aid of a growth-inducing substance called interleukin-2 (which the group had also discovered) had laid the groundwork for both teams' work on HIV. "Without Gallo, there wouldn't have been any work on this [virus] at Pasteur," French immunologist Daniel Zagurey told a *Time* magazine reporter in 1984. "Their research is based on his initial discovery." Gallo's group also was the one that showed clearly that HIV was the cause of AIDS. Because of these achievements, both researchers agreed around 2001 to return to the view that credit for the overall discovery of HIV should be shared equally.

Gallo continued his research during these controversial years. He and his coworkers studied HIV's structure and genetics and the way it infects and damages immune system cells, and they tried unsuccessfully to develop a cure and vaccine for AIDS. In 1986, Gallo's laboratory identified a new virus belonging to a common group called herpesviruses and named it HHV-6. This virus was later shown to cause roseola infantum, a skin disease that affects babies.

Gallo left NIH in 1995. The following year he became a professor of medicine, microbiology, and immunology at the University of Maryland in Baltimore and director of the university's new Institute of Human Virology, positions he still holds. The institute investigates a number of chronic virus-caused diseases and cancers linked to viruses, but its chief focus is on AIDS. In 2001, Gallo and his coworkers announced that they had discovered a group of natural chemicals that appear able to block HIV infection. They have also found a substance in the urine of pregnant women that both reduces HIV levels and attacks Kaposi's sarcoma, an AIDS-related disease. In addition, one of their AIDS vaccines has produced promising results in monkeys. Gallo's group is also trying to develop inexpensive treatments for AIDS that can be used worldwide.

In early 2002, Gallo and former rival Luc Montagnier agreed to work together in attempts to develop an effective vaccine for AIDS as codirectors of the Program for International Viral Collaboration, sponsored by UNESCO. "We are friends and collaborators and we look forward to this new chapter in which we both strive for new solutions . . . in halting the destructive path of HIV/AIDS," Gallo said.

Although Gallo has sometimes been criticized for his aggressive, competitive nature ("Everything I do is competitive, in science and in my life," he admitted to *People* magazine in 1984), he has also been praised for his scientific work. In 1991, for instance, Samuel Broder, then director of the National Cancer Institute, called him "one of the greatest people in American science." He has won numerous awards, including an unheard-of two from the Albert and Mary Lasker Foundation (1982 and 1986), considered the most prestigious private prize for biomedical research in the United States. He has also received the General Motors Prize for Cancer Research (1984), the American Cancer Society's Medal of Honor, the World Health Award (2001), and 15 honorary degrees. Even if he was not the first to discover the AIDS virus, Gallo's work with HTLV unquestionably laid the groundwork for that discovery, and he has played a central role in AIDS research ever since.

Further Reading

"Biographical Sketch." Institute of Human Virology. Available online. URL: www.ihv.org/bios/gallo.html. Accessed 2001.

Cohen, Jon. "Longtime Rivalry Ends in Collaboration." *Science*, February 22, 2002.

Crewdson, John. *Science Fictions: A Scientific Mystery, a Massive Cover-up, and the Dark Legacy of Robert Gallo*. Boston: Little, Brown, 2002.

Gallo, Robert. "The Early Years of HIV/AIDS." *Science*, November 29, 2002.

———. *Virus Hunting*. New York: HarperCollins, 1991.

⊠ Galvani, Luigi

(1737–1798)
Italian (Papal States)
Physiologist, Anatomist

Luigi Galvani showed that nerves transmit electricity and muscles respond to it. He was born on September 9, 1737, in Bologna, Italy (then called the Papal States). He studied medicine at the University of Bologna, graduating in 1759. In 1762, he obtained an advanced degree and became a lecturer in anatomy at the university. He married Lucia Galleazzi, the daughter of one of his fellow professors, in the same year.

At first, Galvani did research on comparative anatomy. In the late 1770s, however, he became interested in the effects that the newly discovered force of electricity, then thought to be a kind of invisible fluid, had on the bodies of living things. He showed that static electricity stored in a device called a Leyden jar could make muscles twitch. In 1786, he happened to touch the leg of a dissected frog with a pair of scissors during a thunderstorm, when the air was charged with electricity, and he noticed that the leg twitched then, too. Furthermore, when he hung frog's legs outdoors from an iron railing on brass hooks, the muscles twitched every time they touched the iron, even in good weather.

Galvani concluded, correctly, that electricity made the frog muscles contract. He thought that the electric current came from the nerves and muscles themselves. He claimed that living things possessed a special form of electricity, which he called "animal electricity." He summarized his findings in a paper called "On the Effects of Electricity on Muscular Motion," published in 1791.

Most scientists of the time accepted Galvani's work, but a physics professor named Alessandro Volta refused to do so. The two men's feud over their differing interpretations of the phenomena Galvani had demonstrated became notorious in scientific circles during the mid-1790s. In 1800, two years after Galvani's death, Volta proved that electric current could be produced by any two dissimilar metals in a salty solution. In Galvani's experiment, the frog's body had provided the salty liquid, not the electricity itself. Volta's efforts to disprove Galvani's theory led him to invent the storage battery.

Galvani enjoyed considerable prestige during the first part of his career: He was president of the Bologna Academy of Science in 1772. He became a professor of anatomy and gynecology (the treatment of women's ailments) in 1775. His later life, however, was unhappy. His beloved wife died in 1790, and in 1797 he also lost his position at the university because he refused to swear allegiance to Napoleon as ruler of what was then called the Cisalpine Republic. Galvani retired to his family home and died in poverty on December 4, 1798.

Although Galvani was wrong in some of his beliefs, he was correct that electricity plays a role in the actions of nerves and muscles. For instance, he demonstrated that a dead frog's leg would contract if a nerve from another frog's leg was touched to the muscle, even when no metal was present. He was even correct that the body can generate electricity, though the currents involved are much smaller than the ones he observed, and the electricity is no different from that generated in any other way. Galvani's work helped to lay the foundation for electrophysiology, the study of the functions of electricity in body tissues.

Further Reading

Pera, Marcello. *The Ambiguous Frog: The Galvani-Volta Controversy on Animal Electricity*. Translated by Jonathan Mandelbaum. Princeton, N.J.: Princeton University Press, 1992.

Piccolino, Marco. "Galvani and Animal Electricity: Two Centuries After the Foundation of Electrophysiology." *Trends in Neurosciences*, October 1997.

Gilbert, Walter

(1932–)
American
Physicist, Molecular Biologist

Walter Gilbert began his career as a physicist, but he made his greatest scientific contributions in molecular biology. His father, Richard V. Gilbert, was a professor of economics at Harvard University. His mother, the former Emma Cohen, was a child psychologist. Walter was born in Boston, Massachusetts, on March 21, 1932, but his family moved to Washington, D.C., when he was seven years old, and he grew up there. He was interested in science from an early age. In high school, he made his own telescopes and frequently cut classes to read about nuclear physics at the Library of Congress.

Gilbert returned to Harvard as an undergraduate, majoring in physics and chemistry. In 1953, he graduated and also married Celia Stone, a poet he had known since high school. They later had a son and a daughter. Gilbert earned a master's degree in physics from Harvard in 1954 and then studied theoretical physics and mathematics at Britain's Cambridge University, from which he earned a Ph.D. in 1957. After taking his degree, he went back to Harvard, where he became a postdoctoral fellow, a lecturer, and finally, in 1959, an assistant professor in physics. He kept this last post until 1964.

Gilbert's career began to change in 1960, when he renewed an acquaintance with JAMES WATSON, the codiscoverer of the structure of DNA. The two men had met when Gilbert was at Cambridge. Watson, who was doing research at Harvard, persuaded Gilbert to join a project to isolate messenger RNA, a short-lived molecule related to DNA that carries DNA's information from the nucleus to the outer part of the cell. By the time the group achieved its aim in the early 1960s, Gilbert had decided to abandon physics and devote himself to the relatively new field of molecular biology. He became an associate professor of biophysics at Harvard in 1964, a full professor of biochemistry in 1968, and the American Cancer Society Professor of Molecular Biology in 1972.

During the mid-1960s, Gilbert first did research on how cells use genetic information to make proteins. He then searched for another elusive molecule called a repressor, which French geneticists François Jacob and JACQUES MONOD had proposed in 1961 as a controller of the action of genes in cells. All cells in the body contain the same genes, Jacob and Monod had pointed out, yet different kinds of cells make different proteins. Most genes in any particular cell therefore must be "turned off." They had suggested that repressors perform this task by binding to parts of DNA molecules. No one had been able to isolate a repressor, however, partly because these molecules are present in cells only in extremely tiny amounts.

Gilbert and his chief coworker, Benno Müller-Hill, studied a common intestinal bacterium called *Escherichia coli*, which makes an enzyme called beta-galactosidase only when lactose, its chief food, is present. They expected the *lac* repressor, the predicted substance that prevented formation of beta-galactosidase at other times, to detach itself from the bacterium's DNA and bind to lactose when the sugar was present, leaving the beta-galactosidase gene free to become active. In 1966, using lactose-like molecules labeled with a radioactive substance and a technique called equilibrium dialysis that they invented, they succeeded in purifying the *lac* repressor, the first genetic control element to be identified. Later, Gilbert proved that the *lac* repressor bound to DNA at the beginning of the gene for beta-galactosidase.

While doing this work, Gilbert broke DNA into pieces by treating it with enzymes that cut it at spots where particular sequences of bases (the smaller molecules within DNA that carry the genetic code) appear. This gave him an idea for a way to determine the sequence of bases in DNA, which scientists must know in order to find out which proteins genes make. In the 1970s, Gilbert and graduate student Allan Maxam broke up radioactively labeled DNA molecules with enzymes that attacked each base separately, creating fragments of different sizes. They then sorted the fragments by size, using a process called gel electrophoresis. When a photographic film is exposed to the gel, the fragments, which are labeled with radioactive atoms, create a pattern of dots that provides information about the base sequence.

This rapid sequencing method, which Gilbert first described in 1975, made it possible for the first time to work out the sequence of an entire gene. In 1980, he shared the Nobel Prize in chemistry with FREDERICK SANGER, who had developed a similar process at about the same time, and PAUL BERG, who had created a way to combine genes from different living things, the beginning of genetic engineering.

Genetic engineering also interested Gilbert. In 1978, he and 10 other scientists joined with venture capitalists to form Biogen N.V., one of the first businesses to use genetic engineering technology. Gilbert became chairman of the Swiss-based company's scientific board of directors. He left Harvard in 1981 to work full time for Biogen as its chief executive officer, but the company did not make the profit its investors had hoped for, and some criticized Gilbert's business abilities. He resigned from Biogen in 1985 and returned to Harvard. In 1986, he became the H. H. Timken Professor of Science in the university's department of cellular and developmental biology, and in 1987 he became head of that department and the Carl M. Loeb University Professor, a member of the highest rank of faculty at the university.

In the late 1980s and early 1990s, Gilbert devoted much of his attention to the Human Genome Project, an ambitious effort to sequence all the genes in a human cell. (The project announced success in this effort in June 2000.) In the 1990s and beyond, his laboratory has investigated the relationship between the parts of the DNA molecule that carry the code for proteins, which are called exons, and other stretches of DNA that occur in between these segments, which are called introns. Introns are sometimes termed "junk DNA" because they have no apparent function; they may be relics of an earlier stage of evolution.

Gilbert has received many awards for his work in addition to the Nobel Prize. They include the Institut de France's Prix Charles-Leopold Mayer (1977), the Albert Lasker Medical Research Award (1979), the Louisa Gross Horwitz Prize from Columbia University (1979), and the Biotechnology Heritage Award from the Chemical Heritage Foundation and the Biotechnology Industry Organization (2002). He has also continued to be involved in the biotechnology industry. For instance, he and Columbia University neurobiologist Eric Kandel have formed a company to develop drugs that enhance memory. In October 2001, Gilbert announced that he was taking a leave of absence from Harvard to work for BioVentures Investors, which provides venture capital to help new companies in this industry.

Further Reading

Gilbert, Walter. "Walter Gilbert—Autobiography." *Les Prix Nobel 1980*. Available online. URL: www.nobel.se/chemistry/laureates/1980/gilbert-autobio.html. Last updated 2002.

"Gilbert, Walter." *Current Biography Yearbook 1992*. New York: H. W. Wilson, 1992.

Hall, Stephen S. *Invisible Frontiers: The Race to Synthesize a Human Gene*. New York: Oxford University Press, 2002.

Golgi, Camillo
(1843–1926)
Italian
Neurobiologist, Histologist

Camillo Golgi developed a way of dyeing nerve cells that allowed them to be seen clearly for the first time. Using this method, he made numerous discoveries about the histology, or microscopic structure, of the nerves and brain.

Golgi was born in Corteno, a mountain village near Brescia, northern Italy, on July 7, 1843. Following in the footsteps of his father,

Alessandro, a physician, he studied medicine at the University of Pavia, earning his medical degree in 1865. During his medical studies and subsequent work at the Hospital of San Matteo, he became interested in the brain. He studied psychiatry briefly, but he believed that too little was known about mental illness to allow it to be approached in a scientific way. He therefore turned to examining the "hidden structure" of brain and nerve tissue.

In 1872, Golgi became chief medical officer at the Hospital for the Incurably Ill in Abbiategrasso, a town near Pavia. He converted the hospital kitchen into a makeshift laboratory for his histological research. There, he found, almost by accident, that compounds containing the metal silver randomly stain some nerve cells black, making them stand out against the background of other, unstained cells. Before this discovery, scientists looking through a microscope at brain or nerve tissue had seen only a confused tangle of fibers, but Golgi's staining method, which came to be called the black reaction, showed the spiderlike structure of individual cells clearly. He described it in a paper in 1873, and it is still used.

Golgi returned to the University of Pavia in 1875 as a lecturer in histology and became a professor the following year. He stayed at the university for the rest of his career, except for about a year around 1879, when he worked for the University of Siena. In 1877, he married Lina Aletti, the niece of the professor who had taught him histology at Pavia. Golgi became head of Pavia's general pathology department in 1881 and, later, the dean of the faculty of medicine and rector (president) of the university. He also was active in public life, becoming a senator in 1900 and crusading for better public health measures.

Through much of his career, Golgi continued to study nerve tissue with his stain. At the time, no one was sure whether nerves were made up of separate cells, and Golgi was never convinced that they were. Instead, he believed that nervous tissue consisted of a network of cells joined together to make a single unit. He described several kinds of nerve cells and proposed that there are two types of nerves, those that convey messages inward from the sense organs to the brain (sensory nerves) and those that carry messages outward from the brain to the muscles (motor nerves). He showed that nerve fibers are separated from one another by tiny gaps that came to be called synapses.

Between about 1885 and 1893, Golgi made important contributions to a quite different field, the study of malaria. The microscopic parasite that causes this disease had been identified in 1880, and Golgi helped to verify the stages of its complicated life cycle. He showed that the several forms of malaria, which are marked by attacks of chills and fever that recur at different intervals, are caused by different species of parasites. He also demonstrated that a fever attack occurs when a new generation of parasites breaks free of the blood cells in which it has been developing. He recommended that quinine, the only drug available to treat the disease at the time, be given a few hours before a bout of fever was expected.

In 1898, Golgi also discovered a tiny structure, consisting of many folded or stacked layers, inside nerve cells. He and others later found that this structure, called the Golgi body or Golgi apparatus, exists in cells of almost all types. Its purpose is still somewhat mysterious, but scientists now believe that it is involved in the making and distribution of proteins, the large group of chemicals that do most of the work in cells.

Golgi shared the 1906 Nobel Prize in physiology or medicine with Spanish researcher SANTIAGO RAMÓN Y CAJAL, a fact that probably irritated both men because they were bitter rivals, for their discoveries about the structure of the nervous system. Golgi, in fact, made a point of criticizing Cajal during his prize acceptance speech. Golgi retired in 1918 but continued to work at Pavia as a professor emeritus until close to his death on January 21, 1926.

Further Reading

Bentivoglio, Marina. "Life and Discoveries of Camillo Golgi." Nobel Foundation. Available online. URL: www.nobel.se/medicine/articles/golgi/. Posted 1998.

Mazzarello, Paolo. *The Hidden Structure: A Scientific Biography of Camillo Golgi*. Translated by Henry Buchtel and Aldo Baldiani. New York: Oxford University Press, 2000.

Goodall, Jane

(1934–)
British/Tanzanian
Primatologist

Jane Goodall's research on chimpanzees in Africa is one of the longest continuous studies of animals in the wild and, according to naturalist and science writer STEPHEN JAY GOULD, is "one of the Western world's great scientific achievements." For it, she was awarded the Kyoto Prize, Japan's equivalent of the Nobel Prize (1990), named a commander of the British Empire (1995), and given the U.S. National Geographic Society's Hubbard Medal (1995), among many other awards.

Jane, the older of Mortimer and Vanne Morris-Goodall's two daughters, was born in London on April 3, 1934. Her father was an engineer, her mother a housewife and writer. Her favorite toy as a baby was a stuffed chimpanzee named Jubilee, which she still owns.

An incident that happened when Jane was just four years old showed her patience and determination as well as her interest in animals. One day, while on a farm, she vanished for almost five hours. Vanne Goodall called the police, but before a search could be launched, Jane reappeared. She explained that she had been sitting in the henhouse, waiting for a hen to lay an egg. "I had always wondered where on a hen was an opening big enough for an egg to come out," Goodall recalled later. "I hid in the straw at the back of the stuffy little hen house. And I waited and waited."

The Goodalls moved to the seaside town of Bournemouth at about that time, and Jane stayed there with her mother and sister after her parents divorced several years later. They had no money for college, so she went to work as a secretary. Then, in 1957, a former school friend invited Goodall to visit her in Kenya. This opened the door to fulfilling a dream of going to Africa that Goodall had formed in childhood after reading books such as Hugh Lofting's fantasies about Dr. Doolittle, who lived in Africa and could talk with animals. She began saving her money and left as soon as she could pay the fare.

While in Kenya, Goodall met famed British anthropologist LOUIS S. B. LEAKEY, who had made pioneering discoveries about early humans, and Leakey hired her as his assistant. He told her about his belief that the best way to learn how human ancestors might have lived was to study the natural behavior of their closest cousins, the great apes—chimpanzees, gorillas, and orangutans—over long periods of time. He wanted to start with chimpanzees, and he asked Goodall if she would like to do the research. "Of course I accepted," she says.

Following Leakey's recommendation, Goodall decided to work at Gombe Stream, a protected area on the shore of Lake Tanganyika in what is now Tanzania. When British officials insisted that they could not let her live alone in the wilderness, her mother agreed to stay with her for a few months. The Goodalls and their African assistants set up camp at Gombe in July 1960.

The chimpanzees ran away from Goodall at first, but as the months passed, they grew used to the "peculiar, white-skinned ape." She, in turn, learned to recognize them as individuals. She gave them names such as Flo and David Graybeard. In the first year of her research, Goodall made several observations that overturned long-held beliefs about chimpanzees, including the discoveries that they ate meat and could make and use tools. For instance, she saw David Graybeard lower a grass stem into the open tower of a termite mound, pull it out a few minutes later with several of the antlike insects clinging to it, and eat them. She also observed chimpanzees stripping leaves off grass stems to make them more useful for "termite fishing"—in short, making tools as well as using them. Most scientists had believed that only human beings could make tools. Goodall obtained a Ph.D. in primatology (the study of humans, apes, monkeys, and lemurs) from Britain's prestigious Cambridge University on the basis of this research in 1965.

In 1962, the National Geographic Society sent a Dutch photographer, Baron Hugo van Lawick, to take pictures of Goodall at work. Van Lawick and Goodall fell in love and married on March 28, 1964. In 1967, they had a son, whom they named Hugo after his father, but everyone called the blond youngster Grub, Swahili for "bush baby." Goodall and van Lawick divorced in 1974, and a year later Goodall married Derek Bryceson, who was in charge of Kenya's national parks. Unfortunately, Bryceson developed cancer in 1980 and died within a few months.

As Goodall's observations continued, she discovered dark sides to chimpanzee behavior. For instance, she saw the animals wage war. One group repeatedly attacked a neighboring group over a period of four years, eventually wiping them out. "When I first started at Gombe, I thought the chimps were nicer than we are," she said in a 1995 *National Geographic* article. "But time has revealed that they are not."

Around 1975, Goodall decided that "I had to use the knowledge the chimps gave me in the fight to save them." She left the continuing observation of the Gombe chimps to other scientists, students, and assistants and began to travel the world as a spokesperson for the animals. In 1977, she formed a nonprofit organization, the Jane Goodall Institute, to help with her work and "advance the power of individuals to take informed and compassionate action to improve the environment of all living things." Its United States headquarters are in Silver Spring, Maryland.

The greatest dangers to wild chimpanzees, Goodall says, are loss of their forest habitat, which is shrinking their population greatly, and poaching (illegal killing). She has set up sanctuaries in Africa for orphan chimpanzees whose mothers have been killed by poachers. She is equally concerned about captive chimpanzees, which are often confined under miserable conditions. The use of chimpanzees in medical research is a third source of distress for Goodall, who believes that such use should be minimized and that the conditions under which the animals are kept should be improved.

The ultimate way to help chimpanzees and other animals, Goodall believes, is to teach people—especially children—to respect them. In 1991, she developed a program called Roots & Shoots, now implemented worldwide, which encourages and supports students, from preschool to university age, in programs benefiting people, animals, and the environment. "Teaching [children] to care for the earth, and each other, is our hope for the future," she says.

Goodall herself, meanwhile, travels around the world speaking not only about chimpanzees but about threats to humanity and nature in general. She urges compassion and tolerance among people and stresses her reasons for hope that humankind will solve the problems it has imposed on the earth. In April 2002, she became the 10th United Nations Messenger of Peace.

Further Reading

Goodall, Jane. *Beyond Innocence: An Autobiography in Letters, the Later Years*. Edited by Dale Peterson. Boston: Houghton Mifflin, 2001.

———. *Reason for Hope: A Spiritual Journey*. New York: Warner Books, 2000.

———. *Through a Window: My Thirty Years with the Chimpanzees of Gombe*. Boston: Houghton Mifflin, 1987.

"Jane Goodall." Jane Goodall Institute. Available online. URL: www.janegoodall.org/jane/index.html. Accessed 2001.

Meachum, Virginia. *Jane Goodall, Protector of Chimpanzees*. Berkeley Heights, N.J.: Enslow Publishers, 1997.

Gould, Stephen Jay

(1941–2002)
American
Paleontologist, Evolutionary Biologist, Philosopher of Science

Stephen Jay Gould is known for defending CHARLES ROBERT DARWIN's theory of evolution by natural selection, revising it, and, above all, explaining it and other complex scientific topics to the public in hundreds of highly acclaimed essays. Gould first decided to be a paleontologist when his parents, Leonard and Eleanor (Rosenberg) Gould, took him to the American Museum of Natural History in New York City and he stood under the fossil bones of a gigantic *Tyrannosaurus rex*. He had been born in the same city about five years before, on September 10, 1941.

When Gould was in junior high school, other students called him "Fossilface" because of his seemingly odd interest, but his parents and teachers encouraged him. He attended Antioch College in Ohio, where he majored in geology and graduated in 1963. He then returned to New York for graduate studies at Columbia University, from which he obtained a Ph.D. in 1967.

As soon as he earned his doctorate, Gould was hired as an assistant professor of geology at Harvard University. He became an associate professor in 1971 and a full professor in 1973. He was a professor of geology there, as well as (since 1982) the Agassiz Professor of Zoology and curator of invertebrate paleontology at the university's Museum of Comparative Zoology. He was married to the former Deborah Lee, an artist and writer, in 1965, and they had two sons.

Gould made his best known scientific contribution in 1972, when he and American Museum of Natural History paleontologist Niles Eldredge described a modification of Darwin's theory that they called "punctuated equilibrium." Darwin had maintained that changes in species occur slowly, gradually, and at a steady rate throughout geologic time. Gould and Eldredge, however, proposed that species exist for long periods without significant change. When change does come, they said, it occurs rapidly, potentially creating a new species in as little as a few thousand years. Many evolutionists now accept this theory.

From the 1980s on, Gould was best known for his writing, especially the essays that he began composing for *Natural History* magazine in January 1974 and continued until January 2001 in a monthly column called "This View of Life." All his essays were connected in some way with evolution, which Gould called "the story of all of us . . . where we came from, how we got here, and perhaps where we are going."

Gould's essays describe people and concepts in biology by using facts and comparisons from fields ranging from choral music to baseball. He himself said, "If I have one special ability, it is as a tangential thinker. I can make unusual connections." His column won the National Magazine Award for Essays and Criticism in 1980, and some of the numerous books into which his essays were collected have also won awards. *The Panda's Thumb*, for instance, received the American Book Award in 1981. In that year, Gould received a MacArthur Foundation "genius grant" as well.

In addition to his essays, Gould wrote books and gave talks on several controversial topics. In 1981, his testimony helped to convince a jury in Little Rock, Arkansas, that "creation science," which is based on ideas in the Bible, was an aspect of religion rather than science and that a state law requiring that it be given equal time with evolution in biology classes was therefore unconstitutional. In *The Mismeasure of Man*, published in 1981, he claimed that standardized intelligence tests are inaccurate and often misused to reinforce social biases. Not everyone agreed with this conclusion, but the book won a National Book Critics Circle Award in 1982.

Gould died of cancer on May 20, 2002, in New York City. In one of many tributes in a 25th-anniversary article about him in the November 1999 *Natural History*, Niles Eldredge wrote, "The guy has eyes in the back of his head. He sees stuff—fossils, ideas, whatever. He can sense the gist of an important issue and cut to the chase faster than anyone else I've ever met."

Further Reading

Golden, Frederic. "Bones, Baseball, and Evolution: Stephen Jay Gould Turns a Musty Discipline into a Joy." *Time*, May 30, 1983.

Gould, Stephen Jay. *The Book of Life: An Illustrated History of the Evolution of Life on Earth*. 2nd ed. New York: W. W. Norton, 2001.

———. *The Mismeasure of Man*. Rev. ed. New York: W. W. Norton, 1996.

"Gould, Stephen Jay." *Current Biography Yearbook 1982*. New York: H. W. Wilson, 1982.

"This View of Stephen Jay Gould." *Natural History*, November 1999.

H

⊠ Haldane, J. B. S.

(1892–1964)
British
Biochemist, Geneticist, Evolutionary Biologist

John Burdon Sanderson Haldane contributed to fields in biology ranging from population genetics to biochemistry. He also wrote highly acclaimed popular articles and books on science and involved himself in numerous social controversies. Nobel Prize-winning biologist PETER BRIAN MEDAWAR called him "the cleverest man I ever knew."

Haldane's connection to biology began almost from his birth on November 5, 1892, in North Oxford, England. His father, John Scott Haldane, was an eminent Oxford University researcher on the physiology of the respiratory system. The elder Haldane tested the effects of various gases on himself, and his son took part in these experiments from childhood on. Haldane was also exposed to politics from an early age, hearing opposing views from his liberal father and conservative mother, Louisa.

Haldane won a scholarship to study mathematics at New College in Oxford in 1911. He began making his mark on science while still a student. Around 1912, based partly on a study of 300 mice belonging to his sister, he concluded that characteristics usually inherited together were "linked," which meant that the genes transmitting them were probably close together on the same chromosome. THOMAS HUNT MORGAN and others reached a similar conclusion at about the same time.

Haldane earned his M.A. from Oxford in 1914, just as World War I was starting. He served overseas in a Scottish regiment, the Black Watch, and was seriously wounded twice, yet claimed later that he had enjoyed his war experience. While recovering from his wounds in India, he developed a procedure for calculating the relative distance between genes based on the frequency with which linked characteristics were inherited together.

Haldane returned to Oxford as a graduate student after the war. He continued his research on genetics until 1922, when biochemist Frederick Hopkins established a new laboratory at Britain's other most famous university, Cambridge, and invited Haldane to become his second-in-command. Haldane remained at Cambridge until 1932, chiefly studying enzymes. He showed in 1924 that enzymes obey the laws of thermodynamics. His description of these chemicals' action, elaborated in a 1930 book titled *Enzymes*, is still considered basically correct.

Haldane was one of the first people to connect biochemistry with genetics. During his

Cambridge years, he oversaw genetic experiments at the John Innes Horticultural Institute at Merton, where he held a part-time position from 1927 to 1937. Work there supported the idea, which Haldane had first suggested in 1920, that genetic information is translated into visible characteristics by means of chemicals, especially enzymes.

While at Cambridge, Haldane developed mathematical models that allowed geneticists to predict the effects of natural selection on specific inherited traits under different circumstances. Perhaps even more important, he described the features that any such model should have. He also wrote a book published in 1932, called *The Causes of Evolution*, which summarized what was known about genetics and evolution at the time. With RONALD AYLMER FISHER and Sewall Wright, Haldane is considered to have founded the field of theoretical population genetics, which linked CHARLES ROBERT DARWIN's theory of evolution by natural selection with GREGOR MENDEL's rules of genetic inheritance.

Eventually deciding to devote his full time to genetics, Haldane became a professor in this field at University College, London, in 1933. There, he investigated human genetics, discovering, for example, a link between color blindness and an inherited blood disease called hemophilia. He became Weldon Professor of Biometry at the college in 1937.

In the late 1930s, Haldane became increasingly disturbed by what he saw as Britain's weak stand against the warlike Fascist governments in Germany, Italy, and Spain. He was drawn to the Communist Party, which opposed the Fascists and, he believed, strongly supported scientific research. He joined the party in 1942 and headed the editorial board of the *Daily Worker*, its English-language newspaper, from 1940 to 1949.

At the urging of his first wife, the former Charlotte Burghes, a journalist whom he had married in 1926, Haldane wrote numerous popular articles on science for the *Daily Worker*, many of which were later collected into books. (Haldane and Burghes later divorced, and he married Helen Spurway, a fellow geneticist, in 1945.) Science fiction writer Arthur C. Clarke has called him "perhaps the most brilliant scientific popularizer of his generation." During his long career, Haldane also wrote many books connecting science and social issues, such as *Science and Ethics* (1928), *Heredity and Politics* (1938), and *The Marxist Philosophy and the Sciences* (1939).

Haldane left the Communist Party in 1949, after the Soviet Union's government banned the study of genetics. Disillusion with the Communists did not make him any happier with British politics, however. In 1957, Haldane and his wife moved to India in protest of Britain's invasion of Egypt after the Egyptian government took control of the Suez Canal. They became Indian citizens in 1960. After working briefly for the Indian Statistical Institute in Calcutta, Haldane left in 1962 to head a new genetics and biometry laboratory in the state of Bubaneshwar, now known as Orissa. There, he continued to investigate and write about the interaction of genetic variation and natural selection.

In 1963, Haldane learned that he had cancer. He responded, typically, by writing a humorous poem about the disease. Treatment was unsuccessful, and he died in Orissa on December 1, 1964.

Although Haldane made no single overwhelmingly important contribution to biology, the breadth of his scientific work and his vivid, if not always pleasant, personality made him memorable both in his own day and later. French geneticist Boris Ephrussi said of him, "He is not merely a man. He is a force of nature." A hundred years after Haldane's birth, Sahotra Sarkar wrote that "few biologists have exerted [so] much influence on new research so long after their death."

Further Reading

Haldane, Alexander. *Lives of Robert and James Haldane*. Carlisle, Pa.: Banner of Truth, 1991.

Sarkar, Sahotra. "A Centenary Reassessment of J. B. S. Haldane, 1892–1964." *BioScience*, November 1992.

Selected Genetic Papers of J. B. S. Haldane. Edited by Krishna R. Dronamraju. New York: Garland Publishing, 1991.

Hales, Stephen
(1677–1761)
British
Naturalist, Biophysicist

Although he was primarily an Anglican minister all his life, Stephen Hales found time to found the science of plant physiology, the study of the functions of various parts of plants. He also contributed to animal physiology, made numerous inventions, and helped to introduce the idea that careful measurement is essential in scientific research.

Hales was born in Beakesbourne, Kent, England, on September 17, 1677. His father, Thomas, was wealthy. Hales trained for the ministry at what is now Corpus Christi College, part of Cambridge University, where he also studied sciences such as chemistry, physics, and botany. He became a fellow of the college in 1703 and remained there until 1709, when he became perpetual curate at Teddington (now a suburb of London), a position he kept for the rest of his life. He died there on January 4, 1761.

Hales admired physicist Isaac Newton and applied the principles and methods of physics to the study of plants and animals. He did most of his experiments on plants between 1719 and 1725 and described them in a book called *Vegetable Staticks*, published in 1727. He included his animal experiments in an expanded edition called *Statical Essays* (1733).

After seeing sap ooze upward out of a cut on a plant, Hales concluded that the liquid inside plants is under pressure. He measured this pressure by cutting off a vine at ground level and attaching a glass tube to the stump. Sap rose 8.3 feet (7.6 meters) into the tube. Hales went on to show that the liquid flows only in one direction, calculate how fast it flows, and determine that this speed is different in different kinds of plants.

Scientists later learned that this "root pressure" draws water from the ground and pulls it up through the plant. The water eventually evaporates through the plant's leaves and returns to the atmosphere, and this evaporation also draws liquid upward in the plant. Hales measured pressure in cut side-shoots of plants at different times of day and found that, when evaporation is occurring, the pressure is greatest when sunlight shines on the plant most intensely. The pressure is also affected by temperature.

Having found that the liquid in plants is under pressure, Hales thought that the same might be true of animals. He measured the blood pressure of a horse—the first time animal blood pressure had been measured—by connecting one of its blood vessels to a giant glass tube, much as he had done with plants. The blood shot nine feet up into the tube. He found that different kinds of animals had different blood pressures.

Hales also studied gases and invented a device for collecting and measuring them that many later scientists used. His work in this field was hindered by the fact that, like other scientists of his time, he believed that air was a single gas rather than the mixture it is. Nonetheless, he concluded, correctly, that plants absorb some part of air and require it for their nutrition. Later scientists showed that the gas plants use is carbon dioxide.

Hales's work on gases led him to conclude that the "spent," foul-smelling air in hospitals, prisons, and similar places contributed to disease. He invented a ventilator to increase air circulation in such places, and it did improve occupants' health, although Hales was mistaken in the belief (common in his day) that the "bad air" itself was the problem. He presented his ideas about ventilation in a 1743 book, *A Description of Ventilation*.

Hales seems to have been an endlessly creative man. He invented processes for preserving meat, drying and cleaning grain, and preventing the spread of fires, to name just a few. While trying to find an effective treatment for the painful stones, or mineral deposits, that sometimes accumulate in human kidneys or bladder, he invented the tweezerlike surgical tool called a forceps. Britain's premier organization of scientists, the Royal Society, awarded him its Copley Medal in 1739.

Further Reading

Allan, D. G. C., and Robert E. Schofield. *Stephen Hales, Scientist and Philanthropist*. London: Scolar Press, 1980.

Westfall, Richard S. "Hales, Stephen." Rice University Galileo Project. Available online. URL: http://es.rice.edu/ES/humsoc/Galileo/Catalog/Files/hales.html. Posted 1995.

Haller, Albrecht von
(1708–1777)
Swiss
Physiologist, Botanist

Viktor Albrecht von Haller helped to found the modern science of physiology. He was born on October 16, 1708, in Bern, Switzerland, and grew up in a village in the Alps. He was a child prodigy—for instance, compiling a Greek dictionary before he was 10 years old. His father, an attorney, made sure that his sickly but brilliant son obtained the best possible education.

Haller studied medicine at the Universities of Tübingen, Germany, and Leiden, Holland, earning his degree from Leiden in 1727. He then traveled around Europe, learning about medicine and literature in different countries. (He himself wrote poetry and, later, novels and other literary works.) He returned to Bern to begin a medical practice in 1729, in which year he also married Marianne Wyss. They eventually had three children.

In 1736, Haller moved to what is now Germany to become professor of medicine, anatomy, surgery, and botany at the new University of Göttingen. Haller, a moody man, seems never to have been happy away from his home city, however, and his sadness intensified when his wife died soon after the move. He returned to Bern in 1753 and went to work as a minor official in the city government. He also ran a salt plant, worked as a physician, and continued his research on physiology. He produced an eight-volume encyclopedia on the subject, *Elementa Physiologiae Corporis Humani* (*The Physiological Elements of the Human Body*, 1757–1766), and a textbook, *Basic Beginnings in Human Physiology* (1759), which continued to be used for about a hundred years. He died in Bern on December 12, 1777.

Haller's most important discovery in physiology concerned the nerves. At the time, most physiologists believed that nerves were hollow tubes that carried some sort of liquid, much as blood vessels carry blood. Haller rejected this idea because he could find no traces of the liquid. He showed that if animal tissue was touched or otherwise stimulated, only the nerves responded. He found that nerves always connected to the brain and spinal cord and therefore concluded that those organs were probably the ones that registered awareness of sensations and sent out responses to them, such as commands for muscles to move. Because of these and other discoveries, Haller is considered the founder of neurology, the study of the nervous system.

Haller's research on muscles was also valuable. He showed that they bunch together, or contract, in response to either a direct stimulus or a stimulus applied to nerves in the muscle, a property he called irritability. He made discoveries about other parts of the body as well, for instance, showing that bile, a yellowish green substance made in the liver, helps to digest fats.

Haller was interested in plants as well as animals. He collected plants, wrote a book about the plants of Switzerland, and developed a system of classifying plants, although this system never became as popular as that of CAROLUS LINNAEUS. In spite of his attempt to bury himself among the musty city records of Bern, Haller's many talents earned him a lasting reputation.

Further Reading

Roe, Shirley A. *The Natural Philosophy of Albrecht von Haller*. New York: Arno Press, 1981.

Harvey, William
(1578–1657)
British
Anatomist, Physician

William Harvey proved that the blood flows through the body in circles, pumped by the heart. He was born in the coastal town of Folkestone, in the part of southern England called Kent, on April 1, 1578. His father, Thomas, was a well-to-do merchant. Thomas Harvey and his second wife, Joan, had seven sons, of whom William was the oldest, and two daughters.

Harvey studied medicine first at Gonville and Caius College, part of Cambridge University. After his graduation in 1597, he went to the University of Padua, in Italy, for advanced studies. One of his professors at Padua, Fabricius (Girolamo Fabrizio), was unusual in stressing that students should learn chiefly by observing the dissection of animal and human bodies rather than by memorizing the works of ancient authorities, such as the Greco-Roman physician GALEN.

Harvey took his medical degree from Padua in 1602 and returned to England, where he quickly became a respected physician. In 1604, he married Elizabeth Browne, whose father had been a doctor at the court of Elizabeth I. Harvey's father-in-law and one of Harvey's brothers, John, introduced him to people at the court of the current monarch, James I, including Charles, the heir to the throne, who became Harvey's lifelong friend. In 1618, Harvey, too, was hired as a court physician.

In the year he married, Harvey became a member of the Royal College of Physicians, an organization of the most important doctors in London. Harvey helped the group with administrative affairs and also gave twice-weekly lectures on anatomy to other physicians from 1616 to 1643. From 1609 to 1629, he was chief physician at St. Bartholomew's, a large London hospital, as well.

In spite of all these duties, Harvey found time to perform extensive dissections of and experiments on different kinds of animals. For instance, when Prince Charles hunted deer, a common pastime for the wealthy, he let Harvey cut up the animals' bodies; indeed, he sometimes watched. Harvey also dissected snakes because

British physician William Harvey described the circulation of the blood and function of the heart in an influential book published in 1628. *(National Library of Medicine)*

their hearts, the organs that interested him most, beat more slowly than those of mammals, so he could see what happened during each beat. These dissections led Harvey to his most important work, which he described in a small book first published in Germany in 1628, titled *De Motu Cordis et Sanguinis in Animalibus* (*On the Movement of the Heart and Blood in Animals*).

Physicians of Harvey's time accepted Galen's belief that the heart's only function was to warm the blood. Harvey wrote that his dissections and experiments had shown, however, that the heart pumps blood through the body. It fills with blood, then squeezes or contracts to force the blood out of its two lower chambers, the ventricles, and into blood vessels called arteries. This squeezing creates the heartbeat. Harvey described the actions of the heart in the first half of his book, which many historians think was written as much as 10 years before the second half.

The second half of Harvey's book was even more revolutionary than the first half. In it, he stated that "it is absolutely necessary to conclude that the blood in the animal body is impelled [pushed] in a circle" rather than being constantly created and used up as Galen had claimed. In fact, Harvey wrote, the blood makes two circles, one from the right ventricle of the heart through the lungs to the left atrium (upper chamber of the heart), and the other from the left ventricle through the body and then back to the right atrium. The blood flows into the body and lungs through the arteries and flows back to the heart through another group of blood vessels, the veins. Back in Padua, Fabricius had shown his students that the veins contained valves, which he called "little doors." Harvey proved that these "doors" opened in only one direction—toward the heart—and claimed that their purpose was to force the blood in the vessels to flow in that direction.

Harvey did not fill in quite all the pieces of the circulation puzzle. For instance, he could not tell exactly how blood moved from arteries to veins. Capillaries, the tiny vessels that transport it, can be seen only with a microscope, and he did not use one. (Italian anatomist MARCELLO MALPIGHI first saw capillaries in 1660, three years after Harvey's death.) Nonetheless, Harvey laid out the basic pattern of the blood circulation completely for the first time.

Just as important as Harvey's conclusions was the way he reached them—by careful observation, experimentation, and measurement—techniques that scientists still use. For instance, he measured the amount of blood that the heart ventricles can hold, which is the most that it can pump in a single beat. He then counted the average number of heartbeats in a single minute. From these figures, he determined that the heart could pump in just half an hour "a larger quantity [of blood] than is contained in the whole body." He said that the body could not possibly make or use so much blood so quickly.

Harvey's ideas caused controversy at first. John Aubrey, who met Harvey when Harvey was an old man and left a description that provides most of what is known of him as a person, wrote that Harvey told him that "after his book on the circulation of the blood came out . . . he fell mightily in his practice [lost many patients]. . . . All the physicians were against his opinion." Harvey, or perhaps Aubrey, seems to have exaggerated these problems, however. Most physicians apparently just ignored Harvey's book.

If Harvey's list of patients grew shorter, it was probably because he was spending most of his time at court. His friend had become King Charles I in 1725, and around 1630, Harvey gave up his other duties to stay with the king. He became Charles's chief physician in 1639 and remained loyal to him even after civil war broke out between the king and his legislature, the Parliament, in 1642. Harvey lived with Charles's court in semi-exile at Oxford from 1642 to 1646, serving as warden of Merton College for the later part of that time. He paid a steep price for his loyalty when a mob broke into his deserted rooms in London and destroyed notes and other papers he had left there, a loss he later said was the worst of his life.

The king's party lost the war, and the new British government, the Commonwealth, executed the former monarch in 1649. Harvey, by then an old man, escaped with a heavy fine. He retired to live with his two surviving brothers, Eliab and Daniel, who had become wealthy merchants. He published a small book that year, answering some criticisms that had been made of his circulation theory. On the whole, however, he was pleased to note, "I perceive that the wonderful circulation of the blood, first found out by me, is consented to by almost all."

Harvey published a third book, *Exercitationes de Generatione Animalium* (Essays on the generation [reproduction] of animals), in 1651. Drawing on experiments he had done at Oxford and probably earlier, this book described the way animals, especially chickens, develop before birth. For instance, he pinpointed the spot at which an unborn chicken's heart first appears. Unlike his book on the circulation, however, this one contained no major new discoveries. Harvey died of a stroke on June 3, 1657.

Harvey's discovery of the blood circulation had little impact on the way physicians of his time treated patients, but medical historians agree that it was one of the most important events in the history of medicine. It not only produced the first accurate description of a major organ system but established that observation, experimentation, and measurement were the proper ways to learn about the body.

Further Reading

Harvey, William. *The Anatomical Exercises: De Motu Cordis* and *De Circulatione Sanguinis, in English Translation*. Edited by Geoffrey Keynes. 1628. Reprint, Mineola, N.Y.: Dover Publications, 1995.

Nuland, Sherwin B. *Doctors: The Biography of Medicine*. New York: Alfred A. Knopf, 1988.

Helmholtz, Hermann von

(1821–1894)
Prussian
Physiologist, Physicist

Hermann Ludwig Ferdinand von Helmholtz made major contributions to two very different branches of science, physiology and physics. He was born Hermann Helmholtz in Potsdam, Prussia (one of the small states that later became part of Germany), on August 31, 1821; the "von" was added to his name in 1882, after the German emperor made him a member of the hereditary nobility in recognition of his scientific achievements.

Helmholtz, as he is usually known, discovered his first scientific love, physics, when he found some textbooks that his father, Ferdinand, a high school teacher, had left on a shelf. He would have liked to study physics at a university, but his family could not afford to pay for an advanced education. At age 17, therefore, he signed up for a government program that paid for his training at the Royal Friedrich-Wilhelm Institute for Medicine and Surgery in Berlin in return for his promise to work for eight years as an army surgeon afterward. While in Berlin, he also took classes in chemistry, physics, physiology, and mathematics.

Helmholtz earned his medical degree in 1843 and then began his military service at the army barracks in Potsdam. He set up a small laboratory there and did physiology experiments in his spare time. One of his first projects was a set of mathematical calculations showing that the amount of physical energy releasable by chemical combination of all the food and oxygen a living thing consumed was enough to produce all the muscular work that the organism did and all the heat it gave off. In short, as Canadian psychologist Raymond Fancher wrote in *Pioneers of Psychology*, a history of the field, "he showed that it was feasible to analyze a living body as if it were a machine, in terms of fuel input, work output, and overall efficiency."

Helmholtz also did research in physics. He developed in detail the very important concept of conservation of energy, which states that energy cannot be created or destroyed but rather is simply changed from one form to another—from chemical energy to heat, for instance. He wrote a groundbreaking paper, "The Conservation of Force," to describe this idea in 1847. As his physiology experiments had shown, energy is conserved in living things as well as in the nonliving environment.

Helmholtz's energy research impressed Prussian government officials so much that they excused him from his remaining three years of military duty and found him a position as professor of physiology at the University of Königsberg, which he took up in 1849. Now that he could afford to start a family, he married Olga von Velten, a young woman to whom he had been engaged for several years. They later had two children. (Olga died in 1859, and Helmholtz married Anna von Mohl in 1861. They had three additional children.)

In 1851, while preparing a lecture on the workings of the eye, Helmholtz happened to shine a mirror into a person's eye at a certain angle and found that he could see inside the eyeball. By placing a mirror on a band that a doctor could wear around the forehead, Helmholtz invented the ophthalmoscope, which allows physicians to examine the inside of the eye. He showed the device to eminent physicians and researchers all over Europe, and most quickly adopted it as a useful tool. A modified form of it is still used.

For Helmholtz's next major experiments, he devised an ingenious way to turn a galvanometer, a device that measures electric current, into a kind of stopwatch that could measure fractions of a second, which no conventional timepiece of the era could do. He used his creation to determine the speed at which electrical signals travel along a nerve, a task other scientists had thought impossible. After measuring and comparing the tiny time intervals between stimulation of a nerve at various points in a dissected frog's leg and the twitch of the frog's foot, he concluded that signals moved down the nerve at about 83 feet (30 meters) per second.

Helmholtz left Königsberg in 1855 and continued his teaching and research at the Universities of Bonn (1855–58) and Heidelberg (1858–71). His most important physiology studies, carried out between 1852 and 1868, involved the way mammals see and hear. His study of vision refined a theory of color vision first proposed by English researcher Thomas Young, which said that the retina (the light-sensitive part of the eye) contains three kinds of color-sensitive receptors. Each kind generates a nerve signal in response to a different one of the primary colors: red, blue, and green. When light stimulates more than one kind of receptor at once, the brain perceives a color made up of a combination of these primaries. For instance, when both red and blue receptors are stimulated, the brain sees purple. Later scientists found this theory to be basically correct. Helmholtz described this and other ideas about vision in *Handbook of Physiological Optics*, which appeared in three volumes between 1856 and 1867.

Just as Helmholtz had drawn on discoveries about the physics of light to explain vision, he drew on information about the physics of sound to explain hearing in another book, published in 1863. He held that sound waves make an organ of the inner ear called the cochlea resonate, or vibrate, thereby stimulating nerve endings in the organ. A nerve transmits messages from these endings to the brain, just as the optic nerve that ends in the retina transmits messages from the eye.

In addition to describing how the eye and ear detect information, Helmholtz considered how the brain assembles this raw data, which he called sensations, into patterns and gives them interpretation or meaning, producing what he termed perceptions. Sensations of different tones from the ear, for instance, can be assembled into perceptions of words or music. Helmholtz believed that all perceptions are learned, but scientists now think that some are inborn, or genetically determined.

Helmholtz's discoveries about vision and hearing made him one of the most famous scientists in Europe. He received awards such as the Copley Medal of Britain's leading science organization, the Royal Society (1873). This fame, ironically, finally won him the chance to abandon physiology and return to physics. In 1871, he became a professor of physics at the University of Berlin, and thereafter he devoted all his time to this science, eventually heading a research institute that the new German government built for him. He worked there until his death on September 8, 1894.

Further Reading

Fancher, Raymond E. *Pioneers of Psychology*. New York: W. W. Norton, 1979.

O'Connor, J. J., and E. F. Robertson. "Hermann Ludwig Ferdinand von Helmholtz." University of St. Andrew's (Scotland) School of Mathematics and Statistics. Available online. URL: www-groups.dcs.st-andrews.ac.uk/~history/Mathematicians/Helmholtz.html. Accessed 2003.

Hershey, Alfred Day

(1908–1997)
American
Molecular Biologist, Microbiologist

Along with MAX DELBRÜCK and SALVADOR LURIA, with whom he shared the Nobel Prize in physiology or medicine in 1969, Alfred Hershey established basic facts about genes by studying viruses called bacteriophages. Most importantly, Hershey proved that genes are made of DNA rather than protein.

Hershey was born in Owosso, Michigan, on December 4, 1908, and grew up in the nearby town of Lansing. His father, Robert Hershey, was an auto worker. His mother was the former Alma Wilbur. Hershey attended Michigan State College, from which he obtained a B.A. in 1930 and a Ph.D. in 1934. His specialties were bacteriology and chemistry. In 1945, he married Harriet Davidson, with whom he later had a son.

Hershey joined the Washington University School of Medicine in St. Louis as an assistant bacteriologist in 1934, rising to the rank of associate professor by 1942. Most of his work there concerned the immune system, but another member of the faculty also interested him in bacteriophages, viruses that infect bacteria. Hershey's bacteriophage research brought him to the attention of Delbrück, a charismatic German-born scientist then working at Vanderbilt University in Tennessee, who was interested in the same subject. At Delbrück's urging, he, Hershey, and Luria, a researcher at Indiana University, set up an informal team called the Phage Group in 1943. Its members, which eventually included other scientists, worked independently but shared results and ideas.

During the next several years, Hershey worked out many details of bacteriophage genetics, for instance, showing (as Luria also did) that two viruses infecting the same bacterial cell can exchange genes. His research intensified after 1950, when he joined the genetics research unit (then called the department of genetics) of the Carnegie Institute of Washington's Cold Spring Harbor Laboratory on Long Island, New York. Hershey became head of the unit in 1962 and kept this post until his retirement in 1974.

Hershey's best-known bacteriophage experiment was aimed at settling an issue that researchers had debated for a decade. Chromosomes, bodies in the cell nucleus that had been shown to carry genetic information, contain both nucleic acid and protein, and no one was sure which of these two types of chemical actually transmitted the information. OSWALD THEODORE AVERY had provided strong evidence in 1944 that the information resides in DNA, but his experiments had not convinced everyone. Hershey realized that bacteriophages were well suited to answer this question because they are simply nucleic acid cores wrapped in protein coats.

In 1952, Hershey and coworker Martha Chase labeled the nucleic acid in bacteriophages with radioactive phosphorus and the protein with radioactive sulfur. They then allowed the viruses to infect bacteria, a process in which the virus's protein shell attaches itself to the surface of the bacterial cells. After a few minutes, they spun the infected bacteria in a kitchen blender at a speed that, they had determined, would knock loose the viral proteins but would not break open the bacterial cell walls. By checking for the two radioactive labels, they showed that the bacteria contained viral nucleic acid but no viral protein. New viruses still emerged from the cells, however, which showed that the nucleic acid must be providing the information that allowed the viruses to reproduce.

Hershey shared the Nobel Prize with Luria and Delbrück for their work on bacteriophages, which established the basis for molecular biology. Earlier, Hershey had received the American Public Health Association's Albert Lasker Medical Research Award (1958) and the (U.S.) National Academy of Science's Kimber Genetics Award (1965) as well. Unlike the more outgoing Delbrück and Luria, however, Hershey was what his friends called "a bit of a hermit." His idea of "Hershey Heaven" was not receiving prizes but working quietly in his laboratory. His later work contributed to the development of vaccines against several virus-caused diseases. Hershey died in Syosset, New York, on May 22, 1997.

Further Reading

"Hershey, Alfred (Day)." *Current Biography Yearbook 1970*. New York: H. W. Wilson, 1970.

Stahl, Franklin W., ed. *We Can Sleep Later: Alfred D. Hershey and the Origins of Molecular Biology*. Long Island, N.Y.: Cold Spring Harbor Laboratory Press, 2000.

Hess, Walter Rudolf

(1881–1973)
Swiss
Neurobiologist

Walter Hess showed that different parts of the brain perform different functions and began to determine what those functions are. He also created a bridge between neurology (the study of the physical brain) and psychology by showing that stimulation of some brain areas could produce changes in behavior. His work earned a share of the Nobel Prize in physiology or medicine in 1949.

Hess was born in Frauenfeld, Switzerland, on March 17, 1881, to Clemens Hess and the former Gertrud Saxon. His father was a physics teacher and interested him in science at an early age. He studied medicine at five Swiss and German universities, finally receiving a medical degree from the University of Zurich in 1906. In 1908, he married Louise Sandmeyer; they later had a son and a daughter.

After working for 11 years as an ophthalmologist (physician who treats the eyes), Hess gave up this high-paying career to devote himself to physiological research. He joined the Physiological Institute, part of the University of Zurich, and became the institute's director, as well as a professor of physiology at the university, in 1917. He later became head of the university's physiology department and remained so until his retirement in 1951.

After studying the regulation of blood pressure, pulse rate, and breathing, Hess turned in 1925 to investigating the way the brain controls these processes. He anesthetized cats and implanted tiny electrodes into their brains. After the animals recovered, he stimulated different brain areas with small amounts of current sent through the electrodes and recorded changes in the animals' body temperature, blood pressure, breathing and pulse rates, digestion, and behavior. The brain region that controls automatic functions proved to be part of the diencephalon, at the base of the brain. Stimulation of this area also affected the animals' balance and movement.

When Hess stimulated some parts of the diencephalon, he noticed that the cats' behavior as well as their automatic functions were sometimes affected. The cats might act frightened, for instance, as if facing a fierce dog, even though no dog was present. With other stimulation, they might fly into a rage. By showing that brain stimulation could produce behavior associated with different emotions, Hess provided evidence that emotions are produced by activity in the physical brain, which some scientists had doubted.

Hess published numerous articles and books about his research, including *The Functional Organization of the Diencephalon* (1948) and *The Biology of the Mind* (1964). In addition to the Nobel Prize, he won the Swiss government's Marcel Benorst Prize in 1933 and the Ludwig Medal of the German Society for Circulation Research in 1938. He died in Locarno, Switzerland, on August 12, 1973.

Further Reading

"Walter Rudolf Hess—Biography." *Nobel Lectures, Physiology or Medicine, 1942–1962*. Available online. URL: www.nobel.se/medicine/laureates/1949/hess-bio.html. Last updated 2001.

⊠ Hippocrates

(ca. 460–370 B.C.)
Greek
Physician

The ancient Greek physician Hippocrates, often called "the Father of Medicine," is an almost legendary figure. Little is known about him except that he practiced medicine on the island of Cos, off the western coast of Asia Minor, around the late fifth and early fourth century B.C. and was probably born there. Tradition has it that he learned medicine from his father, a physician named Heraclides, and that he in turn established a school for physicians on Cos.

Hippocrates' teachings are preserved in a group of 60 to 70 writings called the Hippocratic Collection, which were most likely part of a medical library. They seem to have been written by a number of different people, and no one knows which ones, if any, Hippocrates himself authored. Some of the best known books in the collection are *Aphorisms* (short sayings); *Epidemics I* and *III*; *Airs, Waters, and Places;* and *The Nature of Man*.

The life of Hippocrates may be shadowy, but the impact of his ideas on Western medicine was not. They were part of the great revolution in thinking taking place in Greece at this time, spearheaded by such famous philosophers as Socrates and Plato. Most physicians of Hippocrates' era and earlier saw disease as punishment from the gods or possession by evil spirits, so they treated it with religious or magical rites. Hippocrates and his followers, however, insisted that health and disease are strictly part of nature. They stressed that physicians could understand illness, predict its progress, and choose treatment (where treatment was possible) by observing sick people and their surroundings and using reason to interpret what they saw. These ideas began to change medicine from an aspect of religion to a part of science.

Lacking the knowledge of the interior structure and functions of the body that later scientists built up, Hippocratic physicians accepted the common Greek belief that health represented a balance among four liquids, or humors, in the body: blood, black bile, yellow bile, and phlegm (mucus). According to this theory, disease appeared when some factor, such as unusual weather or poor diet, caused one humor to be produced in excess or to accumulate in some part of the body. Physicians treated illness by removing the excess humor (by taking blood from the patient, for instance) and restoring an interior and exterior environment that kept the humors in balance.

Although the four humors theory seemed so logical that most physicians believed it until well into the 19th century, it was mistaken. Other Hippocratic ideas, however, are still valued. One is that study of the relationship between patients and their environment is vital to understanding disease and restoring health. The Hippocratic writings state that observation of "airs [winds], waters, and places" can tell physicians much about the general health of people who live in a particular location.

Careful observation of individual patients is even more important. One document recommends that doctors observe a patient's "diet; customs; the age of the patient; speech; manners; fashion; even his silence; his thoughts; if he sleeps or is suffering from lack of sleep; the content and origin of his dreams." Hippocratic teachings told physicians to write down their observations so that other physicians could learn from them, and several books in the collection consist of such case notes.

Another important Hippocratic idea was that nature tries to restore health on its own and is usually better able to do so than physicians are. The physician's art lies in recognizing when and how to help nature. Unlike many other physicians of their time and later, Hippocratic doctors usually recommended mild treatments such as changes in diet, cleaner or quieter surroundings, exercise, and relaxation rather than drugs or surgery. Their philosophy was summarized in a statement from the Hippocratic writings usually translated as "First, do no harm." Even though modern physicians can treat disease far more effectively than those of Hippocrates' time, many still recognize the wisdom of avoiding harsh or drastic treatments whenever they can.

Hippocratic physicians tried to distinguish themselves from other medical practitioners by their high moral code. This code became enshrined in the Hippocratic Oath, which some physicians still recite when they graduate from medical school. Takers of this oath promise, among other things, to keep patients' personal information private and not to give out dangerous drugs even if asked to do so. Although physicians in Hippocrates' time and since have argued about the value of honoring particular parts of the oath, such as a promise not to provide abortions, most respect the oath's basic idea that the physician should put the patient's needs before everything else.

Further Reading

Cantor, David, ed. *Reinventing Hippocrates*. Aldershot, Hampshire, England: Ashgate Publishing, 2002.

Hippocrates. Translated by W. H. S. Jones. Loeb Classical Library, no. 147. Cambridge, Mass.: Harvard University Press, 1923.

Nuland, Sherwin B. *Doctors: The Biography of Medicine*. New York: Alfred A. Knopf, 1988.

Riley, Mark T. *Great Thinkers of the Western World Annual 1999*. New York: HarperCollins, 1999.

Hitchings, George Herbert

(1905–1998)
American
Chemist, Pharmacologist

George Hitchings and his chief coworker, GERTRUDE BELLE ELION, used their understanding of basic chemical processes in cells to create drugs. For this new, "rational" approach to drug design, they shared the 1988 Nobel Prize in physiology or medicine with JAMES WHYTE BLACK, a British scientist who used similar methods.

Hitchings was born in Hoquiam, Washington, in 1905 to George Herbert Hitchings Sr., a master shipbuilder, and his wife, Lillian (Mathews) Hitchings. Hitchings Senior died after a long illness when his son was just 12 years old, and, Hitchings wrote later in his autobiographical sketch for the Nobel Foundation, "the deep impression made by this event turned my thoughts toward medicine."

Hitchings attended the University of Washington, beginning as a premedicine major but eventually changing to chemistry. He earned a bachelor's degree in 1927 and a master's the following year. He then did advanced studies at Harvard, moving from chemistry to biochemistry, and obtained his Ph.D. in 1933. In that same year, he married Beverly Reimer, an artist, writer, and teacher. They later had a daughter and a son. Reimer died in 1985, and Hitchings married Joyce Shaver, a physician, in 1989, shortly after he received the Nobel Prize.

The uncertain employment conditions of the depression forced Hitchings into what he called "a nine-year period of impermanence, both financial and intellectual." That difficult time ended when the Wellcome Research Laboratories, a pharmaceutical company then headquartered in Tuckahoe, New York, hired him in 1942 as head—indeed, at first, the only member—of its biochemistry department. Hitchings acquired several coworkers over the next few years, including Elion in 1944.

Researchers at the time normally developed drugs more or less by trial and error, but Hitchings thought that an approach based on cell chemistry could produce better results. He wanted to look for or create compounds that interfered with essential processes taking place only (or at least more often) in cells that cause disease, such as cancer cells or bacteria.

Scientists had not yet learned that DNA carries the inherited information on which all cells depend, but they did know that DNA molecules must somehow reproduce themselves each time a cell divides. They also knew that the large DNA molecule includes several smaller molecules, some of which belong to families of compounds called purines and pyrimidines. DNA must take up these compounds from the cell in order to reproduce. Hitchings reasoned that if DNA in, say, cancer cells could be made to take up substances that were similar to purines and pyrimidines, yet also slightly different, these chemicals might block DNA reproduction and ultimately kill the cell, much as an ill-fitting part can jam and even destroy a machine.

Around 1947, Hitchings and his team began to search for such compounds and send them to the Sloan Kettering Institute for testing as possible anticancer drugs. Hitchings set Elion to work synthesizing "almost"-purines, while other laboratory workers did the same for pyrimidines.

The laboratory's first major success was 6-mercaptopurine (6-MP), which Elion created in 1950. It was one of the first drugs to fight cancer by interfering with cancer cells' DNA. It worked especially well against childhood leukemia, a blood cell cancer that had formerly killed its victims within a few months. When combined with other anticancer drugs, 6-MP now cures about 80 percent of children with some forms of leukemia.

6-MP's effects on leukemia proved to be just the beginning of its powers—and those of Hitchings's approach. Scientists in other laboratories discovered that, by acting on the same cells that are overproduced in leukemia, the drug halts some actions of the immune system. The immune system's attacks on "foreign" substances protect the body against invaders such as bacteria, but they also destroy transplanted organs, except in identical twins. At Hitchings and Elion's recommendation, researchers eventually tested not only 6-MP but azathioprine, a related compound that proved able to suppress the immune system even more effectively than 6-MP did, as possible drugs to prevent the rejection of organ transplants. Boston surgeon Joseph Murray used azathioprine in the first successful kidney transplant between unrelated humans in the early 1960s. It was the breakthrough drug that made organ transplants practical.

The list of drugs produced by Hitchings's laboratory grew longer and more amazing as time went on. It came to include allopurinol, used to treat gout, a painful joint disease; acyclovir, which attacks a dangerous group of viruses called herpesviruses; pyrimethamine, used against malaria; trimethoprim, used to fight bacterial infections; and zidovudine, or AZT, the first drug used to treat AIDS.

Hitchings became Burroughs Wellcome's vice president in charge of research in 1967 and kept this post until 1976, when he retired to gain more time for research, travel, and philanthropy. He had become director of the Burroughs Wellcome Fund, a nonprofit foundation that provides grants for biomedical research, in 1968 and its president in 1971. He also founded (in 1983) and was director for life of the Greater Triangle Community Foundation, which provides social services for needy people in Research Triangle Park, North Carolina, to which Burroughs Wellcome (now GlaxoSmithKline) had moved in 1968. He gave all his Nobel Prize money to this foundation. He once wrote that "my greatest satisfaction has come from knowing that our efforts helped to save lives and relieve suffering." Hitchings died on February 27, 1998.

Further Reading

"Curing Childhood Leukemia: A Leap of Faith." National Academy of Sciences Beyond Discovery. Available online. URL: www.beyonddiscovery.org/content/view.page.asp?I=292. 2001.

Hitchings, George. "George H. Hitchings—Autobiography." *Les Prix Nobel 1988*. Nobel Foundation. Available online. URL: www.nobel.se/medicine/laureates/1988/hitchings-autobio.html. Last updated 2002.

Ho, David (Ho Da-i)
(1952–)
Taiwanese/American
Virologist

In the mid-1990s, David Ho changed the way physicians treated AIDS and, for the first time, gave people hope that it could become a manageable illness rather than an immediate death sentence. Because of his discoveries about the disease, *Time* magazine chose him as its Man of the Year in 1996.

Ho Da-i (his given name means "the Great One") was born near Taichung, on the Chinese island of Taiwan, on November 3, 1952. When he was a child, his father moved to the United States, hoping to establish a better life there. Nine years later, after he found good work as an engineer, he sent for his wife and two sons. When they arrived in Los Angeles, the whole family took new names, and Da-i became David.

David's classmates called him stupid at first because he knew no English, but he learned quickly. Determined to shine academically, he soon excelled in all his subjects. He attended the Massachusetts Institute of Technology for a year, then transferred to the California Institute of Technology (Caltech). He also changed his major from physics to premedicine. After earning a B.S. with high honors in 1974, he went to Harvard Medical School, from which he took his M.D. in 1979. In 1976, he married artist Susan Kuo, whom he had met while at Caltech, and they later had two daughters and a son.

While doing his residency (part of his postmedical training) at the Cedars-Sinai Medical Center in Los Angeles in 1981, Ho treated several homosexual men who had multiple infections of types that the immune system normally fights off. These men proved to be some of the first people identified as having the condition later known as AIDS. "I had a long-standing interest in infectious diseases, and I love puzzles," Ho told *Esquire* interviewer Alec Wilkinson in 1999, "and I . . . decided to make this new disease the focus of my research." From the beginning, Ho suspected that the illness was caused by a virus that damaged the immune system.

In 1982, Ho moved to Massachusetts General Hospital in Boston to study AIDS in the laboratory of Martin Hirsch. Hirsch's laboratory became the fourth in the world to isolate HIV, the virus that causes the disease. They were the first to find it in a type of immune cells called macrophages and among the first to show that it was present in semen and could infect the nervous system. Ho played a leading role in many of these discoveries. "David had the Midas touch," Hirsch said later. "Whatever he did worked."

Ho and his family moved back to California and Cedars-Sinai in 1986. He also joined the faculty of the University of California at Los Angeles (UCLA) Medical School, where he was promoted to associate professor in 1989. While in Los Angeles, Ho proved that the first stage of HIV infection was marked by a flulike illness.

In 1990, philanthropist Irene Diamond chose Ho to head the new Aaron Diamond Center for AIDS Research in New York City. He was relatively unknown and very young for such a high honor, but, Diamond said later, "I [didn't] want a star, I want[ed] a wonderful scientist." Ho is still the scientific director and CEO of the center, which is the largest privately supported AIDS research center in the world. He is also a professor at Rockefeller University, with which the Diamond Center is now affiliated.

In the mid-1990s, David Ho revealed new facts about the behavior of the AIDS virus that suggested that a combination of drugs given soon after infection offered the best hope for controlling the virus's progress. *(David D. Ho)*

Scientists in the early 1990s knew that people infected by HIV often remain apparently healthy for up to 10 years after the virus enters the body. Most believed that during this "quiet time" the virus hid inside certain immune system cells, not reproducing until some stress triggered it and produced full-blown AIDS. Ho, however, suspected that this might not be the case because the number of T cells, the type of immune cell that HIV usually infects, slowly declined during that time. He thought that HIV in fact might reproduce rapidly throughout the period, but the immune system might send equally large numbers of cells to destroy the new viruses, leaving only a few virus particles in the blood. AIDS would appear only when the viruses finally overwhelmed the exhausted immune system. Ho has compared the situation to that of a person on a treadmill, who stays in the same place but is nonetheless walking quickly.

Ho realized that the only way to test his theory would be to stop the virus from reproducing temporarily and see whether the number of immune system cells increased. A new group of anti-AIDS drugs called protease inhibitors, developed in the early 1990s, finally made such a test possible. Ho and George Shaw at the University of Alabama at Birmingham, working independently, verified this idea at about the same time and published their work in 1995.

Ho and Shaw's discovery changed the way physicians treated people with HIV infection. Before then, doctors had seen little reason to give such people antiviral drugs before symptoms of AIDS appeared because the number of viruses in their bodies had seemed to be so low. After the two scientists found that viruses in fact were being produced rapidly, however, Ho proposed that giving drugs as soon as infection was detected might allow virus numbers to be kept small, delaying or perhaps even preventing AIDS from developing. Hitting HIV "early and hard," as he put it, also decreased the chances that the virus would develop resistance to the medications.

Doctors had concluded by the mid-1990s that a combination of protease inhibitors and other anti-AIDS drugs such as AZT, each of which affects HIV at a different stage of its reproductive cycle, worked better than any one drug because the virus was less likely to develop resistance to the mixture than to a single drug. Ho and his coworkers combined this "cocktail" approach with their early treatment idea in a small number of HIV-positive volunteers. They showed that, in many cases, the number of viruses in the volunteers' blood dropped to undetectable levels, the number of immune cells increased, and the people's health improved considerably. Mathematical models even suggested that two or three years of such treatment might wipe out the infection entirely.

Ho announced his findings and recommendations at the 11th International Conference on AIDS, held in Vancouver, Canada, in July 1996. By "provid[ing] concrete evidence that HIV is not insurmountable," he created a sensation, leading to his selection as *Time*'s Man of the Year. (He has also received other awards, including the Ernst Jung Prize in Medicine and the Scientific Award of the Chinese-American Medical Society.) Ho and others were quick to point out that many scientists had contributed to this success. They also stressed that it did not represent a cure for AIDS. Nonetheless, wide application of early treatment with drug "cocktails" caused the death rate from AIDS in the United States in the late 1990s to drop to a fifth of what it had been in the 1980s, and similar encouraging results have occurred in western Europe.

As Ho is the first to admit, many problems remain in treating AIDS. The drugs now in use do not work for everyone; some strains of virus resist them, and they often produce side effects that range from unpleasant to life-threatening. They must be taken according to a complicated schedule that some patients are unable or unwilling to follow. Furthermore, the drugs are extremely expensive, placing them out of reach of most of the millions of infected people in developing countries. Even if drug manufacturers gave out the medicines for free, Ho says, many countries lack the medical infrastructure to administer them properly.

Today, Ho's laboratory at the Diamond Research Center continues to study HIV and its effects on immune cells in the hope of finding better drugs to control it. His researchers know that small amounts of virus hide inside cells where drugs cannot reach them, making Ho's dream of curing the disease so difficult to achieve that he is one of the few AIDS specialists who still thinks it possible. They are trying to learn more about these hiding places and possible ways of forcing the viruses to emerge from them. In addition, they are working to develop a vaccine against the disease, which Ho thinks "is our only real hope to avert a disaster unparalleled in medical history." In 2002, Ho and other researchers reported isolating proteins called defensins, produced by certain white blood cells, which attack viruses and may protect some people infected with HIV against developing AIDS. They hope that these proteins can be incorporated into a vaccine.

Further Reading

"Ho, David." *Current Biography Yearbook 1997*. New York: H. W. Wilson, 1997.

Park, Alice, and Dick Thompson. "The Disease Detective." *Time*, December 30, 1996–January 6, 1997.
Wilkinson, Alec. "Please Leave David Ho Alone." *Esquire*, March 1999.

Hodgkin, Alan Lloyd
(1914–1998)
British
Neurobiologist

Alan Hodgkin and his coworkers showed how changes in nerve cell membranes that affect the flow of electrically charged atoms, or ions, allow the cells to send messages by means of electrical signals. He was born on February 5, 1914, in Banbury, Oxfordshire, England. His father, George, died overseas in World War I while Hodgkin was still a young child, and his mother, the former Mary Wilson, raised him.

Hodgkin attended Trinity College at Cambridge University, earning a B.S. degree in 1936. He never earned a Ph.D. He spent essentially all of his career at Cambridge, as lecturer and assistant research director of the physiology department from 1945 to 1952, Foulerton Research Professor of the Royal Society from 1952 to 1969, and Plummer Professor of Biophysics from 1970 to 1984. He was also Master of Trinity College from 1978 to 1984 and Chancellor of the University of Leicester.

Since the end of the 19th century, scientists had known that nerve cells transmit messages by means of tiny amounts of electric current sent down long fibers, called axons, that extend from the cell bodies. They knew that the transmission process involved temporary changes in the electrical qualities of the nerve cell membranes, but they could not determine exactly what these changes were or how they were produced. The subject was hard to study because most animals' nerves are delicate and thin as hairs and because changes in nerve cells take place within thousandths of a second.

Hodgkin discovered a possible solution to the first problem when he visited the United States on a Rockefeller Foundation fellowship in 1937 and 1938. He spent some time at the famed marine biology laboratory in Woods Hole, Massachusetts, and scientists there pointed out that the giant squid has nerve fibers up to a millimeter across—40 times as large as the biggest nerves in vertebrates. Hodgkin realized that these large fibers might allow him to perform experiments that could not have been done before. On this same trip he met Marion Rous, daughter of cancer virologist PEYTON ROUS. Hodgkin and Rous, a children's book editor, were married when Hodgkin revisited the United States briefly in 1944, and they had three daughters and a son.

Back at Cambridge in 1939, Hodgkin began working on squid nerve fibers with Andrew Huxley. Their research had to halt during World War II, when both did military service (Hodgkin helped to develop radar for planes). After they resumed their studies in 1945, they developed a method called the voltage clamp technique, which involved placing one microscopic electrode inside the squid nerve and another outside, resting on the nerve cell's membrane. This technique allowed them to make the first electrical recordings taken from inside a nerve cell. They also used the clamp to set the voltage at the cell membrane to different fixed levels so they could study the effect of different voltages on the membrane.

Hodgkin and Huxley found that changes in voltage open and close microscopic pores in the cell membrane, later called channels, thus permitting or blocking the flow of sodium and potassium ions. Because ions carry an electrical charge, their movement, in turn, affects the electrical state of the cell. These changes move down the fiber as a pulse of electrical activity called the action potential. Hodgkin and Huxley described the interaction between electricity and membrane changes that characterize the nerve impulse mathematically in a paper published in 1952.

Beginning in the late 1940s, Hodgkin also did research with a different coworker, BERNHARD KATZ. Hodgkin and Katz showed that a change in the nerve cell membrane's openness to sodium ions, allowing these positively charged ions to flow into the cell, is the key difference that gives the inside of the cell a temporary positive charge as the action potential passes. They also did research on the means by which the cell pumps these ions back out through the membrane afterward and on how the movements of sodium and potassium ions interact with each other.

Hodgkin and Huxley were awarded shares of the Nobel Prize in physiology or medicine in 1963 for their work on nerve transmission, which became the basis for later studies of electrical changes in the membranes of other cells, such as muscle cells. Hodgkin was knighted in 1972 and was president of Britain's top science organization, the Royal Society, from 1970 to 1975. His other awards included the Royal Society's Royal Medal (1958) and Copley Medal (1965).

Around 1970, Hodgkin turned to the study of vision. He showed that a series of chemical reactions occurs when certain cells in primitive sea creatures called horseshoe crabs react to light. He also investigated more complex cells in the retina, the light-sensitive part of the eye in higher animals. Hodgkin retired in 1984 and died in 1998.

Further Reading

"Alan Lloyd Hodgkin—Biography." *Nobel Lectures, Physiology or Medicine 1963–1970.* Available online. URL: www.nobel.se/medicine/laureates/1963/hodgkin-bio.html. Last updated 2001.

Hodgkin, Alan. *Chance and Design: Reminiscences of Science in Peace and War.* Cambridge, England: Cambridge University Press, 1992.

"Wiring up Biology." *The Economist*, December 12, 1992.

Hodgkin, Dorothy Crowfoot

(1910–1994)
British
Chemist

By interpreting X-ray photographs of crystals, Dorothy Hodgkin worked out the three-dimensional structure of complex biological molecules. She won the Nobel Prize in chemistry in 1964.

Dorothy Mary Crowfoot was born on May 12, 1910, in Cairo, Egypt, where her father, John Crowfoot, worked for the Ministry of Education, part of the British government that controlled Egypt at the time. Her mother, Molly, was an expert on ancient weaving. Dorothy and her sisters lived with relatives in England during most of their childhood, but they often visited their parents in the Middle East during the summers, helping to excavate archaeological sites and meeting intellectuals and diplomats.

Crowfoot became interested in chemistry, especially the study of crystals, while still a teenager. She learned about the new science of X-ray crystallography, in which information from photographs made by shining a beam of X rays through a crystal is used to calculate the three-dimensional structure in which atoms are arranged within the molecules that make up the crystal. A course Crowfoot took while at Somerville, a women's college at Oxford University, cemented her decision to devote her career to this field of chemistry.

Crowfoot earned a bachelor's degree in chemistry from Somerville in 1932. She then began working at Cambridge, Britain's other most famous university, with J. D. Bernal, who was among the first to use X-ray crystallography to study the complex molecules in the bodies of living things. While "clearing Bernal's desk," as she called it, Crowfoot tried the technique on a variety of biological molecules.

Somerville persuaded Crowfoot to return there as a researcher and teacher in 1934, and she stayed at that institution for the rest of her career. She became a reader (the equivalent of a full professor in the United States) in X-ray crystallography in 1956 and the Wolfson Research Professor in 1960. She kept this position until 1977.

In her late 20s, Crowfoot developed rheumatoid arthritis, which severely deformed the joints of her hands and feet, but she refused to let this painful disease keep her from her work. She became famous for developing new crystallographic techniques that allowed her to tackle molecules that had defeated others. For instance, to study cholesterol, a substance now best known for its contribution to heart disease, she made artificial cholesterol crystals that contained an extra atom of a heavy element such as mercury. Differences between X-ray photos of these crystals and natural ones allowed her to determine features of the cholesterol molecule that could not be seen in any other way. Cholesterol was the most complex molecule analyzed by X-ray crystallography up to that time and the first to have its structure worked out by X-ray studies alone. Crowfoot's research on cholesterol became the thesis for which she earned her Ph.D. in 1937.

In that same year, Crowfoot met Thomas L. Hodgkin, and they married on December 16. Theirs was often a long-distance marriage, since Thomas Hodgkin, an expert on African history, first taught at a different university than Dorothy and later lived in Africa. Nonetheless, the union was happy, and the couple had three children.

During World War II, HOWARD WALTER FLOREY's team at Cambridge University was trying to find ways to mass-produce penicillin, a possible new germ-killing drug. ERNST BORIS CHAIN, a biochemist in Florey's laboratory, asked Hodgkin in 1942 to try to determine the penicillin molecule's structure, which he hoped would help him make the drug artificially. This was a tremendous challenge, since only tiny amounts of penicillin were available for analysis and its chemical formula was unknown.

Hodgkin and graduate student Barbara Rogers-Low finally solved the penicillin puzzle in 1946. They used one of the earliest IBM computers to help them do so, the first time a computer had been used to solve a biochemical problem. This work helped chemists create synthetic penicillins that were better than the natural form at attacking certain kinds of bacteria. It also made Hodgkin internationally famous.

Hodgkin's next challenge was vitamin B_{12}, a compound essential for healthy blood. Some people could not extract the vitamin from their food and therefore needed to take it as a drug, and in 1948 the pharmaceutical company Glaxo asked her to determine the vitamin's molecular structure to help them in manufacturing it. Even less was known about B_{12} than had been understood about penicillin, and its molecule was four times as large. Hodgkin's team finished its analysis of vitamin B_{12} in 1956, a feat that J. D. Bernal called "the greatest triumph of crystallographic technique that has yet occurred." It was this work that earned Hodgkin the Nobel Prize.

The analysis of insulin, a hormone that helps cells turn sugar into energy, was perhaps Hodgkin's masterwork. Hodgkin had worked on the insulin molecule from time to time since the start of her career, but its complex structure defeated even her until 1969. Her work in deciphering insulin's structure helped scientists learn how the hormone functions.

The Nobel Prize was Hodgkin's greatest award, but it was far from her only one. She won the Royal Medal of Britain's Royal Society in 1957. In 1965, she received the Order of Merit, Britain's highest royal order. Only one other woman, Florence Nightingale, had won this award. She also won the Royal Society's Copley Medal in 1976 and the Longstaff Medal of the British Association for the Advancement of Science in 1978.

Throughout her life, Hodgkin was as renowned for her teaching, humanitarian work, and personal kindness as she was for her scientific achievements. She worked toward world peace, international cooperation among scientists, and independence and advancement of developing countries. Hodgkin retired in 1977 and died of a stroke on July 30, 1994, at her home in Stour, England. All those who knew her mourned the woman whom a scientist friend, Max Perutz, had called the "gentle genius."

Further Reading

Cohen, Linda Juliana. "Dr. Dorothy Crowfoot Hodgkin: Chemist, Crystallographer, Humanitarian (1910–1994)." Almaz.com. Available online. URL: http://almaz.com/nobel/chemistry/dch.html. 1996.

Ferry, Georgina. *Dorothy Hodgkin: A Life*. Long Island, N.Y.: Cold Spring Harbor Press, 2000.

McGrayne, Sharon Bertsch. *Nobel Prize Women in Science: Their Lives, Struggles, and Momentous Discoveries*. New York: Birch Lane Press, 1993.

Hood, Leroy

(1938–)
American
Molecular Biologist

In the 1970s and 1980s, Leroy Hood and his coworkers at the California Institute of Technology (Caltech) invented four machines that made much of biotechnology, as well as the Human Genome Project, possible. "Lee," as he is known, was born in Montana in 1938 and grew up there. In high school, he excelled in everything from science to football. He earned a bachelor's degree from Caltech, followed by an M.D. from Johns Hopkins University in Baltimore in 1964. He took a position at the National Cancer Institute, part of the National Institutes of Health, in 1967, then came back to Caltech for his Ph.D. in biochemistry in 1968. The university hired him as an assistant professor in 1970. Hood is married to Valerie Logan, and they have two children.

By the time Hood joined the Caltech faculty, scientists had learned that the order, or sequence, of bases—four kinds of small molecules within the long molecules of DNA—is the "code" that transmits inherited information and tells cells how to make proteins. Proteins are also large molecules made up of combinations of smaller ones, in this case called amino acids. The order of bases in the DNA tells the cell which amino acids, in which order, make up a particular protein. Learning the sequence of bases in a gene (which normally contains the instructions for making one protein) therefore can help scientists figure out which protein the gene makes, and learning the sequence of amino acids in a protein can lead them to the gene that makes it as well as help them understand how the protein works in the cell.

Unfortunately, determining either the sequence of bases in a gene or the sequence of amino acids in a protein was a complex, tedious process that often took months. To speed up these tasks and make them less labor-intensive, Hood and his coworkers set out to develop automatic gene and protein sequencers. They also invented gene and protein synthesizers, machines that can assemble stretches of DNA or protein from known sequences of bases or amino acids. Making artificial genes and proteins helps scientists study these substances and drug companies mass produce them. The group completed these "four instruments that would change the world," as Hood puts it, in the late 1980s.

From the beginning, Hood was interested in the commercial as well as the academic world. An enthusiastic individual whose favorite activity seems to be starting new projects or helping others do so, he has advised numerous biotechnology companies, including Amgen, which in 2001 was the country's largest biotech company.

Hood also encouraged an even larger project, this one sponsored by the federal government, and he and his genome center took an active part in it. First proposed in the late 1980s, this program, eventually called the Human Genome Project, had the ambitious goal of determining the base sequence of all the genes in a human cell. Some scientists called the project a waste of money, but Hood insisted that having the "genetics parts list for humans" in computer databases could revolutionize biology and medicine by greatly increasing understanding of the way genes and proteins work. Those involved in the project, which published a "rough draft" of the human genome in June 2000, agree that it would have been impossible without the machines that Hood and his coworkers invented.

In 1992, lured partly by a grant of $12 million from Microsoft founder Bill Gates, Hood moved to the University of Washington to set up a new molecular biotechnology department. At the university, as throughout his career, Hood and his researchers studied the immune system, uncovering new facts about the way the system produces millions of different antibodies, each of which can attach itself to a different kind of microbe or other invader. (Hood had received the prestigious Albert Lasker Medical Research Award in 1987 for earlier studies of immune diversity.) Other projects included autoimmune diseases, in which the immune system mistakenly attacks the body, and prostate cancer. The group also continued to invent new research tools, such as a type of "gene chip" made with inkjet printer technology. A gene chip allows thousands of genes to be analyzed at once so that scientists can, for instance, determine how the pattern of genes active in a cancer cell differs from the pattern in a normal cell.

Hood ultimately concluded, however, that no university could provide the kind of interdisciplinary laboratory he dreamed of, in which chemists, molecular biologists, computer experts, mathematicians, physicists, and engineers would work together to study the cell. In 1999, therefore, he left the University of Washington to found his own research organization, the Institute for Systems Biology, in Seattle. Instead of focusing on particular genes or chemicals, Hood and the institute's other scientists examine the "complex interactions of numerous gene, protein, and cell elements that form informational networks and systems" and try to develop mathematical models that describe and predict the systems' behavior. They are currently studying such diverse subjects as the immune system, prostate cancer, the genome of puffer fish, and science education for children. Hood believes that the systems approach will ultimately revolutionize medicine, allowing physicians to identify the genes that make individuals likely to develop

Leroy Hood developed automatic gene and protein sequencers and synthesizers in the late 1980s. *(Leroy Hood)*

particular diseases and then prescribe drugs, lifestyle changes, or other treatments to keep these illnesses from developing.

Further Reading

Menosky, Joseph A. "The Gene Machine." *Science 81*, July–August 1981.

Zacks, Rebecca. "Under Biology's Hood." *Technology Review*, September 2001.

Hooke, Robert

(1635–1703)
British
Naturalist

Robert Hooke contributed to an immense range of sciences, from astronomy to zoology, at a time when science was just starting to take its modern form. He was born at Freshwater, on the Isle of Wight (a small island off the southern coast of Britain), on July 18, 1635. His father, John, was the curate (minister) at a church there.

Hooke showed artistic skill from childhood on, so after his father died when Robert was about 12, he was sent to London to be an assistant to a famous portrait painter. The paint fumes made him ill, however, so he soon left. Friends apparently helped him enter Westminster School, where he astonished his teachers by the speed at which he learned such subjects as ancient languages and mathematics, not to mention playing the organ and devising "thirty . . . methods of flying."

Hooke entered Christ Church, one of the colleges of Oxford University, in 1653. While earning his B.A. and M.A. degrees, he met many of the influential men who would be his colleagues and sometimes—though not always—his friends for the rest of his life. He earned money by assisting some of them, such as chemist Robert Boyle. In 1662, when this group formed a scientific organization called the Royal Society of London, they hired Hooke to set up apparatuses for experiments demonstrated at the society's weekly meetings. By 1665, he had advanced to being a full member of the society. He was also appointed professor of geometry at Gresham College in London in that year, which gave him permanent living quarters in the building where the society met.

A modern essay on Hooke's life calls him "England's Leonardo," comparing him to the versatile Italian Renaissance scientist and artist Leonardo da Vinci. Like Leonardo, Hooke was a prolific inventor, creating tools to extend scientists' senses, such as the compound microscope (a microscope with two or more lenses) and the reflecting telescope. He was also an artist, skilled at both drawing and architecture. He was so well known as an architect that in 1666 the British government put him in charge of surveying London, preparatory to rebuilding the city after a great fire had destroyed much of it.

Hooke's chief contribution to biology was *Micrographia*, a book containing his descriptions and meticulous drawings of common objects and creatures viewed under the compound microscope in the early 1660s. Published in 1665, *Micrographia* was the first scientific "coffee-table book" and a best-seller among scientists and educated nonscientists alike. Gentleman diarist Samuel Pepys called it "the most ingenious booke that ever I read in my life" and wrote that he had stayed up until 2 A.M. to finish reading it.

Hooke's microscope revealed an amazing world. Fleas and other minute insects emerged as monsters with complex, strangely beautiful bodies. (For instance, Hooke described a flea as being "adorn'd with a curiously polish'd suite of sable [black] Armour.") A piece of cork (a woody material) proved to contain rows of squarish chambers that reminded Hooke of the tiny rooms in which monks lived. He therefore called them *cells*, the name given to those rooms. The name was later extended to refer to the microscopic units of which all living things are composed, although what Hooke saw was not living cells but the fibrous walls of plant cells, all that was left after the cells had died.

Hooke's investigations in geology had implications for biology as well. Among the objects he studied with his microscope were fossils, which some scientists thought were simply unusual rocks with designs that happened to resemble parts of plants and animals. Hooke, however, said that they were the remains of actual plants and animals that had lived long ago, turned to stone by being filled with water that contained mineral deposits. He wrote that "there have been many other Species of Creatures in former Ages, of which we can find none at present; and . . . 'tis not unlikely . . . that there may be divers [various] new kinds now, which have not been from the beginning." This recognition that the kinds of plants and animals on Earth had changed during the geologic past was far ahead of his time and foreshadowed CHARLES ROBERT DARWIN's discoveries about evolution.

Hooke's last years were embittered by poor health and ongoing feuds with famed physicist Sir Isaac Newton and some other Royal Society members about which of them had thought of certain concepts and devices first. Still, a friend wrote that he "was of an active, restless, indefatigable [tireless] Genius even almost to the last" of his life. Hooke died in London on March 3, 1703, probably from complications of heart disease and diabetes.

Further Reading

Chapman, Allan. "England's Leonardo: Robert Hooke (1635–1703) and the Art of Experiment in Restoration England." *Proceedings of the Royal Institution of Great Britain*, 1996. Available online. URL: www.roberthooke.org.uk/leonardo.htm.

Hooke, Robert. *Micrographia*. 1665. Reprint, Lincolnwood, Ill.: Science Heritage Ltd., 1987.

Nichols, Richard. *Robert Hooke and the Royal Society*. Philadelphia, Pa.: Trans-Atlantic Publications, 1999.

Horner, John R.

(1946–)
American
Paleontologist

John (Jack) R. Horner flunked out of college seven times, yet he went on to make discoveries that revolutionized the way scientists think about dinosaurs. Horner was born in Shelby, Montana, in 1946 to John Horner, who owned a sand-and-gravel business, and Miriam Horner.

Throughout Horner's school years, everyone (including himself) thought he lacked intelligence because he read and took in information very slowly. He enrolled in the University of

John R. Horner of Montana State University's Museum of the Rockies has uncovered collections of baby dinosaur fossils that suggest that some dinosaurs took care of their young. *(Museum of the Rockies/Bruce Selyem)*

Montana at Missoula in 1964 but kept getting grades too poor to let him stay in school. Each time he flunked out, he simply reapplied. (After his first failure, he was drafted into the marines and spent two years in Vietnam.) Only when he was 31 years old did he learn that he suffered from dyslexia, a brain problem that makes reading hard but does not affect intelligence.

Horner might have felt lost in school, but out in Montana's desertlike badlands, looking for fossil bones, he was happy. He found his first dinosaur bone during a walk with his father when he was just eight years old. He pursued his interest by taking paleontology classes in college. After he and his brother took over the family business in 1973, he would stop during his long drives across the state in a tractor-trailer truck whenever he came to "what seemed like a fossily area" and hunt bones.

"I [w]ould look for fossils whether I got paid for it or not," Horner told Prehistoric Planet interviewer Tony Campagna in 1999. Nonetheless, in spite of his lack of academic credentials, he kept hoping that someone *would* pay him to do what he loved. He applied to museums all over the English-speaking world, seeking any kind of job with them. Finally, in 1975, the Museum of Natural History at Princeton University hired him as a preparator (someone who cleans and puts together fossil fragments sent in by others).

Fortunately, Horner's supervisor, Don Baird, recognized his skill and encouraged him to bring in fossils that he found during his summer visits to Montana. One day in 1977, Horner showed Baird what looked at first like a "smashed turtle." When Horner compared it with some other specimens in the museum, however, he realized that it was a slightly damaged dinosaur egg—one of the very few ever found. (Dinosaurs hatched from eggs, just as modern reptiles do.)

Horner made another major find the following summer—not in the ground but in a rock shop. The shop's owner showed him a handful of tiny bone fragments, and he realized with a shock that they belonged to baby hadrosaurs, or duck-billed dinosaurs, a type of large, plant-eating dinosaur that became Horner's specialty. Encouraged, Horner and a friend went back into the badlands near where the bones had been found. A few days later, they spotted a collection of baby dinosaur fossils, mixed with the remains of eggshells, in a bowl-shaped depression about six feet across. It appeared to be a dinosaur nest, the first ever discovered.

Horner went on excavating the area around this find, called the Willow Creek anticline, throughout the 1980s. His revolutionary discoveries there made headlines—and his reputation. In 1982, he became a curator of the Museum of the Rockies in Bozeman, a paleontological museum associated with Montana State University. In the field, substantial funding from the National Science Foundation transformed the former college dropout into the leader of North America's largest paleontological research team. He also won a MacArthur Foundation "genius grant" in 1986. During this decade, he found not only more nests but the remains of thousands of young and adult hadrosaurs in all stages of development.

After analyzing his fossils, Horner concluded that hadrosaur babies had remained in their nests until they were about three feet long and three months old, being weak and relatively helpless before that time, as some kinds of birds are at birth. This, in turn, led him to the startling theory that, unlike most living reptiles, which abandon their eggs after laying them, the parents of these babies must have remained with their young and brought them food. No one had thought dinosaurs capable of such parental care. Horner named this new genus of hadrosaurs *Maiasaura,* which means "good mother lizard."

Horner's ideas about the "good mothers" are only some of his startling theories. His studies of hadrosaur growth rates have convinced him that at least some dinosaurs were probably warm-blooded, because they seem to have grown faster than cold-blooded creatures usually do. Similarly, his spectacular find of eggs containing fully formed unborn dinosaurs—the first dinosaur fetuses ever discovered—suggest to him that dinosaurs developed more the way birds do than the way reptiles do. Not all paleontologists agree with Horner's conclusions, but his discoveries and theories have given the field much food for thought.

Unlike some other paleontologists, who travel the world seeking new kinds of fossils, Horner has usually confined his work to Montana's bountiful horde of dinosaur remains. Indeed, he is the official state paleontologist. He is said to have found more dinosaur fossils than any other paleontologist in history. He looks chiefly for hadrosaurs and related dinosaurs because, he says, his main interest is trying to work out dinosaur behavior and ecology, and for that purpose he needs as many fossils of a single kind of dinosaur as possible so he can compare them. He also studies other dinosaurs, however, both very large and very small: He has found eight *Tyrannosaurus rex* skeletons, including one that in 2002 was said to be the largest ever found, and he studies micro-sites, collections of minute fossils that may offer clues to such mysteries as what made dinosaurs become extinct. Horner is still a curator at the Museum of the Rockies as well as an adjunct professor of geology at Montana State University.

In addition to studying dinosaurs himself, Horner enjoys teaching the public about these popular animals. For instance, he was a scientific consultant for the movies *Jurassic Park* and *The Lost World* (though he says that the basic idea of the films, that living dinosaurs could be produced from tiny amounts of fossil DNA, is impossible). He also frequently speaks to school classes, explaining how science is done. He likes to emphasize that "science [is] a process rather than a body of knowledge."

Further Reading

Dobb, Edwin. "What Wiped Out the Dinosaurs?" *Discover*, June 2002.

Horner, John R. *Digging Dinosaurs*. New York: Workman Publishing, 1988.

———. *Dinosaurs: Under the Big Sky*. Missoula, Mont.: Mountain Press Publishing, 2001.

"Horner, John R." *Current Biography Yearbook 1992*. New York: H. W. Wilson, 1992.

Lessem, Don. *Jack Horner: Living with Dinosaurs*. New York: W. H. Freeman, 1994.

Humboldt, Alexander von

(1769–1859)
Prussian
Naturalist

Friedrich Wilhelm Heinrich Alexander von Humboldt's exploration of South America contributed to many branches of science, including biology. Among other things, he was a founder of biogeography, the study of the distribution of living things on the Earth's surface.

Humboldt was born in Berlin, then part of the German state of Prussia, in 1769. His parents were both wealthy. His father, Baron Alexander Georg von Humboldt, had been a major in the army of Prussia's emperor, Frederick the Great, and an adviser to the emperor. He died when Alexander was about nine years old.

Humboldt studied at Göttingen and other universities but did not obtain a degree. He then entered the Prussian Academy of Mines at Freiberg in 1791 and began working as inspector of mines in Bayreuth the following year. A long trip around Europe with a friend in 1790 had interested him in travel, but he had no way to fulfill this urge until his mother died in 1796, leaving him some property. Selling it gave him enough money to quit his mining job.

European countries at the time were just beginning to sponsor voyages of exploration that included scientific investigations, which the

In 1799, young Prussian naturalist Alexander von Humboldt and a French fellow scientist, Aimé Bonpland, began what was to become a famous voyage of exploration through South America. *(National Library of Medicine)*

countries' governments hoped to put to economic use. Humboldt tried to attach himself to some of these voyages, but political changes canceled them before they started. Frustrated, he finally decided to finance his own expedition. He wanted to go to South America, then chiefly controlled by Spain, so he and French botanist Aimé Bonpland, whom he had met while preparing for one of the aborted expeditions, visited the Spanish court in 1799. Foreigners had almost never been allowed inside South America, but the two young men obtained the king's permission for their journey by promising to give him reports about such potentially useful subjects as deposits of precious metals.

Humboldt's first view of South America thrilled him. "What magnificent vegetation! How brilliant the plumage [feathers] of the birds, the colors of the fish!" he wrote to his brother when he arrived in Venezuela. "We have been running about like a couple of fools. . . . Bonpland declares he'll go crazy if these wonders don't let up."

Humboldt wrote to a friend that he hoped to "investigate the interaction of all the forces of nature" during his trip. He had plenty of opportunity. His journey with Bonpland, which lasted from 1799 to 1804, took them 6,000 miles (10,000 kilometers) around South and Central America by foot, canoe, and horseback, covering the present-day countries of Venezuela, Colombia, Ecuador, Peru, Mexico, and Cuba. They explored rivers, including the Amazon and Orinoco, which they proved were connected, and climbed mountains such as the 20,702-foot (6,269-meter)-high Chimborazo in Ecuador, then thought to be the tallest mountain in the world (though they did not quite reach the top). They collected some 60,000 plant specimens, including 3,500 new species, as well as samples of rocks, animals, and native artifacts.

Humboldt's observations during this long trip had implications for biology as well as geology, climatology, and many other sciences. For instance, his temperature measurements of the Pacific Ocean revealed a current of cold water flowing past the Peruvian coast, which has major effects on sea life there. Humboldt also pointed out that altitude as well as latitude (distance from the equator) affect the types of plants that grow in an area; for example, plants high on a mountain in Ecuador may resemble those at sea level in northern Europe.

When the two explorers returned to Europe, they found themselves famous. The Prussian government made Humboldt a member of its delegation to the court of Napoleon, the emperor of France, so he took up residence in Paris. Humboldt liked the worldly French capital much better than the then-backward Berlin and managed to stay there for almost 25 years. During that time, he became friends with influential people such as Simón Bolívar, who eventually freed Colombia, Peru, Venezuela, and Bolivia from Spanish rule, a change Humboldt strongly favored.

Humboldt spent much of his time in Paris preparing an extensive account of his South American voyage, which was published in 30 volumes between 1805 and 1834. Printing these large, expensive books exhausted his fortune, forcing him to rely on his government salary, so when the Prussian ruler insisted that he return to Berlin in 1827, he had to obey. Like his father, he became an imperial adviser, and he also gave popular lectures at the University of Berlin. What he really wanted to do, however, was return to traveling.

Humboldt finally had another chance to go exploring in 1829, when the czar (emperor) of Russia asked him to study mining and minerals in the Ural Mountains. During a six-month trip, he not only traveled through the Urals but journeyed through Siberia to the Chinese border, then returned to Moscow by way of the Caspian Sea. He did not write about this journey, but two scientists who accompanied him wrote a geography of Central Asia based on it.

On his return to Germany, Humboldt began collecting his earlier lectures into an encyclopedia called *Cosmos*, four volumes of which were published during his lifetime and a fifth after his death (between 1845 and 1862). This monumental work attempted to draw together all areas of science and stressed the unity of nature. Humboldt also worked for political unity, or at least cooperation, in scientific matters, persuading the governments of Russia and Britain to set up stations throughout their possessions to make and share observations of Earth's weather and magnetic field—the first large-scale international cooperative effort in science. Humboldt, whom *National Geographic* writer Loren McIntyre calls "perhaps the most widely admired man of the nineteenth century," died on September 14, 1859, just short of his 90th birthday.

Further Reading

de Terra, Helmut. *Humboldt: The Life and Times of Alexander von Humboldt, 1769–1859*. New York: Alfred A. Knopf, 1955.

Humboldt, Alexander von. *Personal Narrative*. Translated by Jason Wilson. Reprint, New York: Penguin Books, 1996.

Lee, Jeff. "Alexander von Humboldt." *Focus*, Summer 2001.

McIntyre, Loren. "Humboldt's Way." *National Geographic*, September 1985.

J

Jenner, Edward

(1749–1823)
British
Physician

Edward Jenner discovered a way to protect people against smallpox, a highly contagious disease that had killed, blinded, or disfigured large numbers of people, especially children, throughout human history. In doing so, he helped to lay the foundation for immunology, the study of the body's defense system (which his protection process, called vaccination, activated), and virology (he coined the term *virus* for the still-unknown agent that caused smallpox).

Jenner was born on May 17, 1749, in the small town of Berkeley, part of the area of western England called Gloucestershire. His father, Stephen Jenner, was a minister. His parents died when he was about five years old, and an older brother raised him (Edward was the youngest of six children). At age 13, at his own request, he was sent to a nearby town to assist and be trained by a surgeon. He went to London in 1770 to continue his training under John Hunter, a famous Scottish surgeon. He then returned to Berkeley in 1773 and established a medical practice there. He married Catherine Kingscote in 1788, and they had four children. He watched birds in his spare time and wrote a paper about the cuckoo that earned him membership in the Royal Society, Britain's top science organization, in 1789.

At the time, the only way of avoiding smallpox was variolation, a procedure used for centuries in Africa, the Middle East, and Asia. Lady Mary Wortley Montagu, the wife of the British ambassador to Turkey, had introduced it to England around 1722. Based on the observation that smallpox survivors almost never caught the disease a second time, variolation involved scratching the skin and then rubbing in matter from the sores of a person who had a mild case of smallpox. The person receiving this treatment usually suffered a mild attack of the disease and was thereafter protected against catching a more serious infection during an epidemic. Sometimes, however, the induced disease proved serious or even set off a new epidemic.

Jenner often performed variolation as part of his medical duties, and he noticed that people who had had a disease called cowpox seldom developed smallpox after the treatment. Cowpox, similar to smallpox but much milder, was common among cows in that dairy farming region, and people who milked or otherwise handled cows often caught it. It produced sores on their hands but few other signs of illness. Jenner remembered hearing a folk belief that people who had had cowpox could not catch smallpox, and he began to wonder if cowpox could protect against smallpox without the risk of serious illness that variolation carried.

On May 14, 1796, after some 25 years of observation and investigation, Jenner decided to test his cowpox theory with an experiment. He gave an eight-year-old boy, James Phipps, a treatment like variolation except that he used material from a dairywoman's cowpox sore instead of from a smallpox sore. The boy experienced only a very mild illness. Then, on July 1, Jenner gave the boy a standard variolation—in other words, exposed him to smallpox. Phipps remained healthy. Jenner tested the procedure on about 15 other people, including his own baby son, during the next two years and obtained similar results.

Jenner sent an article about his procedure, which he called vaccination after the Latin word for *cow*, to the Royal Society in 1797. The president of the society refused to publish it, saying that Jenner "ought not to risk his reputation by presenting to the learned body anything which appeared so much at variance with established knowledge." Jenner then published the article at his own expense as a pamphlet in September 1798. Several prominent London physicians, apparently braver than the Royal Society, tried the new treatment, and after they, too, reported success, the practice spread rapidly. Within a year, more than 1,000 people had been vaccinated.

Jenner devoted most of the rest of his life (which ended on January 26, 1823, when he died of a stroke) to promoting vaccination and trying to make it reliable. He refused to take any money for the process itself, although the British legislature voted him substantial sums in the early 1800s as a reward for his discovery. Vaccination had its critics and its failures, partly because other physicians sometimes failed to follow Jenner's instructions for preparing and preserving the vaccine. The procedure was successful in most cases, however, and saved the lives and health of millions. As Jenner wrote in 1801, "The numbers who have partaken of [vaccination's] benefits throughout Europe and other parts of the Globe [vaccination quickly spread to North America and Asia as well as Europe] are incalculable." The final triumph of his invention came in 1977, when an intensive vaccination campaign by the World Health Organization wiped out natural smallpox worldwide.

Further Reading

Bardell, David. "Nestling Cuckoos to Vaccination—A Commemoration of Edward Jenner." *BioScience*, December 1996.

Bazins, Herve. *The Eradication of Smallpox: Edward Jenner and the First and Only Eradication of a Human Infectious Disease*. Translated by Andrew Morgan and Glenise Morgan. San Diego, Calif.: Academic Press, 2000.

Fisher, R. B. *Edward Jenner*. London: Andre Deutsch Ltd., 1991.

Jenner, Edward. *Vaccination Against Smallpox*. New York: Prometheus Books reprint, 1996.

Johanson, Donald C.

(1943–)
American/Ethiopian
Paleontologist, Anthropologist

In 1974, Donald Carl Johanson and his coworkers made history—or perhaps one should say prehistory—by discovering what was then the oldest fairly complete skeleton of a human ancestor known to have walked upright. He has continued to challenge concepts in paleoanthropology, the study of human origins and evolution.

Johanson was born on June 28, 1943, in Chicago, the son of Carl Johanson, a barber, and the former Sally Johnson. Both his parents had immigrated from Sweden. He became interested in paleoanthropology in high school, partly as a result of reading about the discoveries of prehuman fossils that LOUIS S. B. LEAKEY and his family had made in Africa. "It became an obsession," he said later.

At the University of Illinois, Johanson first studied chemistry because of the field's good job prospects, but he found it boring and changed his major to anthropology. He received a B.A. in 1966, then did graduate work in paleoanthropology at the University of Chicago, obtaining an M.A. in 1970 and a Ph.D. in 1974. In 1972, he moved to Cleveland, Ohio, to become an associate curator of anthropology at the Cleveland Museum of Natural History and an assistant professor of anthropology at Case Western Reserve University.

Most of Johanson's searches for prehuman remains have been in the East African country of Ethiopia, especially in the Hadar Valley, an area in the northeastern part of the country where the Afar people live. On November 30, 1974, during his team's third dig, they noticed part of a humanlike arm bone sticking out of the ground. They explored further and found other fragments of what appeared to be the same fossilized skeleton. Ultimately, the skeleton proved to be 40 percent complete. The shape of the pelvis, or hipbone, showed that the bones had belonged to a female. That night, as the group celebrated around their campfire to the tune of a taped Beatles song, "Lucy in the Sky with Diamonds," they decided to call the skeleton Lucy.

The discovery of Lucy made Johanson world famous. In 1975, he was promoted to curator of physical anthropology and director of scientific research at the Cleveland Museum and adjunct professor at Case Western. Returning to Hadar that year, Johanson's team, now much larger, made a second spectacular find, the skeletons of at least thirteen individuals of the same type as Lucy, buried close together. Johanson believed that this group, which was nicknamed the First Family, had been killed suddenly and together, probably by a flash flood, then buried quickly under river silt, which kept the bodies intact. The existence of such a large group, Johanson said, suggested that these ancestors of humans had already learned to cooperate.

Lucy and the First Family were hominids, creatures with both apelike and humanlike features. Johanson announced in 1978 that he considered them to belong to a new species, which he named *Australopithecus afarensis* (the southern ape from Afar). He claimed that this species was a direct ancestor of humans. He and others eventually determined that Lucy had lived about 3.18 million years ago and the First Family slightly earlier.

The most striking fact that the skeletons of Lucy and the First Family revealed was that they had walked upright, on two legs. The Leakeys and most other paleoanthropologists had believed that human ancestors developed upright walking, large brains, and the ability to make and use tools at about the same time. Johanson suggested, however, that bipedal (two-legged) walking had come almost two million years before the other two changes and perhaps had even caused them. "Lucy proved that . . . bipedalism was the thing that separated us from the apes," he said in the early 1980s. He proposed that humans had developed through "mosaic evolution," with some parts of the body becoming humanlike while others were still apelike.

RICHARD LEAKEY and MARY LEAKEY disagreed with Johanson's interpretation of his skeletons, and a widely publicized dispute ensued. The Leakeys, for instance, claimed that the skeletons represented two species, not one. They pointed out that Lucy was delicate in build, only about 3 feet 6 inches tall and weighing 60 pounds or so, but some of the other skeletons seemed to belong to creatures much taller and sturdier. Johanson thought that this was because *A. afarensis* males differed considerably in size from the females, as is true in gorillas and some other living apes. The Leakeys also doubted that Johanson's supposed new species was an ancestor of the human genus, *Homo*. These questions are still being debated.

Political unrest kept Johanson out of Ethiopia during most of the late 1970s, so he concentrated on developing the physical anthropology laboratory at the Cleveland Museum, analyzing his earlier finds, and writing a book about the discovery of Lucy. *Lucy: The Beginnings of Humankind*, which Johanson coauthored with Maitland A. Edey, was published in 1981 and won that year's American Book Award in science. The year 1981, in fact, was a pivotal one for Johanson. In May, he married Susan Rannigan (a second marriage, since he had been married briefly as a graduate student), with whom he later had two children. He also moved to Berkeley, California, where he set up his own paleoanthropological research organization, the Institute of Human Origins.

Johanson continued to make important discoveries during the 1980s. In 1986, for instance, he found a partial skeleton of *Homo habilis* ("Handy Man"), the first toolmaker, at Olduvai Gorge in Tanzania, the site of the Leakeys' spectacular discoveries. Only skulls of this species, which lived about two million years ago, had been known before. Johanson claimed that this skeleton, also a female, bore strong resemblances to that of Lucy, being both smaller and more apelike than the Leakeys had expected *Homo habilis* to be. This suggested to Johanson that the modern human body pattern did not appear until the species *Homo erectus* developed about 1.6 million years ago. Awards Johanson won during this decade included the American Humanist Association's Distinguished Service Award (1983).

In 1992, back in Ethiopia, Johanson's group discovered fragments that they pieced together into the first more or less complete skull of *Australopithecus afarensis*. The skull, which belonged to a male, reinforced Johanson's belief that males and females of this species were of greatly different sizes.

Johanson has a "dramatic personality," as one fellow scientist put it, and he has often been a controversial figure. Some colleagues have criticized him for spending so much time on public-oriented activities such as writing popular books, giving speeches, and hosting television shows. He has also been at the center of several scientific conflicts. In addition to his well-known disagreements with the Leakeys, he had quarrels with some other members of the Institute of Human Origins in the early 1990s that essentially split the organization in 1994.

In 1997, the part of the institute that kept the name, with Johanson still at its head, moved to Tempe, Arizona, and became affiliated with Arizona State University. Johanson continues to direct the institute and is also a professor of anthropology at the university. In addition to doing research, he writes and gives talks on such subjects as humans' responsibility to take care of the rest of the Earth. "If we are the guardians of the past," he said in 1995, "we must also be guardians of the future."

Further Reading

Johanson, Donald C. *From Lucy to Language*. New York: Simon & Schuster, 1996.

———, and Maitland Edey. *Lucy: The Beginnings of Humankind*. 1981. Reprint, New York: Touchstone Books, 1990.

"Johanson, Donald C(arl)." *Current Biography Yearbook 1984*. New York: H. W. Wilson, 1984.

Katz, Bernhard

(1911–)
German/British
Neurobiologist

Bernhard Katz refined biologists' understanding of the way nerves transmit messages by showing how acetylcholine, one of the chemicals by which these messages are sent, is released from the endings of nerve fibers. He was born in Leipzig, Germany, on March 26, 1911. His parents were Max Katz and the former Eugenie Rabinowitz, both of Russian Jewish descent.

Katz studied medicine at the University of Leipzig and earned his degree in 1934. He went to University College, part of London University, for postgraduate studies, completing his Ph.D. in 1938 and taking an additional doctor of science degree in 1942. He did research at the Sydney Hospital in Australia from 1939 to 1942, after which he served as a radar officer in the Australian air force until the end of the war (he became a naturalized British citizen in 1941). He married an Australian, Marguerite Penly, in 1945; they later had two sons.

Katz returned to University College with his new bride in 1946 and stayed there for the rest of his career. He was assistant director of research at the biophysics research unit and Henry Head Research Fellow until 1950, reader in physiology in 1950 and 1951, and professor and head of the biophysics department from 1952 until he retired in 1978.

Beginning in the late 1940s, Katz did research with ALAN LLOYD HODGKIN of Cambridge University that continued Hodgkin's studies of the way the electric current in nerve cells (which the cells use to transmit messages) affects the movement of ions, or atoms with an electric charge, in and out of the cell membranes. Katz and Hodgkin showed that a change in the nerve cell membrane's openness to sodium ions, which allows these positively charged ions to flow into the cell, is the key difference that gives spots on the inside of the cell a temporary positive charge as the electrical signal flowing along the nerve fiber passes them. Katz and Hodgkin also studied the means by which the cell pumps these ions back out through the membrane after the signal goes by.

Katz started the research for which he was most famous in the early 1950s. Researchers had shown that nerves transmit messages from one cell to another across tiny gaps between the cells, called synapses, by releasing chemicals termed neurotransmitters. An electric current passing down a nerve cell fiber causes the release of neurotransmitter molecules at the fiber's end. The molecules float across the synapse and are taken up by the ending of the cell on the other side of the gap. Once inside the cell, they produce a current in its fiber, and the message proceeds. There are several kinds of neurotransmitters, and different types of nerves release different ones. Katz studied acetylcholine, the first neurotransmitter to be discovered (by HENRY HALLETT DALE in 1914).

Scientists had thought that nerve endings release neurotransmitter molecules only when electrical (message) signals stimulate them to do so. Katz found, however, that the endings of the nerves he studied released tiny amounts of acetylcholine all the time, whether the cells were transmitting a signal or not. When a signal did arrive, it greatly increased the amount of acetylcholine released. Other researchers extended these findings to other neurotransmitters.

Katz also found that the amount of electric charge at the nerve ending, which is related to the amount of neurotransmitter being released, is always a multiple of a certain minimum number. He therefore suggested that neurotransmitter is released in "packets," or "quanta," each consisting of a more or less fixed number of molecules. The packets are released slowly when the nerve is at rest and much more quickly when an electrical signal arrives at the nerve ending. Support for this idea came when other scientists, using electron microscopes, saw nerve endings budding off tiny hollow spheres that could contain the packets of molecules. Katz's discoveries about nerve transmission earned him a share of the 1970 Nobel Prize in physiology or medicine, as well as the Copley Medal of the Royal Society in 1967 and a knighthood in 1969.

Further Reading

"Sir Bernard Katz—Biography." *Nobel Lectures, Physiology or Medicine 1963–1970*. Available online. URL: www.nobel.se/medicine/laureates/1970/katz-bio.html. Last updated 2001.

Khorana, Har Gobind

(1922–)
Indian/American
Biochemist, Molecular Biologist

Har Gobind Khorana won a share of the 1968 Nobel Prize in physiology or medicine for helping to decipher the "genetic code" that tells cells how to make proteins. He also created the first artificial gene.

Khorana was born around January 9, 1922, in the village of Raipur, in the Punjab, India,

Indian-born Har Gobind Khorana helped to decipher the genetic code in the early 1960s and created the first artificial gene in the 1970s. *(Donna Coveney/MIT)*

now part of West Pakistan. His father, Ganpat Rai, was the village tax collector. Although poor, Rai and his wife, Krishna Devi, believed strongly in education. They and their five children, of which Har Gobind was the youngest, were almost the only family in the village who could read and write.

Khorana attended Punjab University in Lahore, majoring in chemistry and graduating in 1943. After earning a master's degree in 1945, he won a government scholarship to continue his studies in Britain. He earned a Ph.D. in biochemistry from the University of Liverpool in 1948.

After postgraduate study in Zurich, Switzerland, and at Britain's Cambridge University, Khorana moved to the University of British Columbia in Vancouver, Canada, in 1952 to direct the university's organic chemistry section. He married Esther Elizabeth Sibler, a Swiss woman, in that same year. They later had three children. In 1960, Khorana went to the United States and became codirector of the Institute for Enzyme Research at the University of Wisconsin at Madison. He was named a professor of biochemistry in 1962 and the Conrad A. Elvehjem Professor of Life Sciences in 1964. He became a naturalized U.S. citizen in 1966.

Khorana's early research involved the synthesis of enzymes and other biological chemicals. Making these substances "from scratch" was often cheaper than extracting them from animal or human tissue. At the University of Wisconsin, however, he turned to work on nucleic acids (DNA and RNA) and the genetic code. Scientists had discovered that inherited information, consisting chiefly of instructions for making proteins, is carried in the order, or sequence, in which four kinds of small molecules called bases are arranged inside the large molecules of nucleic acid. Proteins, like nucleic acids, are made up of smaller molecules, in this case called amino acids. There are 20 different kinds of amino acids, and FRANCIS CRICK, the codiscoverer of DNA, had proposed in the 1950s that each amino acid is represented in the nucleic acid code by one or more sequences of three bases. The four kinds of bases, he pointed out, can be combined in 64 ($4 \times 4 \times 4$) different "triplet" sequences, more than enough to specify all 20 amino acids.

By the mid-1960s, biochemists and molecular biologists were poised to "crack the code" by finding out which triplet sequences stood for which amino acids, and Khorana became a leader in this work. He synthesized all 64 sequences, inserted them in cells, and found out which amino acids they attached to during the protein-making process. He confirmed that some amino acids are represented by more than one triplet sequence and also showed that some sequences, rather than standing for amino acids, tell the cell when to start or stop assembling a protein molecule.

Thanks to Khorana and other key researchers, such as MARSHALL NIRENBERG of the National Institutes of Health in Bethesda, Maryland, and Robert W. Holley of New York's Cornell University, all 64 base triplets were "translated" by 1966. These three men shared the 1968 Nobel Prize for this work. Khorana and Nirenberg also won the Albert Lasker Medical Research Award and Columbia University's Louisa Gross Horwitz Prize in that year.

The work that won the Nobel Prize by no means ended Khorana's achievements. In 1970, he announced that he and his coworkers had made the first artificial gene. It was a simple gene from yeast, with only 144 bases—far smaller than most human genes, which often contain millions of bases. In that same year, Khorana moved to the Massachusetts Institute of Technology (MIT), where he became the Alfred P. Sloan Professor of Biology and Chemistry.

At MIT, Khorana went on to manufacture a gene normally found in the bacterium *Escherichia coli*. He first synthesized this gene in 1973, but he was unable to prove that his gene could actually make its protein until 1976, when he determined the sequences of the natural "start" and "stop" signals for that gene and added them to his artificial one. He then used early genetic engineering techniques to insert his synthetic gene into the genome of *E. coli* cells that lacked the natural form of the gene and showed that the cells became able to make the protein for which the gene carried the code. Although synthesizing most genes has proved impractical because they are so complex, Khorana's work on artificial genes laid part of the foundation for today's biotechnology industry.

In recent years, Khorana, in 2002 an emeritus professor but still working at MIT, has changed his focus to biochemical studies of vision, especially of rhodopsin, the pigment that allows the eye to sense dim light. Rhodopsin is found in rod cells, one of two types of light-sensitive cells in the retina. Khorana's laboratory uses both biochemical techniques and genetic ones, such as creating artificial mutations in the gene that carries the instructions for making rhodopsin, to learn about this chemical's structure and the way it changes in response to light.

Further Reading

"Har Gobind Khorana—Biography." *Nobel Lectures, Physiology or Medicine 1963–1970*. Available online. URL: www.nobel.se/medicine/laureates/1968/khorana-bio.html. Last updated 2001.

"Khorana, Har Gobind." *Current Biography Yearbook 1970*. New York: H. W. Wilson, 1970.

Yount, Lisa. *Asian-American Scientists*. New York: Facts On File, 1998.

Kimura, Motoo
(1924–1994)
Japanese
Geneticist, Evolutionary Biologist

Japanese population geneticist Motoo Kimura challenged the conventional view of evolution by proposing that, at least at the molecular level, blind chance governs changes in genes rather than natural selection, as CHARLES ROBERT DARWIN had proposed. Kimura was born on November 13, 1924, in Okazaki, Japan. He earned a master's degree in botany from Kyoto University in 1947 and worked there as an assistant for two years afterward. He also studied in the United States at Iowa State College and the University of Wisconsin, obtaining his doctorate from the latter in 1956. He spent most of his career at the National Institute of Genetics in Mishima, Japan. He was a research member there from 1949 to 1957, laboratory head of the department of population genetics from 1957 to 1964, and overall head of the department after that. He died of a stroke in Mishima in November 1994.

Kimura's special interest was the mathematics of genetics and evolution. He became a leader in theoretical population genetics, developing mathematical models that are still standards in the field. His most famous contribution was the neutral theory of molecular evolution, which he first propounded in 1968.

According to the most widely accepted modern understanding of Darwin's theory of evolution, changes in living things begin as mutations in individual genes. These mutations occur randomly (by chance). Interaction between living things and their environment determines which mutations will be passed on to increasing numbers of a species and which ones will die out. Only mutations that give a living thing an advantage in a particular environment, increasing the chances that it will survive long enough to pass on its genes to offspring, will survive in the long run.

Techniques developed in the second half of the 20th century made possible the study of evolution not only at the level of whole organisms but at the molecular level, where genes do their basic job of telling cells how to make proteins. Mutations in genes produce changes in proteins, and scientists can compare the composition of genes and proteins in related types of living things and note how often changes in both occur. Kimura drew on this information when formulating his theory.

Kimura said that natural selection applies to features that can be seen in whole living things, but he claimed that the rules of evolution are different at the molecular level, just as the laws of physics that apply to the everyday world are different from those that govern behavior inside the atom (at the quantum level). Changes in the composition of particular proteins, he maintained, occur at about the same rate in numerous, widely separated groups of living things, amounting to a sort of "molecular clock." This steady rate of change suggested to him that, on average, individual changes neither help nor harm organisms' survival—they are neutral as far as natural selection is concerned—and their likelihood of being passed on, like their original occurrence, is governed solely by chance. Michael Nachman wrote in the March 1996 issue of *BioScience* that Kimura's neutral theory "revolutionized the way we think about molecular evolution."

Far more population geneticists accept Kimura's theory now than did so when it was first proposed, although aspects of it are still controversial. Kimura won several awards for his work, including the Order of Culture, Japan's highest cultural award, and the Darwin Medal of Britain's Royal Society (1992).

Further Reading

Kimura, Motoo. *Population Genetics, Molecular Evolution, and the Neutral Theory: Selected Papers.* Edited by Naoyuki Takahata. Chicago: University of Chicago Press, 1994.

King, Mary-Claire
(1946–)
American
Geneticist

Some geneticists merely analyze DNA in test tubes, but Mary-Claire King sees genetics as closely tied to the needs of living people. *New York Times* reporter Natalie Angier wrote in 1993, "Nearly everything [King] has ever chosen to work on has had, at its core, a deep sense of humanity."

Mary-Claire King was born in Wilmette, Illinois, a suburb of Chicago, on February 27, 1946. Her father, Harvey, was head of personnel for Standard Oil of Indiana. Clarice, her mother, was a housewife. A childhood love of solving puzzles drew King to mathematics, which she studied at Carleton College in Minnesota. A friend's death from cancer during King's teenage years also interested her in medicine.

After graduating from Carleton in 1966, King went to the University of California at Berkeley to learn biostatistics, or statistics related to living things. While there she took a class in genetics that appealed to her by presenting aspects of heredity as puzzles and also by showing how information about genes could be used to help people. King changed her study plans accordingly.

In protest against government actions in Vietnam and in Berkeley during student demonstrations, King dropped out of graduate school to work for consumer advocate Ralph Nader. After about a year, however, Allan Wilson, one of her favorite Berkeley professors, persuaded her to return and join his molecular biology laboratory. As part of a project to trace human evolution, Wilson asked King to compare the genes of humans and chimpanzees. To her own and everyone else's amazement, she found that the two species share more than 99 percent of their genes. King's research not only earned a Ph.D. in genetics from Berkeley in 1973 but was featured on the cover of *Science* magazine in 1975.

In 1974, King turned to a quite different aspect of genetics: the possibility that some women inherit a susceptibility to breast cancer. Scientists at the time were discovering that all cancer results from damaging changes in genes, but usually those changes occur during an individual's lifetime. Only a few rare cancers were known to be inherited, and many researchers doubted that inheritance could play a role in common cancers such as breast cancer. King eventually proved, however, that about 5 percent of breast cancers are inherited. Women with the inherited form of the disease usually develop it at an earlier age than other breast cancer patients and have mothers, sisters, or other close female relatives who also suffer the illness.

When King began her hunt for the breast cancer gene, finding the location of a particular gene on one of humans' 23 pairs of chromosomes, she says, was like looking for an address at night in an unfamiliar city that has only one streetlight for every 10 blocks. Technology for finding genes improved during the 1980s, however, and in 1990, Jeff Hall, a researcher in King's laboratory, finally determined that the gene was about halfway down the lower arm of chromosome 17. By then, King had become a professor of epidemiology at UC Berkeley's School of Public Health (in 1984) and a professor of genetics in the university's Department of Molecular and Cell Biology (in 1989).

Mary-Claire King helped to isolate a gene that predisposes some women to develop breast cancer and has also used genetic techniques to identify children in Argentina so they can be reunited with their families. *(University of Washington/Mary Levin)*

King's laboratory lost the race to pinpoint BRCA1, the breast cancer gene; Mark Skolnick and other scientists at the University of Utah Medical Center identified it in September 1994. However, King continues to study this and other genes involved in breast cancer (several have now been found) to learn how both the cancerous and the normal forms of the genes function. Her work on breast cancer has earned awards such as the Susan G. Komen Foundation Award (1992), the Clowes Award of the American Association for Cancer Research (1994), and a lifetime grant from the American Cancer Society (1994).

King's most unusual genetic project, tied to her lifelong concern for human rights, was helping to reunite families torn apart during the "dirty war" waged in Argentina between 1976 and 1983. Beginning in 1984, she helped a group called the Abuelas de Plaza de Mayo (Grandmothers of the Plaza of May) identify their grandchildren, who had been orphaned when the military government that formerly controlled the country killed their parents. The children, born in prison or captured with their mothers, had been sold or given away, and many of the families who acquired them refused to give them up unless the grandmothers could prove their kinship.

To demonstrate the relationship, King adapted a technique that her old mentor, Allan Wilson, had recently developed. Most human genes are carried on DNA in the nucleus of each cell, but small bodies called mitochondria, which help cells use energy, also contain DNA. Unlike the genes in the nucleus, which come from both parents, mitochondrial DNA is passed on only through the mother. It therefore is especially useful in showing the relationship between a child and female relatives. King's work has reunited more than 50 Argentinian children with their birth families, and her technique has also been used to identify the remains of people killed in wars or murdered by criminals.

In 1995, King moved to Seattle to head a laboratory at the University of Washington. She is currently American Cancer Society Professor of Medicine and Genetics and an adjunct professor of molecular biotechnology at the university. She writes that "the focus of [her] lab is the identification and characterization of genes responsible for complex, common human conditions." The laboratory's chief area of study continues to be BRCA1 and other genes involved in breast and ovarian cancers, including noninherited forms of the disease. King's researchers are also investigating the genetics of inherited deafness and have isolated several genes associated with deafness in different families. Study of these genes may add to knowledge about normal hearing as well as possibly helping the affected families.

A third project in King's laboratory involves using changes in mitochondrial DNA (inherited only through females) and DNA on the Y chromosome (inherited only through males) to track migrations of human populations and answer other questions about human evolution and history. Her group's studies have shown, for instance, that migration patterns of females are different from those of males. They also suggest that differences among the so-called races of humanity come from variation in only a handful of genes, chiefly those that determine skin color. "The myth of major genetic differences across 'races' is . . . worth dismissing with genetic evidence," King wrote in a 1999 article in *Science* magazine.

Mary-Claire King believes that women bring a special gift to science. "We're more inclined to pull together threads from different areas, to be more integrative in our thinking," she says. She hopes to apply her share of this gift to "to improve the lives of people."

Further Reading

"King, Mary-Claire." *Current Biography Yearbook 1995*. New York: H. W. Wilson, 1995.

McHale, Laurie. "Putting the Puzzle Together." 1996. University of Washington. Available online. URL: www3.cac.washington.edu/alumni/columns/sept96/king1.html. Accessed 2003.

Waldholz, Michael. *Curing Cancer: Solving One of the Greatest Medical Mysteries of Our Time*. New York: Simon & Schuster, 1997.

Yount, Lisa. *Disease Detectives*. San Diego, Calif.: Lucent Books, 2001.

Kitasato, Shibasaburo
(1852–1931)
Japanese
Bacteriologist

Shibasaburo Kitasato was the codeveloper of antitoxin, an immune system-based treatment for certain bacterial diseases, and codiscoverer of the bacteria that cause plague. He was born in Oguni, a village in the mountains of Kyushu, Japan, in 1852. His father was the village mayor. Kitasato studied at the Kumamoto Medical College and then the Imperial (later Tokyo) University, graduating from the latter in 1883. He worked for the government's Public Health Department for two years, then was sent to study with famed bacteriologist ROBERT KOCH in Berlin, Germany.

Kitasato made several important discoveries during his six years at Koch's institute. In 1889, he became the first person to grow a pure culture of the bacteria that cause tetanus, a disease that produces fatal muscle spasms. A year later, he and German researcher EMIL VON BEHRING discovered that if they injected nonfatal doses of the tetanus bacteria's toxin into an animal, the animal's immune system formed substances that made the toxin harmless and protected the animal against the disease. The substances were in the serum, or liquid part of the blood. If this serum was injected into another animal, it protected that animal as well. Kitasato, von Behring, and PAUL EHRLICH developed this discovery of what was called antitoxic immunity or serum immunity into treatments for the deadly bacterial diseases tetanus, anthrax, and diphtheria.

By the time Kitasato returned to Japan in 1891, he was famous. The following year he started his own research institute, the Institute for Infectious Disease. Here, he identified the bacterium that causes dysentery, a serious intestinal disease, and studied tuberculosis.

In 1904, an epidemic of bubonic plague that had been raging on the Chinese mainland reached Hong Kong, an area on the southern Chinese coast then controlled by Britain. Research teams from Japan and France came to Hong Kong to attempt to identify the microbe that caused the disease. Kitasato headed the Japanese team, while the leader of the French group was Swiss-born bacteriologist Alexandre Yersin. Lacking a language in common and considering themselves to be rivals, the teams worked separately. Both found a bacillus, or rod-shaped bacterium, in the blood of people with the disease. Kitasato found the bacillus first and published his results first, but Yersin carried out tests that more conclusively proved that this kind of bacteria caused the disease, such as injecting a pure culture of the bacteria into mice and showing that they developed plague. Historians of science still argue about whether Kitasato, Yersin, or both should receive credit for identifying the plague bacillus, now called *Yersinia pestis*.

When the Japanese government, which had provided part of the support for Kitasato's Institute for Infectious Disease after 1899, incorporated it into Tokyo University in 1915, Kitasato, who had opposed the change, resigned as the institute's director and started another organization, the Kitasato Institute. He remained head of this institute until his death. To honor his work, Japan made him a baron in 1923. Kitasato died on June 13, 1931, in Nakanocho.

Further Reading

Solomon, Tom. "Hong Kong, 1894: The Role of James A. Lowson in the Controversial Discovery of the Plague Bacillus." *Lancet*, July 5, 1997.

Koch, Robert

(1843–1910)
Hanoverian/German
Bacteriologist

Along with LOUIS PASTEUR, Robert Koch founded the science of bacteriology. He made major contributions to identifying, studying, and preventing diseases caused by bacteria.

Koch was born on December 11, 1843, in Klausthal, part of the German state of Hanover. His father was a mining official. Koch, who had amazed his parents by teaching himself to read at age five, earned a medical degree from the University of Göttingen in 1866. One of his teachers

In the late 1870s and 1880s, German bacteriologist Robert Koch created methods for proving that a specific microorganism causes a certain disease and identified the bacteria that cause tuberculosis, cholera, and other deadly illnesses. *(National Library of Medicine)*

there was anatomist Jakob Henle, an early believer in the idea that microorganisms could cause disease—the germ theory of disease, which Pasteur helped to prove. Koch came to share Henle's convictions.

After working briefly in several hospitals and in the army during the Franco-Prussian War (1870), Koch became a district medical officer in Wollstein in 1872. By then, he was married to the former Emmy Fraats and had a daughter. (He married a second time, to Hedwig Freiberg, in 1893.) His memories of Henle's teaching revived when his wife gave him a microscope for his 28th birthday.

Koch's first studies were on anthrax, an illness that killed thousands of cattle and sheep yearly and sometimes also affected humans. Several scientists had reported seeing microscopic rod-shaped bodies in the blood of animals that had died of the disease, but others questioned whether the rods were even alive, let alone the cause of anthrax. Indeed, many physicians and medical researchers refused to believe that microorganisms could cause any disease.

Turning part of his consulting room into a makeshift laboratory, Koch used his new microscope to examine blood and tissues. He, too, saw the rods in the blood of animals that had died of anthrax, but they did not appear in the blood of healthy animals. After trying many other materials, he finally persuaded the rods to multiply by placing them in fluid from the inside of a slaughtered ox's eye. The multiplication, which he witnessed with his microscope, proved that they were alive. Koch moved drops containing the rods from one batch of fluid to another until they appeared to be the only kind of microbes in the liquid. He then injected some of the fluid into mice and showed that they developed anthrax. He withdrew some blood from the dying animals, confirmed that it contained the rods, and then used it to produce anthrax in a second batch of mice. This laborious series of experiments not only proved that the rodlike microbes (which were a type of bacteria) caused anthrax but became the basis for what were later called Koch's postulates, the standards for proving that a particular microorganism causes a particular disease.

Koch also found that, when the anthrax bacteria were exposed to harsh conditions, they took on a new form that resisted heat, cold, drying, and other treatments that normally would have killed them. He called these resistant forms spores. He discovered that the spores could be stored for months and still, when exposed to favorable conditions, turn back into the rodlike forms and cause disease once more. The existence of these spores, which had never been reported before, explained why animals often developed anthrax after grazing in fields where corpses of other animals with the disease had been buried as much as a decade earlier. Koch recommended that bodies of anthrax-infected animals be burned, or at least buried very deeply, to stop the spread of the disease.

Koch demonstrated his discoveries to professors in the city of Breslau in 1876, and one was impressed enough to publish Koch's paper about his work in the botanical journal that the professor edited. Koch described similar experiments involving bacteria that infect wounds in 1878. His fame spread rapidly, and in 1880 the new government of Germany (formed in 1870) made him part of the Imperial Health Office in Berlin. Here, he obtained, for the first time, a reasonably well-equipped laboratory and two assistants.

In 1882, Koch announced a discovery even more impressive than his findings about anthrax: his identification of the bacteria that cause tuberculosis. At the time, this disease caused one out of every seven deaths in Europe and North America. Koch had developed a special stain that allowed him to see the tuberculosis bacteria under the microscope.

When an epidemic of cholera broke out in Egypt in 1883, the German government sent Koch and several assistants there to study it because of fears that the deadly disease would spread to Europe. Koch found comma-shaped bacteria that he believed caused the disease and confirmed this identification during another cholera outbreak in India a year later. The bacteria appeared in water supplies containing waste from people who had the disease, and Koch, like British physician JOHN SNOW about 35 years earlier, concluded that contaminated drinking water was an important route by which the illness spread. The German government awarded Koch a prize of 100,000 marks for his discovery.

In addition to providing convincing proof that the germ theory of disease was correct and identifying specific disease-causing microbes, Robert Koch established basic laboratory methods for the new science of bacteriology. He was one of the first researchers to use dyes or stains to make microorganisms more visible and one of the first to photograph them through the microscope. He pioneered the use of solid and semisolid nutrient materials (beginning with humble potato slices) for raising bacteria in the laboratory. These materials were useful for isolating different types of microbes because each kind of germ formed a separate colony on them as it multiplied. Researchers could transfer samples of particular colonies to fresh tubes or plates of nutrient jelly, allowing cultivation of the pure cultures (containing just one type of microbe) that Koch's postulates demanded.

In 1885, the German government made Koch the director of the Institute of Hygiene as well as a professor at Berlin University. It built him his own research institute, the Institute for Infectious Diseases, in 1891, and he headed this organization until 1904. During this period, in addition to overseeing the research of younger bacteriologists, Koch traveled to various parts of the world to identify disease-causing microorganisms and their methods of spreading and advise governments about controlling them. Among other things, he showed (in 1897) that rats spread plague.

Koch generated a great deal of excitement when he announced in 1890 that injections of a substance he called tuberculin could cure tuberculosis. Partly on the strength of this claim, he was awarded one of the first Nobel Prizes, the 1905 prize in physiology or medicine. The German government also gave him numerous honors, including the Order of the Crown and the Grand Cross of the German Order of the Red Eagle. Unfortunately, further research did not bear out Koch's hopes, although tuberculin (an extract of the bacteria that cause the disease) did prove useful in a test to show whether someone had been exposed to tuberculosis bacteria and therefore might have the disease. Koch was criticized after tuberculin's failure, but later historians agree that his mistake in promoting it was far outweighed by his many contributions to the understanding of disease. He died in Baden-Baden on May 27, 1910.

Further Reading

Brock, Thomas D. *Robert Koch: A Life in Medicine and Bacteriology*. Madison, Wis.: Science Tech Publishers, 1988.

de Kruif, Paul. *Microbe Hunters*. New York: Harcourt, Brace, 1926.

"Robert Koch—Biography." *Nobel Lectures, Physiology or Medicine 1901–1921*. Available online. URL: www.nobel.se/medicine/laureates/1905/koch-bio.html. Last updated 2001.

⊠ Kolff, Willem Johan

(1911–)
Dutch/American
Biophysicist

Working under the difficult conditions of wartime, Willem Kolff created the first artificial kidney used in human beings. He later helped to design the first artificial hearts and other artificial organs.

"Pim" Kolff, as he was called, was born in Leiden, the Netherlands, on February 14, 1911.

As a child, he wanted to be a zookeeper or perhaps a carpenter. His father was a physician, but Kolff doubted that he could do that job because he did not think he could stand to watch people die. Nonetheless, his father eventually persuaded him to study medicine.

Kolff took his medical degree from the University of Leiden in 1938. Later that year, when he was an assistant in the hospital at Groningen, he watched a young man die a slow, painful death after his kidneys failed. Wastes that those organs normally would have filtered out built up in the man's blood and poisoned him. Kolff determined to make an artificial filter that could do some of the kidneys' work so that others would not have to suffer in the same way.

Kolff learned that researchers at Johns Hopkins University in the United States had made an artificial kidney, or dialysis machine, as early as 1913. They had tested it in dogs, but it had not worked well enough to be tried on human beings. Part of the problem had been lack of a suitable material for the filter, which needed to have microscopic pores, or holes, that allow small particles or molecules to pass through but exclude larger ones.

When Kolff began looking for filter materials, a friend amazed him by recommending sausage casings. At the time, such casings were made of a stiff plastic called cellophane. In 1939, Kolff filled a cellophane casing with a mixture of blood and urea (one of the main compounds that the kidneys must filter out of the blood), sealed it shut, and rocked it back and forth in a tub of salt water. After half an hour, the urea had moved out into the water, and the blood inside the casing was clean. Kolff concluded that cellophane indeed might be just what he wanted.

Before Kolff had developed his machine to the point where he dared try it on human beings, World War II broke out, and shortly afterward Nazi Germany seized control of the Netherlands. Refusing to work under the Nazi sympathizer who was put in charge of the Groningen Hospital, Kolff moved to the smaller town of Kampen. There, he continued working as a hospital physician, helping the anti-Nazi resistance movement when he could, and doing research on artificial kidneys. Obtaining parts was difficult, since Dutch Nazis controlled all supplies and were suspicious of any work that they did not understand. Kolff used whatever he could find, for instance, obtaining several parts from a local auto dealership.

Kolff finally felt ready to test one of his machines on a human patient in March 1943. The device consisted of a wooden drum wrapped in 65 feet of cellophane tubing, sitting in an enamel tub full of fluid. One end of the tube connected to an artery in the patient's arm and the other to a vein. The blood flowed from the artery into the tube. As an electric motor slowly turned the drum and the blood moved through the tubing, urea and other wastes passed through the cellophane and into the fluid outside. At the same time, substances the body needed moved from the fluid into the blood. The blood, purified and enriched, finally went back into the patient's body through the vein.

The only patients Kolff could obtain permission to test this experimental device on were already dying of kidney failure. All but one of the first 15 did die, and Kolff believed that that one probably would have survived without dialysis. Nonetheless, he was encouraged because the machine brought enough temporary improvement for the patients to emerge from comas and speak with their families. Finally, in September 1945, just months after the Germans had been driven out of the Netherlands, Kolff obtained his first survivor, a 67-year-old woman named Sofia Schafstadt. The artificial kidney kept Schaftstadt alive for several days, long enough for her own kidneys to begin working again. Ironically, she had been a Dutch Nazi and was in prison awaiting trial for war crimes when her kidneys failed.

After the war, Kolff donated prototypes of his artificial kidney to researchers in England,

Canada, and the United States. He himself immigrated to the United States in 1950 and joined the Cleveland Clinic in Ohio, where he set up and headed the world's first artificial organ research center. (He became a U.S. citizen in 1956.) He spent part of his time there improving dialysis machines. By 1960, the machines were able to keep people with permanent kidney failure alive for relatively long periods, but they were cumbersome and expensive. There were too few machines to treat everyone who needed them, and those who could obtain treatment had to spend many hours in a hospital each week. Attempting to make the devices more practical, Kolff and others invented a wearable dialysis machine in the 1970s.

In addition to working on artificial kidneys, Kolff helped to develop some of the first heart-lung machines, which provide oxygen to the blood while the heart is stopped during surgery. These machines came into use in the 1950s, making open heart surgery possible for the first time.

Kolff moved to the University of Utah in Salt Lake City in 1967, becoming director of the division of artificial organs and a new institute of biomedical engineering at the university's medical center. He was also a research professor at the engineering school and a professor of surgery at the medical center. At the institute, Kolff trained other researchers in the new field of artificial organs as well as continued his own inventions.

One of Kolff's chief areas of research at Utah was artificial hearts, on which he had begun working in the late 1950s. He teamed with a younger scientist, Robert Jarvik, and others to design to the first artificial heart to be used in a human. It kept a 61-year-old retired dentist named Barney Clark alive for 112 days in 1982 and 1983. The heart had many problems and was never widely used, but improved devices that do part of the heart's work are used today in people who cannot obtain heart transplants.

Willem Kolff also worked, or guided the work of others, on artificial skin, eyes, ears, lungs, and arms. He received many awards, including induction into the National Inventors Hall of Fame in 1985, the Albert Lasker Award for Clinical Medical Research in 2002, and the National Academy of Engineering's Russ Prize in 2003. He retired in 1998 but was still giving interviews in late 2002. In life as well as in science, Kolff has shown that persistence can conquer almost any obstacle.

Further Reading

Keck, Patricia S., and John J. Meserko. "Willem J. Kolff: Pioneer in Artificial Organ Research." *Proceedings of the American Academy of Cardiovascular Perfusion*, January 1985. Available online. URL: http://members.aol.com/amaccvpe/history/kolff.htm.

"Kolff, Willem Johan." *Current Biography Yearbook 1983*. New York: H. W. Wilson, 1983.

Landers, Peter. "A Father of Invention, Kolff Rarely Saw the Necessity for Patents." *Wall Street Journal*, September 26, 2002.

Noordwijk, Jacob van. *Dialysing for Life: The Development of the Artificial Kidney*. Dordrecht, the Netherlands: Kluwer Academic Publishers, 2001.

⊠ Kornberg, Arthur

(1918–)
American
Biochemist

Arthur Kornberg discovered enzymes that control vital processes inside the cell, including the manufacture of DNA. He also made infectious viral DNA outside a cell for the first time. He received a share of the Nobel Prize in physiology or medicine in 1959 for his early work on DNA.

Kornberg was born on March 3, 1918, in Brooklyn, New York, to Joseph and Lena (Katz) Kornberg, both immigrants from Austria. His father owned a hardware store. Kornberg attended the City College of New York—beginning at age 15—and obtained a bachelor of science degree in biology and chemistry in 1937.

Arthur Kornberg of Stanford University was the first to synthesize DNA outside cells. *(Arthur Kornberg)*

He went on to the University of Rochester (New York), from which he earned an M.D. in 1941.

During World War II, Kornberg served in the Public Health Service, first in the Coast Guard and then at the National Institutes of Health (NIH), where he remained until 1953. Between 1943 and 1945, he worked in the Nutrition Section of the Division of Physiology, and from 1947 to 1953 he was head of the Enzyme and Metabolism Section. He married Sylvy Ruth Levy, a fellow biochemist with whom he often collaborated, in 1943; they have three sons, Roger, Tom, and Kenneth.

At NIH, Kornberg became an expert in the identification, purification, and synthesis of complex biological chemicals, especially enzymes and nucleic acids and their components. He isolated several enzymes and other substances related to cells' use of vitamins. During his NIH years, he also trained and researched briefly at several universities, including the New York University College of Medicine (1946, with SEVERO OCHOA, with whom he later shared the Nobel Prize), the Washington University School of Medicine in St. Louis (1947, with CARL FERDINAND CORI and GERTY THERESA RADNITZ CORI), and the University of California at Berkeley (1951, with H. A. Barker). Kornberg left NIH to become a professor and head of the Microbiology Department at the Washington University School of Medicine in 1953. Since 1959, he has been professor and executive head of the Biochemistry Department of the Stanford University School of Medicine in California.

The work that earned Kornberg the Nobel Prize took place in the 1950s, while he was at Washington University. Even before JAMES WATSON and FRANCIS CRICK worked out the structure of DNA and suggested how the DNA molecule could copy or reproduce itself, Kornberg was planning to "follow the enzymes" to find out how the cell manufactures the building blocks of nucleic acids and assembles them into RNA and DNA. In 1956, he purified from bacteria a DNA polymerase, the key enzyme that the cell uses to assemble DNA from its components. Using this enzyme and adapting methods that Ochoa had developed for synthesizing RNA, Kornberg synthesized DNA outside a living cell for the first time in 1957. This DNA was not biologically active, but in 1967, Kornberg synthesized the DNA of a virus that infected bacteria, put this DNA into bacterial cells, and showed that the bacteria then began manufacturing that type of virus.

Kornberg was involved in the development of genetic engineering and biotechnology during the 1970s, cofounding a company (the DNAX Research Institute) that created genetically engineered products related to the immune system. In the late 1970s and 1980s, his laboratory also studied chemicals that control the start of DNA replication—the "on switch" for cell reproduction. Later, the group isolated DNA's "off switch" as well. Understanding the actions of these enzymes and learning how to add or remove them might pave the way for treatments for cancer, in which cell reproduction is uncontrolled, and AIDS, which depends on the reproduction of viruses' genetic material.

During most of the 1990s and beyond, Kornberg and his coworkers at Stanford have focused on inorganic polyphosphate, or poly-P, a long chain of phosphates (phosphorus compounds) found in all living things. It also appears outside living cells in places such as volcanoes and deep-sea thermal vents, where many scientists think life on Earth may have begun. Kornberg believes that poly-P is involved in several essential cell processes, including regulation of growth and the ability of cells to take up DNA from outside sources. His laboratory is trying to discover the enzymes involved in its synthesis and function.

The Nobel Prize is only one of Kornberg's many awards. Others include the American Chemical Society's Paul-Lewis Laboratories Award in Enzyme Chemistry (1951), the American Medical Association's Scientific Achievement Award (1968), and the National Medal of Science (1980). The achievements that earned these awards, Kornberg has insisted, came more from persistence and unrelenting work than brilliant insight. "The best way to accomplish something," he told *Discover* reporter David Freedman in 1991, "is to find a focus and stick with it."

Further Reading

Freedman, David H. "Life's Off-Switch." *Discover*, July 1991.

Kornberg, Arthur. "Biochemistry at Stanford, Biotechnology at DNAX." University of California Programs in the History of the Biosciences and Biotechnology. Available online. URL: http://sunsite.berkeley.edu:2020/dynaweb/teiproj/oh/science/kornberg/@Generic_BookView. Posted 1997.

———. *For the Love of Enzymes: The Odyssey of a Biochemist*. Cambridge, Mass.: Harvard University Press, 1989.

———. *The Golden Helix: Inside Biotech Ventures*. Sausalito, Calif.: University Science Books, 1995.

Krebs, Hans Adolf

(1900–1981)
German/British
Biochemist

Hans Krebs worked out key parts of the complex processes through which the body breaks down food to obtain energy. His discoveries earned a share of the Nobel Prize in physiology or medicine in 1953.

Krebs was born in Hildesheim, Germany, on August 25, 1900. His father, Georg Krebs, was an ear, nose, and throat surgeon. His mother was the former Alma Davidson. Krebs studied medicine at the Universities of Göttingen, Freiburg, Munich, Berlin, and Hamburg, obtaining his M.D. from Hamburg in 1925. He did postgraduate study in biochemistry at the Kaiser Wilhelm Institute in Berlin, then transferred to the medical clinic of the University of Freiburg in 1930.

Krebs lost his job in 1933 because the Nazis, who took over the German government in that year, discharged everyone of Jewish descent. Fortunately, British biochemist Frederick Hopkins invited him to come to Cambridge University, and he arrived in England with, as he said later, "virtually nothing but a sigh of relief and a few books." He remained at Cambridge for two years, obtaining an M.A. from the university in 1934.

In 1935, Krebs moved to Sheffield University as a lecturer in pharmacology. He became head of the university's biochemistry department in 1938 and professor of biochemistry and director of the Medical Research Council unit for research on cell metabolism in 1945. He married Margaret Fieldhouse, a fellow Sheffield faculty member, in 1938 (they later had two sons and a daughter) and became a naturalized British citizen in 1939. In 1954, he joined Oxford University as Whitley Professor of Biochemistry and a fellow of Trinity College. He stayed at Oxford until he retired in 1967.

Krebs's first work, done while he was still in Germany, concerned the way the body breaks down protein. Scientists knew that this process occurs chiefly in the liver, resulting in production of urea and other wastes. By 1932, Krebs had worked out the process by which nitrogen atoms are removed from amino acids, the smaller molecules of which proteins are made, and combined with other substances to make urea. This process is called the urea cycle or ornithine cycle.

At Sheffield, Krebs went on to investigate the way the body breaks down glucose, a simple sugar, into carbon dioxide and water, releasing energy—the chief process by which the body obtains energy. He studied the second of three major stages in glucose breakdown, which takes place mostly in mitochondria, tiny bodies within the cell. He found that a chemical named citric acid is both the first product of this series of chemical reactions and the last, so the process is a cycle. This cycle, which proved to be involved in the breakdown of fats and proteins as well as sugars, came to be called the citric acid cycle, the tricarboxylic acid cycle, or simply the Krebs cycle. Krebs described it in 1937 and in more detail in a 1957 book, *Energy Transformation in Living Matter*, which he coauthored with H. L. Kornberg.

Krebs shared the 1953 Nobel Prize with Francis Lipmann of Harvard Medical School, who worked out additional details of the Krebs cycle and discovered coenzyme A, a key chemical involved in it. The American Public Health Association also gave Krebs its prestigious Albert Lasker Medical Research Award in 1953, and Britain's Royal Society awarded him its Royal Medal in 1954 and Copley Medal in 1961. He was knighted in 1958 and died on November 22, 1981, in Oxford.

Further Reading

"Hans Krebs—Biography." *Nobel Lectures, Physiology or Medicine 1942–1962*. Available online. URL: www.nobel.se./medicine/laureates/1953/krebs-bio.html. Last updated 2001.

Krebs, Hans. *Reminiscences and Reflections*. Oxford, England, and New York: Clarendon Press/Oxford University Press, 1981.

"Krebs, H(ans) A(dolf)." *Current Biography Yearbook 1954*. New York: H. W. Wilson, 1954.

L

Lamarck, Jean-Baptiste, chevalier de

(1744–1829)
French
Naturalist, Evolutionary Biologist

Although Jean-Baptiste-Pierre-Antoine de Monet de Lamarck is remembered mostly for a now-discarded theory of evolution, his more lasting contribution to science was a better understanding of invertebrates, or animals without backbones. He also introduced the term *biology*.

Lamarck was born on August 1, 1744, in the village of Banzentin in Picardy (northern France), the youngest of the 11 children of Philippe-Jacques de Monet de Lamarck and the former Marie-Françoise de Fontaines de Chuignolles. His parents were aristocrats, although they had little money. The Lamarcks had a long tradition of military service, which Philippe Lamarck himself followed.

Jean-Baptiste's family sent him to a Jesuit seminary to train for the priesthood in 1756, but after his father died around 1760, he left the school and joined the French army. He fought heroically in the Seven Years' War against Germany, holding off an enemy attack singlehandedly, and was made an officer for his bravery. Health problems resulting from an accident forced him to leave the military in 1766, however. He then studied medicine for a while, supporting himself by working as a bank clerk.

Walks in the countryside around Paris turned Lamarck's interest to botany. In 1778, he published a three-volume work called *Flore française* (*The Plants of France*), which made his name in science and brought him to the attention of famed French naturalist GEORGES-LOUIS BUFFON. Buffon saw to it that Lamarck was elected to the Academy of Sciences, France's most highly regarded scientific organization, and helped the young man obtain the king's patronage. Lamarck traveled through Europe in the early 1780s, collecting plants for the royal botanical garden. When he returned, Buffon found him a job in the garden itself. Lamarck also became a professor of botany at the University of Paris.

In 1793, the leaders of the French Revolution reorganized the botanical garden, which was actually a research institution where many scientific subjects were studied. They renamed it the National Museum of Natural History and decided that it should be run by 12 professors who specialized in different fields. Lamarck and another scientist, Geoffroy Saint-Hilaire, were assigned to divide the professorship for zoology, even though neither man was a zoologist. Geoffroy (as he was called), the better known of the two, took charge of mammals, birds, reptiles, and fish, leaving Lamarck with "insects, worms, and microscopic animals."

Most zoologists had neglected these "lower" animals, but Lamarck began to study them intensely and coined the term *invertebrates* to describe them. He greatly improved classification within this catchall group, for instance, distinguishing for the first time the separate classes of arachnids (spiders and their kin), annelids (segmented worms, such as earthworms), and crustaceans (animals such as lobsters and crabs). His much admired seven-volume *Histoire naturelle des animaux sans vertèbres* (*Natural History of the Invertebrates*), published between 1815 and 1822, described and classified fossil invertebrates as well as living ones.

While studying invertebrates, Lamarck developed a theory of evolution that in some ways foreshadowed the later work of CHARLES ROBERT DARWIN and in other ways was very different from it. Paleontologist and science historian STEPHEN JAY GOULD called Lamarck's scheme "the first comprehensive theory of evolution in modern science." Lamarck first proposed it in 1801 in his *Système des animaux sans vertèbres* (*System of Invertebrate Animals*) and described it most fully in *Philosophie zoologique* (*Zoological Philosophy*), a two-volume work published in 1809.

Unlike many scientists of his day, Lamarck believed that the Earth was very old and that species of plants and animals had changed during that long period of time. The main force driving such evolution, he thought at first, was a tendency for living things to become more complex. The need to adapt to changing environments was a secondary force. Lamarck later modified this picture, putting more emphasis on changes caused by environment and showing evolution as a process of branching rather than a linear progression. He may have been the first to diagram evolution as a sort of tree with many branches.

Lamarck believed that, over generations, parts of the body became larger or stronger if a species used them frequently, but they withered away if not used. In his most commonly cited example, he claimed that the long necks of giraffes had developed because changes in the growth of certain trees had forced the animals' antelopelike ancestors to reach higher and higher for the leaves on which they fed. Such repeated reaching made their necks stretch, and this acquired characteristic was passed on to their offspring. Darwin, by contrast, would maintain that long necks arose by chance and then were passed on to future generations because their possessors were more likely than other members of their species to survive and have offspring.

Lamarck's last years were miserable. Always short of money, he became poorer than ever and slowly lost his sight as well. Two daughters, perhaps the only survivors among his eight children by four marriages, took care of him. When he died on December 28, 1829, in Paris, they had to ask his friends for money to bury him.

Lamarck had been respected for his work on invertebrates, but most of his contemporaries, as well as later thinkers, ridiculed his beliefs about evolution. (Darwin called Lamarck's theory "rubbish.") This happened partly because his ideas clashed with the scientific fashion of the time, for instance, by organizing living things into a grand system rather than focusing on details, and partly because Lamarck, a vain and argumentative man, had made many personal enemies among his fellow scientists. After his death, one of those enemies, respected French naturalist GEORGES CUVIER, delivered a vicious "eulogy" that made Lamarck's ideas seem more foolish than they really were and almost ruined what was left of his reputation permanently. The later work of GREGOR MENDEL and others, which explained the mechanism of heredity, showed why Lamarck's ideas about evolution had to be wrong. Nonetheless, modern science historians such as Gould say that Lamarck deserves to be remembered for his successful work rather than for his mistakes.

Further Reading

Burkhardt, Richard W. *The Spirit of System: Lamarck and Evolutionary Biology*. Boston: Belknap Press/Harvard University Press, 1995.

Jordanova, L. J. *Lamarck*. New York: Oxford University Press, 1984.

Waggoner, Ben. "Jean-Baptiste Lamarck (1744–1829)." University of California, Berkeley, Museum of Paleontology. Available online. URL: www.ucmp.berkeley.edu/history/lamarck.html. Updated 1998.

Landsteiner, Karl
(1868–1943)
Austrian/American
Immunologist, Pathologist

Karl Landsteiner's discovery of blood types helped to make safe blood transfusions possible. Landsteiner was born on June 14, 1868, in Vienna, Austria. His father, Leopold, a journalist and newspaper publisher, died when Karl was still a child, and Karl's mother, the former Fanny Hess, raised him.

Landsteiner studied medicine at the University of Vienna, obtaining his medical degree in 1891. He studied chemistry and biochemistry for five years at various universities before becoming a research assistant at the Vienna Pathological Institute in 1898. He stayed there for 10 years, then became professor of pathology at the Royal Imperial Wilhelminen Hospital in Vienna. He also taught at the University of Vienna after 1911.

Landsteiner made his most important discovery around 1900. He and others had already found that when the blood of two different animal species is mixed together, the red blood cells form clumps. Landsteiner discovered that the same thing sometimes happens when blood cells from one person are mixed with the serum (liquid part of the blood) from another. If such clumping occurred in a living body, he realized, it would cause severe illness and probably death.

In a paper published in 1901, Landsteiner divided people into three groups, or blood types—A, B, and O—depending on the way their blood behaved when mixed with that of other people. His coworkers discovered a fourth group, AB, a year later. No clumping occurred when serum and red cells from two people with the same blood type were mixed. Cells from a type A person clumped when mixed with serum from a type B person, and vice versa. Type O people's cells could be mixed with serum from any other type without clumping, but cells of any type except O would clump when mixed with serum from type O. Type AB was exactly the opposite; AB serum did not cause any other cells to clump, but AB cells would clump when mixed with any kind of serum except that from type AB people.

Austrian pathologist Karl Landsteiner discovered blood types in 1900, paving the way for successful blood transfusions. *(Lasker Foundation)*

These reactions, Landsteiner realized, were caused by substances in the serum called antibodies. The immune system uses antibodies to detect "foreign" cells and mark them for destruction. Antibodies mark cells by attaching to other substances on the cell surfaces called antigens. People normally do not have antibodies against antigens on their own cells.

Type A blood cells carry one kind of antigen, and type B cells carry another. Type A people therefore have antibodies against the B antigen, and vice versa. Type O people's cells carry neither antigen, and their serum has antibodies against both; type AB people, conversely, have both antigens on their cells and no antibodies to either in their serum.

By 1907, Landsteiner had developed a technique for finding out whether a blood transfusion would be safe. He simply mixed the serum of the intended recipient with the blood cells of the potential donor on a microscope slide. If the cells did not clump, the transfusion would probably succeed. This discovery did not immediately bring transfusions into widespread use, however, because at the time there was no way to preserve blood. Transfusions were used more often after 1914, when Richard Lewisohn discovered that adding a chemical called sodium citrate to blood and then refrigerating the blood would preserve it for about 10 days. They became a standard medical treatment in the 1930s, when preservation methods were further improved and blood banks, which could store large amounts of blood for long periods, were developed.

Landsteiner also studied disease-causing microorganisms. During 1911–19, he showed that poliomyelitis (polio) could be transferred to monkeys by injecting them with crushed spinal cord tissue from children who had died of the disease. This still occurred when the tissue had been put through a filter with a mesh small enough to remove all known bacteria. This suggested that the disease was caused by a virus, a group of still-smaller microorganisms that scientists had theorized about but would not be able to see until electron microscopes were invented in the 1930s.

Conditions in Austria were grim after World War I, so in 1919 Landsteiner and his family (he had married Helen Wlasto in 1916, and they had one son) moved to Holland. He worked at the RK Hospital in the Hague until 1922, then emigrated to the United States and joined the Rockefeller Institute for Medical Research in New York City. He stayed there for the rest of his career, becoming a naturalized U.S. citizen in 1929.

Landsteiner made additional discoveries about blood groups while in the United States. In 1927, he and his coworkers identified another set of blood cell antigens, the M and N antigens. These antigens have no effect on transfusions, but because they, like other blood group antigens, are inherited, they can be used to determine whether a man could be the father of a certain child.

A more important discovery was the Rh antigen, which Landsteiner and coworkers Alexander Wiener and Phillip Levine found in the blood of rhesus monkeys in 1940. About 85 percent of the humans Landsteiner tested also possessed this antigen, or were Rh positive; the rest lacked it, or were Rh negative. The Rh antigen proved to explain a mysterious illness that killed some babies soon after birth. When a woman is pregnant, a few blood cells from her fetus enter her bloodstream. If an Rh-negative woman has a child by an Rh-positive man, the child will inherit the Rh antigen from its father, and the mother's immune system will form antibodies to the "foreign" antigen on the unborn baby's cells. The antibodies will not harm this first baby, but if the woman has a second Rh-positive child, they will make her immune system attack that baby's blood cells, causing severe illness or death. Treatments can now destroy such antibodies before they cause harm.

Landsteiner won the Nobel Prize in physiology or medicine in 1930 for his discovery of blood types. He retired officially in 1939 but continued to do research until June 24, 1943, when he had a heart attack in his laboratory. He died two days later.

Further Reading

"Karl Landsteiner—Biography." *Nobel Lectures, Physiology or Medicine, 1922–1941*. Available online. URL: www.nobel.se/medicine/laureates/1930/landsteiner-bio.html. Last updated 2001.

Yount, Lisa. *Medical Technology*. New York: Facts On File, 1998.

⊠ Lavoisier, Antoine-Laurent

(1743–1794)
French
Chemist, Physiologist

Antoine Lavoisier is often called the father of modern chemistry. His studies of respiration were also important contributions to physiology. He was born into a wealthy family in Paris on August 26, 1743. His father, Jean-Antoine, was a lawyer. His mother, the former Emilie Punctis, died when he was five years old, and his father, grandmother, and aunt raised him. He was educated at the highly regarded Collège de Quatre Nations, also called Collège Mazarin. Following his family's expectations at first, he studied law, receiving his degree in 1763, but he never practiced. Instead, he turned to science, notably geology and chemistry.

Lavoisier wasted no time in establishing his career. He presented a paper on the mineral gypsum to the Academy of Sciences, France's leading scientific group, in 1765. The following year, he won a medal in a contest that the academy sponsored for designing street lighting for Paris. These achievements earned his election to the academy in 1768, an amazing honor for someone only 25 years old. He also ensured a high income by buying a share in the Ferme Générale, a private company nicknamed the tax farm because the French government "farmed out" to it the right to collect certain taxes. In 1771, when Lavoisier was 28, he married Marie Anne Paulze, the 14-year-old daughter of a fellow tax farm member.

During most of his life, Lavoisier both worked for the government and carried out private research. His government assignments included improving the country's gunpowder and establishing a metric system of weights and measures. He also investigated innumerable subjects for the Academy of Sciences, ranging from ventilation in prisons to the effects of hypnotism. He became head of the academy in 1785.

Lavoisier's contributions to chemistry include naming the element oxygen, recognizing that oxygen is necessary for combustion (burning), and discovering that water is a combination of hydrogen and oxygen. He developed the first logical system of names for chemical compounds, much as CAROLUS LINNAEUS did for names in biology. He stressed the importance of careful weighing and measuring and emphasized that the total amount of matter present at the end of a chemical reaction is always the same as the amount present at the start. He described most of his discoveries and theories in *Traité élémentaire de chimie* (*Elementary Treatise on Chemistry*), published in 1789. His wife drew illustrations for this book and often helped him in his laboratory work.

In the early 1780s, Lavoisier and mathematician Pierre-Simon Laplace investigated respiration. Although this word is often used to mean *breathing*, it can also mean the chemical reactions in the bodies of living things that use the oxygen obtained through breathing. Lavoisier and Laplace studied respiration by means of an ingenious invention of Laplace's called a calorimeter, which consisted of three metal chambers nested inside one another. The men filled the outer and middle chambers with ice and then placed something that produced heat in the innermost chamber. They measured the amount of heat by determining the amount of water created as the heat melted the ice in the middle chamber. (The ice in the outer chamber shielded the ice in the middle chamber from the heat of the room.) In the respiration experiments, the heat producer was a guinea pig.

Lavoisier and others had already shown that both respiration and combustion used oxygen and gave off another gas, later called carbon dioxide. The guinea pig experiments convinced Lavoisier that "respiration is . . . a combustion, very slow, . . . but [otherwise] perfectly similar to that of carbon" and that the heat released by respiration maintained the animal's body temperature. He went on to show in 1775 that animals breathed out water vapor as well as carbon dioxide and that the heat produced by the reactions needed to make the amounts of these two substances in the breath equaled the animals' body heat. In other words, he said, respiration is really two combustions: One "burns" carbon and produces carbon dioxide, and the other "burns" hydrogen and produces water. Lavoisier's calorimeter studies were among the first to measure chemical reactions in the bodies of living things.

Around 1790, Lavoisier also investigated respiration in human beings. He had an assistant, Armand Séguin, wear a mask that was airtight except for a breathing tube so that the gases Séguin breathed in and out could be measured. He determined how much oxygen Séguin breathed in when sitting quietly and compared this with the amount he breathed in when pressing a foot pedal connected to a weight. His found that Séguin breathed faster and used three times as much oxygen when lifting the weight as he did when resting. Lavoisier concluded that energy from respiration allowed the body to do work as well as to heat itself.

Unlike many other members of the Ferme Générale, Lavoisier had never abused his power as a tax collector. On the contrary, he had argued that taxes should be reduced for the poor. At first, he also supported the French Revolution, a revolt against the country's abusive king and nobles, which began in 1789. The leaders of the revolution, however, cared only that Lavoisier was an aristocrat and a tax farmer. He was arrested along with the other tax farmers in November 1793, condemned to death in a hasty trial ("The Republic has no need for scientists," the judge is said to have snapped), and guillotined on May 8, 1794. Mathematician Joseph Lagrange told a friend, "It required only a moment to sever his head, and probably one hundred years will not suffice to produce another like it."

Further Reading

Holmes, R. L. *Lavoisier and the Chemistry of Life: An Exploration of Scientific Creativity*. Madison: University of Wisconsin Press, 1985.

McKie, Douglas. *Antoine Lavoisier: Scientist, Economist, Social Reformer*. New York: Henry Schuman, 1952.

Lavoisier, Antoine Laurent. *Elements of Chemistry*. 1789. Reprint, Mineola, N.Y.: Dover Publications, 1984.

Leakey, Louis S. B.
(1903–1972)
Kenyan/Tanzanian
Paleontologist, Anthropologist

Louis Seymour Bazett Leakey greatly changed scientists' view of humanity's origins and established a family dynasty that has dominated 20th-century paleoanthropology. He was born in Kabete, near Nairobi in the east African country of Kenya, on August 7, 1903. His parents were Harry and Mary (Bazett) Leakey, British missionaries who lived with the Kikuyu, Kenya's largest tribal group. As a child, Leakey learned the Kikuyu language along with English, and he was initiated as a member of the tribe at age 13. At about the same time, he began collecting fossils and ancient stone tools and decided to become an archaeologist or a paleontologist.

Leakey's parents sent him to Britain for formal education when he was 16. He entered St. John's College, Cambridge University, in 1922 but had to leave temporarily a year later because of headaches resulting from a sports injury. This break was perhaps the first example of what both admirers and rivals later called "Leakey's luck," because it left him free to join Canadian archaeologist William Cutler's expedition to East Africa. Leakey wrote later that he learned more about archaeological techniques on this trip than he could have absorbed in years of college. After a few months he returned to Cambridge, from which he earned a bachelor's degree in anthropology and archaeology in 1926 and a Ph.D. in African prehistory in 1930.

Leakey believed that the human species had originated in Africa, as CHARLES ROBERT DARWIN had maintained, rather than in Asia, as most anthropologists of his own time thought. On several expeditions to East Africa that he led in the late 1920s and early 1930s, he discovered stone tools that showed that sophisticated toolmakers had worked in the area much earlier than others had realized. In 1931, he paid his first visit to Olduvai Gorge in Tanganyika (now Tanzania), where a river had cut a 35-mile-long gash through 300 feet of sedimentary rock that had once been an ancient lake bed. Olduvai quickly became his favorite site.

Although Leakey made some important finds on these early expeditions, other scientists criticized him for drawing conclusions about them that sometimes were not supported by sufficient evidence. His personal life raised eyebrows, too. In 1928, he had married Wilfrida (Frida) Avern, and they had two children, but in 1933, Leakey met and fell in love with 20-year-old Mary Nicol. He divorced Frida and married Mary in 1936. MARY LEAKEY became a valuable partner in Leakey's explorations and, in time, a highly respected paleoanthropologist in her own right. The Leakeys later had three sons, Jonathan, Richard, and Phillip, and RICHARD LEAKEY followed his parents into paleoanthropology.

During World War II, Louis Leakey left his fieldwork to head the African section of British military intelligence. He was an organizer and administrator as much as a researcher for the rest of his career. For instance, he was curator of the Coryndon Museum in Nairobi from 1945 to 1961, after which he founded the National Museum Centre for Prehistory and Paleontology in the same city.

On the research side, the Leakeys found the skull of an apelike creature called *Proconsul africanus* on Rusinga Island in Lake Victoria in 1948. *Proconsul* lived between 25 million and 40 million years ago, about the time that apes diverged from monkeys. This find, however, was merely a prelude to a much more important one that Mary Leakey made at Olduvai in 1959: part of the skull of a hominid that the Leakeys called *Zinjanthropus boisei*, or "East African Man." "Zinj," or "Nutcracker Man," as the Leakeys nicknamed the skull because of its big teeth, was later found to be 1.75 million years old. Its species is now placed in the genus *Australopithecus*, which many paleoanthropologists think was ancestral to humans, although this particular species is believed to have died out. The discovery of *Zinjanthropus* brought the Leakeys considerable fame and, for the first time, ample money from organizations such as the National Geographic Society to finance their digs.

In the early 1960s, the Leakeys found a second species of fossil hominid that apparently had lived alongside *Zinjanthropus* but was considerably more advanced. Leakey named this new species *Homo habilis*, or "Handy Man." Because its bones were found in close association with stone tools, Leakey concluded that *Homo habilis* was the earliest known toolmaker as well as the earliest known species in the genus to which humans belong. The Leakeys also found several other new species of hominid during the 1960s, including *Homo erectus*, a still more advanced species that Leakey maintained was a direct ancestor of modern humans, and members of the genus *Kenyapithecus*, which extended the prehuman lineage back to 20 million years ago. The Leakeys' discoveries helped to convince anthropologists that humans had originated in Africa, although some of Leakey's specific conclusions about the relationships among the *Homo* species have remained controversial.

A combination of health problems, changing interests, and strains in his marriage led Louis Leakey to do less research during the late 1960s and spend more time traveling around the world, speaking about his discoveries and raising funds for further investigations. A charismatic man "with a bold sentient face and a wild shock of thick white hair," as *Life* reporter Ken MacLeish described him in 1961, he persuaded donors to found the Leakey Foundation in 1968 to "increase scientific knowledge and public understanding of human origins and evolution." Among the research projects he sponsored were long-term investigations of living ape species by JANE GOODALL (chimpanzees), DIAN FOSSEY (mountain gorillas), and BIRUTÉ GALDIKAS (orangutans). Leakey believed that studies of apes would provide information about how human ancestors might have lived as well as being valuable in themselves.

During his career, Louis Leakey wrote more than 150 articles and 20 books, both popular and scientific, including two volumes of autobiography, *White African* (1937) and *By the Evidence* (1974). His many awards, often shared with his wife, included the Andrée Medal of the Swedish Geographical Society (1933), the Hubbard Medal of the National Geographic Society (1962), and the Founder's Medal of the Royal Geographic Society (1964). Leakey died of a heart attack in London on October 1, 1972.

Further Reading

Leakey, L. S. B. *By the Evidence: Memoirs, 1932–1951*. New York: Harcourt Brace, 1974.

———. *White African*. Cambridge, Mass.: Schenckman Publishing Co., 1937.

Lewin, Roger. "The Old Man of Olduvai Gorge." *Smithsonian*, October 2002.

Morell, Virginia. *Ancestral Passions: The Leakey Family and the Quest for Humankind's Beginnings*. New York: Simon & Schuster, 1996.

Willis, Delta. *The Leakey Family, Leaders in the Search for Human Origins*. New York: Facts On File, 1992.

Leakey, Mary

(1913–1996)
British/Tanzanian
Paleontologist, Anthropologist

Mary Leakey made many of the discoveries of hominid bones for which her better-known husband, LOUIS S. B. LEAKEY, became famous. After his death, she was celebrated in her own right for such finds as the earliest known fossil footprints of human ancestors walking upright. "It was Mary who really gave that team scientific validity," Gilbert Grosvenor, chairman of the National Geographic Society, once said.

Leakey was born Mary Douglas Nicol on February 6, 1913, in London, to Erskine Nicol, a landscape painter, and the former Cecilia Frere. One of Erskine Nicol's favorite places to paint was southwestern France, where beautiful Stone Age paintings had been discovered in caves. Mary loved exploring the caves and decided that she wanted to study early humans.

Mary Nicol had inherited her father's artistic talent, and at age 17 she began trying to obtain work as an illustrator on archaeological digs in Britain. Two women archaeologists, Dorothy Liddell and (later) Gertrude Caton-Thompson, hired her, and her experience with them took the place of a college education.

In 1933, Caton-Thompson invited the 20-year-old Nicol to a dinner party for Leakey, who was starting to become known for his studies of early humans, Leakey asked Nicol to make some drawings for him, and the two fell in love, even though Leakey was 10 years older than Nicol, married, and the father of two children. Leakey and Nicol went to Africa together in 1935, and Nicol gazed "spellbound" for the first time at Leakey's favorite site, Olduvai Gorge in Tanzania. They married on December 24, 1936, as soon as Leakey's divorce became final, and were, in Mary's words, "blissfully happy." They later had three sons, Jonathan, Richard, and Philip. RICHARD LEAKEY, like his parents, would become famous for studies of early hominids.

Louis and Mary Leakey became professional as well as personal partners, excavating several Stone Age sites in Kenya and Tanzania during the 1930s. Mary continued with archaeology during World War II, while Louis worked for British intelligence, and in the early postwar years, when Louis was spending part of his time as curator of the Coryndon Museum in Nairobi. The two were working together on Rusinga Island in Lake Victoria in 1948 when Mary made the couple's first big find, the skull of an apelike creature called *Proconsul africanus*. The finding of this fossil, 25 to 40 million years old, in East Africa gave weight to the idea that humans had originated on that continent, rather than in Asia as had been thought. It also made the Leakeys world famous.

In the 1950s, Mary Leakey explored a site in Tanzania where people living during the Late Stone Age, when the Sahara was a fertile valley, had painted thousands of human and animal figures on rocks, revealing such details as clothing and hairstyles. She copied some 1,600 of these paintings, an achievement she called one of the highlights of her career. The best of her drawings were published in a book called *Africa's Vanishing Art: The Rock Paintings of Tanzania* in 1983.

Leakey made another major discovery in 1959, when she and Louis were excavating together at Olduvai once more. Louis was sick and had stayed in camp on that July 17, but Mary took a walk through the oldest part of the site and spotted a piece of bone in the ground. It proved to be part of an upper jaw, complete with two large, humanlike teeth. She dashed back to the camp and burst into Louis's tent, shouting, "I've got him! I've got him!"

What Leakey eventually had was about 400 bits of bone, which she painstakingly assembled into an almost complete skull of a 1.75-million-year-old hominid, which the Leakeys named *Zinjanthropus boisei*, or "Zinj" for short. (*Zinjanthropus* means "East African Man.") The Leakeys sometimes called the creature "Nutcracker Man" because of its big teeth, which were adapted to chewing tough plant matter. Zinj was the oldest hominid found up to that time and extended the timeline of human evolution back by a million years. Its species was later reclassified as part of the genus *Australopithecus*, which many paleontologists think includes the ancestors of the human genus, *Homo*, although they agree that Zinj's particular species died out.

A few years after the Leakeys discovered Zinj, they found a skull of a new species in the human genus as well. They named it *Homo habilis*, or "handy man," because they believed it had made the many tools found nearby. They concluded that this species had lived alongside *Australopithecus boisei*.

Louis and Mary Leakey drifted apart during the late 1960s, and their marriage was over in all but name by the time Louis died in 1972. Mary continued working, and in 1978 she made what she felt was her most important discovery: three sets of fossil hominid footprints crossing a patch of hardened volcanic ash at Laetoli, Tanzania, about 30 miles south of Olduvai. The footprints were made about 3.6 million years ago, much earlier than human ancestors had been thought to be walking upright. Leakey noted that no tools were found in the area, which suggested that hominids began to walk on two feet before they started to make tools. "This new freedom of forelimbs posed a challenge," Leakey wrote in

National Geographic. "The brain expanded to meet it. And [hu]mankind was formed."

Mary Leakey received many awards for her work, including the Hubbard Medal of the National Geographic Society, the society's highest award, which she shared with Louis in 1962; the Boston Museum of Science's Bradford Washburn Award; and the Gold Medal of the Society of Women Geographers. Age and failing eyesight forced her to retire from fieldwork in 1983, but she continued to write and lecture. She died in Nairobi on December 9, 1996, at age 83. F. Clark Howell, professor emeritus of anthropology at the University of California, Berkeley, said that Mary Leakey left "an unparalleled legacy of research and integrity."

Further Reading

Leakey, Mary. *Disclosing the Past*. Garden City, N.Y.: Doubleday, 1984.

———. *Olduvai Gorge: My Search for Early Man*. New York: Collins, 1979.

"Leakey, Mary (Douglas)." *Current Biography Yearbook 1985*. New York: H. W. Wilson, 1985.

Morell, Virginia. *Ancestral Passions: The Leakey Family and the Quest for Humankind's Beginnings*. New York: Simon & Schuster, 1996.

Poynter, Margaret. *The Leakeys: Uncovering the Origins of Humankind*. Berkeley Heights, N.J.: Enslow, 1997.

Leakey, Richard

(1944–)
Kenyan
Paleontologist, Anthropologist

Before focusing on wildlife conservation and Kenyan politics, Richard Erskine Frere Leakey added to the knowledge of human evolution already enriched by his famous paleoanthropologist parents, LOUIS S. B. LEAKEY and MARY LEAKEY. In 1994, the National Geographic Society awarded him the Hubbard Medal, its highest honor. Although Leakey is respected more for his organizational ability than for personal scientific achievement and has not worked in paleoanthropology since the 1980s, *Current Biography* wrote in 1995 that he "remain[ed] perhaps the best-known living paleoanthropologist."

Born on December 19, 1944, in Nairobi, Kenya, Richard, the middle of Louis and Mary Leakey's three sons, grew up on his parents' digs. He found his first fossil, a bone of an extinct giant pig, when he was just six years old. Determined to be independent, however, he insisted that he did not want to study paleoanthropology. He preferred tracking wild animals, and while still a teenager he supplied animals to research institutions. He left high school at age 17, without graduating, to form a photo safari company.

In 1963, after hominid bones were found on one of his safaris, Leakey decided that his parents' career appealed to him after all. He finished high school and began attending a university in England but, restless again, quit after two years. He joined an expedition to Ethiopia, and on a plane flight from there to Nairobi in 1967, happened to look out the window and spotted an area below that his knowledge of geology told him would be a likely place to find fossils. A brief ground exploration convinced him that his guess was right. When he visited the National Geographic Society in Washington, D.C., with his father a year later, he startled everyone by asking for funds to conduct a dig at this site, Koobi Fora, near Lake Turkana in Kenya. Society officials granted his request, but they warned him that if he found nothing, he should never ask them for money again.

Leakey's choice of the Lake Turkana site demonstrated that he had inherited the famous "Leakey's luck" as well as his father's skill at self-promotion. Both the number (more than 400) and the quality of hominid fossils found there proved exceptionally high, and several made major changes in scientists' understanding of human ancestry. The first, discovered in 1972, was "Skull 1470," which proved that *Homo habilis*, an early human species discovered by Leakey's parents, had existed at least 2 million years ago, much earlier than had been thought.

In 1984, Leakey's group found "Turkana Boy," the skeleton of a 1.6-million-year-old adolescent male *Homo erectus*, a direct ancestor of modern humans. One of the most complete early *Homo* skeletons known, Turkana Boy provided new information about the species' appearance, showing, for instance, that *Homo erectus* males probably grew up to six feet tall. The "Black Skull," a 1985 find, belonged to an unknown hominid that lived about 2.5 million years ago, even earlier than *Homo habilis*.

The Leakey team's discoveries at Lake Turkana showed that at least three types of hominids had existed together in East Africa 2.5 million to one million years ago: *Australopithecus boisei*, the senior Leakeys' "Nutcracker Man"; *Homo habilis*, the earliest known species in the human genus; and one other, unnamed type. This indicated that human evolution was much more complex than paleoanthropologists had believed, and that complexity has generated some bitter disagreements. For instance, Richard Leakey argued vehemently with DONALD C. JOHANSON from the late 1970s to the early 1990s about whether humans were descended from *Australopithecus afarensis* (the species to which Johanson's famous "Lucy" belonged), as Johanson maintained, or evolved separately, as Leakey held. Leakey believes that the human genus is about 3 million years old, much older than some others think, and that human evolution contains at least two parallel lines of development.

Leakey ended his active involvement with the Turkana digs in 1989, when Kenya's president, Daniel arap Moi, chose him to head the country's wildlife service. Leakey had already directed the National Museums of Kenya since 1968 and built them up to world-class status, but preserving Kenya's wildlife was an even more challenging task. It was also a vital one not only from a biological standpoint but for the country's economy, which depended heavily on the tourists that wildlife brought.

At the time, widespread poaching was decimating Kenya's wild animals, especially elephants, which were hunted for their valuable ivory tusks. Leakey, a passionate conservationist, attacked poachers within Kenya and also lent his support to an international movement to ban the ivory trade. Partly because of that support, a worldwide agreement to halt the trade went into effect early 1990. (Kenya's elephant population was reported to have begun slowly rising again in the early 2000s as a result.) At the same time, Leakey tried to help the farmers who had to share land with the animals, for instance, by having more tourism revenues directed to them.

The outspoken Leakey was perhaps more successful with animals than with people. Political enemies in Kenya's Parliament accused him of mismanagement, leading him to resign his Wildlife Service post in 1994. By this time, he had also suffered health problems, including kidney failure (a transplant from his younger brother, Philip, had saved his life in 1979) and the loss of the lower part of both legs after a plane crash in 1993. His friendship with President Moi became increasingly strained as well, and in 1995 Leakey helped to found a political party called Safina ("The Ark"), which ran against Moi's party (unsuccessfully) in the 1997 elections.

To the surprise of many, Moi asked Leakey to head the Wildlife Service again in 1998 and then, in July 1999, to take over the country's civil service and clean up the corruption for which it had become infamous. Leakey accepted these jobs, but in March 2001 he resigned them. In November 2002, he became chairman of the board of the National Museums of Kenya.

In 1970, Leakey had married Meave Epps, a zoologist who had come to work at Lake Turkana the year before. (He had first married Margaret Cropper, who worked at Olduvai, in 1964, but that marriage had ended in divorce.) Meave

Leakey became interested in her husband's work and retrained as a paleoanthropologist. Much as Mary Leakey had done when L. S. B. Leakey began his years as a speaker and fundraiser, Meave took over the family's paleoanthropological site when Richard started his government career in 1989. Since 1969, she has worked for the National Museums of Kenya, the organization that her husband had helped to shape, and she headed its paleontology division from 1982 to 2001.

Meave Leakey has focused on searching for the earliest hominids. In 1994, she found fossils of a new species, *Australopithecus anamensis*, thought to have lived about 4 million years ago. She even established a new genus in 1999 with her team's discovery of a 3.5-million-year-old fossil of a species that she calls *Kenyanthropus platyops* ("Flat-Faced Man from Kenya"). She believes that humans may have descended from this species. Louise, one of Meave and Richard's two daughters, also took part in this find, and she is now training to become the third generation of the Leakey paleoanthropological dynasty.

Further Reading

Leakey, Richard. *One Life: An Autobiography*. Salem, N.H.: Salem House, 1984.

———, and Roger Lewin. *Origins Reconsidered: In Search of What Makes Us Human*. New York: Doubleday, 1992.

———, and Virginia Morell. *Wildlife Wars: My Fight to Save Africa's National Treasures*. New York: St. Martin's Press, 2001.

"Leakey, Richard." *Current Biography Yearbook 1995*. New York: H. W. Wilson, 1995.

Lederberg, Joshua

(1925–)
American
Geneticist

Joshua Lederberg's discoveries about the genetics of bacteria helped to lay the foundations for modern genetic studies and genetic engineering. He was born to Zwi H. Lederberg, a rabbi, and Esther (Goldenbaum) Lederberg, both recent emigrants from Israel, on May 23, 1925, in Montclair, New Jersey. A few years later, his family moved to the Washington Heights district of New York City, and he grew up there. He said in a 1996 interview that he knew at age six that he wanted to be a scientist and had focused on biochemistry by the time he was 12.

Lederberg obtained a bachelor's degree in zoology in 1944 from Columbia College, then began studies at Columbia Medical School. In 1946, however, he took a leave of absence to do research in biochemical genetics with Edward L. Tatum at Yale University, and he never returned to Columbia. He earned a Ph.D. in microbiology from Yale in 1948. He joined the University of Wisconsin in 1947 as an assistant professor of genetics, then became an associate professor in 1950 and a full professor in 1954, when he was only 29 years old. He organized the university's medical genetics department in 1957 and headed it for the next two years.

Lederberg also established and directed the genetics department at Stanford University Medical School in California, beginning in 1959, and was director of the university's Kennedy Laboratories for Molecular Medicine from 1962 to 1978. He was president of Rockefeller University in New York from 1978 to 1990, when he faced mandatory retirement at age 65. He has continued to work at the university as a Sackler Foundation Scholar and an emeritus professor in the laboratory of molecular genetics and informatics, where he is currently investigating how changes in the shape and structure of DNA molecules affect their susceptibility to mutation by outside agents such as chemicals.

In July 1945, while still at Columbia, Lederberg began investigating the genetics of bacteria—at a time when many scientists did not even believe that bacteria possessed genes. At Yale and the University of Wisconsin, he focused on a common bacterium called *Escherichia coli*, which lives, usually harmlessly, in the human intestine. His work helped to establish *E. coli* as a model experimental organism for genetic studies. He also developed or codeveloped many techniques that later genetic researchers would use. "I had a very definite sense [that] I was founding a new field," he has said.

Because bacteria normally reproduce by dividing, there seemed to be no way for them to exchange genes as more complex organisms do during sexual reproduction. Lederberg, however, blended two strains of *E. coli* with different characteristics in a single culture and showed that the culture eventually came to contain some bacteria that possessed characteristics, and therefore presumably genes, from both strains. By 1947, Lederberg had demonstrated the existence of bacterial genetic exchange and recombination, which he called conjugation, and showed that bacterial genes could be mapped by crossbreeding strains, much as earlier geneticists such as THOMAS HUNT MORGAN had mapped genes by crossbreeding fruit flies.

Lederberg made a second important discovery in 1952, while at the University of Wisconsin. With the help of Norton Zinder and Lederberg's first wife, the former Esther Zimmer, who was also a geneticist (they met in Tatum's laboratory, married in 1946, and divorced in 1966), he showed that viruses that infect bacterial cells, called bacteriophages, force the bacteria to copy the viruses' genes and produce new viruses. The new viruses explode out of the cells, destroying them, and go on to infect other bacteria. Lederberg found that the viruses sometimes accidentally acquire genes from the original bacteria during the copying process and carry them into the new bacteria along with their own genes, a process called transduction. The added genes change the bacteria, creating a new strain. This discovery helped to lay the groundwork for genetic engineering, in which viruses are used deliberately to transfer genes from one cell to another.

Lederberg's work in bacterial genetics won part of the Nobel Prize in physiology or medicine in 1958, when he was only 33 years old. He shared the prize with Tatum and GEORGE WELLS BEADLE, who had also done pioneering work on the genetics of simple organisms. Lederberg was elected to the U.S. National Academy of Sciences in 1957 and was one of *Time* magazine's people of the year in 1960.

In addition to doing basic research, Lederberg has worked often as a scientific consultant to government agencies. Beginning in the late 1950s, for instance, he advised the National Aeronautics and Space Administration (NASA) on such subjects as detection of life on Mars and microbial contamination of spacecraft. He has advised other agencies on topics including artificial intelligence and the use of nuclear, chemical, and biological weapons. In 1989, he won the National Medal of Science largely for his service as a government consultant. He has also lent his expertise to the World Health Organization (on biological warfare) and several biotechnology companies.

In 1968, Lederberg married Marguerite Stein, a French-born psychiatrist, and they have two children. He told interviewer Lev Pevzner in 1996, "I hope I've lived a life of science whose style will encourage younger people."

Further Reading

"Lederberg, Joshua." *Current Biography Yearbook 1959*. New York: H. W. Wilson, 1959.

National Library of Medicine. "Profiles in Science: The Joshua Lederberg Papers." Available online. URL: http://profiles.nlm.nih.gov/BB/Views/Exhibit/. Updated 2002.

Pevzner, Lev. "Interview with Prof. Lederberg, Winner of the 1958 Nobel Prize in Physiology or Medicine." 1996. Nobel Prize Internet Archive. Available online. URL: http://almaz.com/nobel/medicine/lederberg-interview.html. Posted 2000.

Leeuwenhoek, Antoni van
(1632–1723)
Dutch
Naturalist, Histologist

Antoni van Leeuwenhoek had no scientific training, yet he made single-lens microscopes that revealed sights no one else had seen, including bacteria, red blood cells, and spermatazoa (male sex cells). He was born in Delft, Holland, on October 24, 1632, to Philip van Leeuwenhoek, a basketmaker, and his wife, the former Margaretha van den Berch, who came from a family of beer brewers. Antoni's father died when he was six years old, and two years later his mother married a painter, Jacob Molijn.

Antoni van Leeuwenhoek, a cloth merchant in the Dutch town of Delft in the late 17th century, made single-lens microscopes through which he discovered bacteria and other microorganisms, spermatazoa, and many other previously hidden features of nature. *(National Library of Medicine)*

Leeuwenhoek went to Amsterdam, the country's capital, in 1648 to learn the trade of selling cloth. In 1654, he returned to Delft, where he remained for the rest of his life. He opened his own store there and married Barbara de Mei, whose father was also a cloth seller. They had several children, but only one, Maria, survived to adulthood. Barbara Leeuwenhoek died in 1666, and Leeuwenhoek married Cornelia Swalmius in 1671; she died in 1694. In addition to running his shop, Leeuwenhoek served as the chamberlain (janitor), land surveyor, and wine gauger (quality control inspector) for the Delft city government at different times in his life. The latter two jobs involved painstakingly accurate measurement, which was also an important part of Leeuwenhoek's scientific work.

Leeuwenhoek may have begun thinking about the very small when he used a magnifying glass to examine the quality of weaving in the cloth he sold. His interest was probably increased when he visited England in 1668—the only time he left Holland—and most likely saw ROBERT HOOKE's *Micrographia*. This book, a best-seller at the time, described and pictured Hooke's discoveries with a compound microscope.

Around 1671, Leeuwenhoek began making his own microscopes. Unlike Hooke's, they were simple microscopes, with a single tiny lens set in a hole in a pair of metal plates just a few inches long. In most cases a single specimen, the position of which could be adjusted by screws, was permanently mounted in front of the lens. In essence, Leeuwenhoek's strange-looking little devices were extremely powerful magnifying glasses, the best of them enlarging objects by 200 to 300 times. Using methods that he never revealed, Leeuwenhoek ground lenses that magnified more accurately than compound microscopes could do until the mid-nineteenth century. He is thought to have created about 500 microscopes, but only nine of them have survived.

Soon after Leeuwenhoek began making and using microscopes, a friend of his, physician

Regnier de Graaf, persuaded him to describe some of what he had seen through them in a letter to the Royal Society of London, one of the first and most respected organizations of scientists. The letter, dated April 28, 1673, described a fungus (mold), bees, and lice. The society responded, opening a correspondence that was to last for the rest of Leeuwenhoek's long life and encompass hundreds of letters. The society published translations of his letters in its journal, *Philosophical Transactions*, and the letters were also issued later in book form.

Leeuwenhoek's letters and the accompanying drawings, made by an artist acquaintance, opened up a microscopic world far more complex than Hooke's. The Dutchman's most important discovery was microscopic living things, the first of which he saw in the cloudy, green-streaked water of a lake and described in a letter dated September 7, 1674. He later found what he called "little animals" in rainwater, seawater, and even in whitish matter scraped from his teeth. Some of the ones from his mouth "leap[ed] about in the fluid like the fish called a jack," while others moved with a "whirling motion," he wrote. Most of these microorganisms were relatively large creatures belonging to a group now called protists, but on April 24, 1676, he described "incredibly small" microbes that later scientists agree were bacteria. No one had seen any of these creatures before, and some scientists doubted whether they really existed, but Hooke confirmed Leeuwenhoek's observations before the Royal Society in 1677.

Leeuwenhoek also used his microscope to work out the life cycles of insects, spiders, and other tiny living things. At the time, most people believed that such creatures could arise spontaneously from such things as spoiling wheat and rotten meat. Leeuwenhoek showed, however, that they were born from parents like themselves. He traced insects' development from eggs to wormlike larvae to adults. He also did experiments to learn how different species of insects lived and how parts of their bodies worked. He used what he learned about some pest insects to work out ways to control them.

The tissues and organs of human and animal bodies attracted Leeuwenhoek's attention as well. He observed the microscopic blood vessels that connect arteries to veins and saw the disc-shaped cells that give the blood of vertebrates its red color. He noted that such cells in fish, but not in mammals, included a small central body. This was the nucleus, which (later scientists would show) contains the cell's inherited information. He also discovered the sex cells of male animals, whose tadpole-like shape and vigorous swimming reminded him of his "little animals." Until Leeuwenhoek saw these spermatazoa in 1677, no one had been sure what males contributed to the process of creating offspring. He meticulously described and measured many tissues, including muscle and bone, and organs, including the eye.

In spite of Leeuwenhoek's isolation and lack of academic credentials, the Royal Society's publication of his letters made him famous. In 1680, the society made him a full member—a great honor—and the University of Louvain, in what is now Belgium, gave him a silver medal in 1716. Titled people, even kings and queens (most notably Peter the Great of Russia, during a trip to Europe in 1698), came to visit the crotchety old cloth merchant, by then living alone with his daughter, and peer through his little microscopes. He never lost the childlike enthusiasm with which he turned his microscopes on everything he could lay his hands on, no matter how humble or even disgusting it might appear to others. Just hours before his death from a lung disease on August 26, 1723, at the age of 90, he asked a friend to translate two more letters for the Royal Society.

Further Reading

Dobell, Clifford. *Antony van Leeuwenhoek and His "Little Animals."* Reprint, Mineola, N.J.: Dover Publications, 1978.

Ford, Brian J. *The Leeuwenhoek Legacy*. Reprint, Monticello, N.Y.: Lubrecht & Kramer Ltd., 1991.

Levi-Montalcini, Rita

(1909–)
Italian/American
Neurobiologist

Drawing on research begun in her bedroom during World War II, Rita Levi-Montalcini discovered a substance that makes nerves grow. She was born in Turin, Italy, on April 22, 1909. Her father, Adamo Levi, was an electrical engineer and factory owner, and her mother, the former Adele Montalcini, was a painter. Levi-Montalcini later combined her parents' last names to create her own.

When Levi-Montalcini was 20, a family friend's painful death from cancer made her decide to become a physician. She first had to overcome the objections of her father, who believed that women should not have careers. She earned her M.D. from the Turin School of Medicine with the highest honors in 1936. While still a student, she became an assistant to one of her professors, histologist Giuseppe Levi (no relation), and she continued in that capacity after graduation.

Italian Fascist leader Benito Mussolini's regime did not persecute Jews as intensely as Nazi Germany did, but in 1938 it passed a law that deprived all Jews of academic jobs. Both Levi and Levi-Montalcini therefore became unemployed. Levi-Montalcini, who had been doing experiments with chick embryos, was discouraged until a friend suggested that she continue her research at home. To do so, she set up what she called "a private laboratory *a la* Robinson Crusoe" in her bedroom in 1939, turning a small heater into an incubator for her eggs and using sharpened sewing needles to cut up the tiny embryos. She went on with this work in the Piedmont Hills, where her family moved to escape the Allied bombing of Turin in 1941.

In her makeshift laboratory, Levi-Montalcini attempted to duplicate the experiments of researcher Viktor Hamburger, who had shown that when a limb was cut from a developing embryo, nerves starting to grow into that limb died. Hamburger thought this happened because the nerve cells were no longer receiving some substance from the limb that they needed in order to mature. Levi-Montalcini found, however, that the cells did mature before they died. She suspected that the unknown substance kept the cells alive and attracted them toward the limb rather than helping them mature. She published the results of her research in a Belgian science journal.

Mussolini fell from power in July 1943, but German troops occupied Italy a month and a half later. Now, for the first time, Italian Jews were in real danger. Using assumed names, Levi-Montalcini and most of her family hid with a friend in Florence. After the Allies entered Italy in August 1944, she worked as a physician in a refugee camp for several months. She and her family returned to Turin in May 1945, and she resumed research with Giuseppe Levi.

In 1946, Levi-Montalcini was startled to receive a letter from Viktor Hamburger, who had read the report of her findings about his research. He was then at Washington University in St. Louis, Missouri, and he invited her to come to the United States and work with him for a semester. She agreed, never suspecting that the "one-semester" visit that began in 1947 would last 30 years. She became an associate professor of zoology at the university in 1951 and a full professor of neurobiology in 1958.

Levi-Montalcini's work took a new direction in 1950, when Hamburger told her about research done by a former student, Elmer Bücker. Bücker had grafted tissue from a mouse cancer onto a chick embryo and found that the tumor made nerve fibers from the embryo multiply and grow toward it, just as a grafted limb would have. Levi-Montalcini could not see why cancer tissue should have this effect, so she repeated Bücker's experiments. She found that some mouse cancers produced a quantity of nerve growth "so extraordinary that I thought I might be hallucinating" when she viewed it under the microscope. Nerves grew not only into the tumors but into nearby organs of the embryos. Levi-Montalcini noted that the nerve cells did not make contact with other cells, however, as they would have if stimulated by an extra limb. She described her discovery to the New York Academy of Sciences in 1951, but other scientists showed little interest at the time.

Levi-Montalcini decided that she would have a better chance of identifying the "nerve-growth promoting agent" if she studied its effects on nerve tissue in laboratory dishes rather than on whole embryos, which contain many substances that might affect growth and development. She therefore spent several months in Rio de Janeiro, Brazil, learning tissue culture methods from Hertha Meyer, a fellow former student of Giuseppe Levi's. At first, Levi-Montalcini had trouble making her new experiments work, but she finally found one tumor that made nerves grow out from a chick embryo ganglion, or nerve bundle, "like rays from the sun."

In January 1953, biochemist STANLEY COHEN joined Hamburger's group and set out to learn the chemical nature of Levi-Montalcini's mystery substance, which the team was beginning to call nerve growth factor (NGF). Levi-Montalcini later called the period when she worked with Cohen "the most intense and productive years of my life." Among other things, Cohen learned that NGF was a protein and that relatively large amounts of it could be obtained from the salivary glands of male mice. Levi-Montalcini showed that several kinds of nerve cells could mature only when NGF was present. The two scientists' teamwork continued until Cohen moved to Vanderbilt University in Tennessee in 1959.

Levi-Montalcini had become a U.S. citizen in 1956, but she also kept her Italian citizenship. In 1959, she decided that she wanted to spend more time with her family in Italy, so she persuaded Washington University to set up a laboratory of cellular biology in Rome. This laboratory was greatly enlarged in 1969 and became part of a new Institute of Cell Biology,

Building on research begun in her bedroom during World War II, Italian-born Rita Levi-Montalcini discovered a vital natural substance called nerve growth factor in 1950 while working at Washington University in St. Louis, Missouri. *(Archives: Washington University, St. Louis)*

run by Italy's National Council of Research. Levi-Montalcini was the institute's director, although she still spent part of her time in the United States. In 1977, she retired from Washington University and moved back to Italy permanently, and a year later she also retired from her directorship of the Roman institute.

After her retirement, Levi-Montalcini continued to do research as a guest professor at the Institute of Cell Biology. ("The moment you stop working, you are dead," she once told an interviewer.) In 1986, she found that NGF can spur the growth of brain cells as well as those from the spinal cord, which suggests that it might someday be used in a treatment for brain-damaging conditions such as Alzheimer's disease and strokes. She has also found that some cells in the immune system both produce and respond to NGF, suggesting a link between this disease-fighting system and the nervous system.

Levi-Montalcini received many awards for her work on NGF, including election to the U.S. National Academy of Sciences in 1968 and the prestigious Albert Lasker Medical Research Award in 1986. Later in 1986, she and Stanley Cohen shared the Nobel Prize in physiology or medicine. Levi-Montalcini also won the U.S. National Medal of Science in 1987.

In a brief memoir published in 2000, Levi-Montalcini attributed her success to "the absence of psychological complexes, tenacity in following the path I reputed to be right, and the habit of underestimating obstacles." She was still active in 2002, persuading the Italian government to give more money to medical research and sponsoring scholarships for women scientists.

Further Reading

Levi-Montalcini, Rita. "From Turin to Stockholm via St. Louis and Rio de Janeiro." *Science*, February 4, 2000.

———. *In Praise of Imperfection: My Life and Work.* New York: Basic Books, 1988.

———. *Saga of the Nerve Growth Factor.* River Edge, N.J.: World Scientific Publications, 1997.

McGrayne, Sharon Bertsch. *Nobel Prize Women in Science: Their Lives, Struggles, and Momentous Discoveries*. New York: Birch Lane Press, 1993.

Li, Choh Hao

(1913–1987)
Chinese/American
Biochemist

Choh Hao Li discovered and synthesized hormones made by the pituitary, a tiny structure deep in the brain that has often been called the body's master gland. Li was born in Canton, China, on April 21, 1913. His father, Kan-chi Li, was an industrialist, and his mother was the former Mew-ching Tsin.

Li obtained a bachelor's degree in chemistry from Nanjing (Nanking) University in 1933 and was an instructor there for two years, then moved to the University of California, Berkeley, for graduate studies. He earned a Ph.D. in biochemistry in 1938. He remained at the university's Berkeley and San Francisco campuses for the rest of his career, working first at Berkeley's Institute for Experimental Biology. He became an assistant professor in 1944, an associate professor in 1947, and a full professor of biochemistry and experimental endocrinology in 1950. He was director of the Hormone Research Laboratory, located first in Berkeley and then (after 1967) in San Francisco, from 1950 until his retirement in 1983. He was then in charge of the Laboratory of Molecular Endocrinology at San Francisco until his death.

Li devoted his career to hormones made by the front part of the pituitary, which govern growth, reproduction, and other vital processes. He purified growth hormone and adrenocorticotropic hormone (ACTH) from sheep and pig pituitaries in 1943. ACTH stimulates the adrenals, two tiny glands on top of the kidneys, to make other hormones. In 1956, Li and his coworkers determined the chemical composition and structure of ACTH, the first time this had been done for a hormone. ACTH proved to be a protein consisting of 39 amino acids, but Li theorized that only part of the molecule, which he called the active core, was necessary for the hormone's activity. His laboratory supported this assertion in 1960 by synthesizing a molecule only 19 amino acids long that produced most of the hormone's effects. ACTH had been found to be an effective treatment for rheumatoid arthritis and certain other medical problems, and Li's discovery opened the possibility of synthesizing a form of the hormone that would be cheaper than the natural version and also might have fewer side effects.

An even more valuable discovery was Li's isolation of human growth hormone, or somatotropin, in 1956. This hormone provided the first treatment for children who had stunted arms and legs because of a condition called pituitary dwarfism. Unlike the case with ACTH and some other hormones, the growth hormone molecule is different in different species, so humans do not respond to growth hormones from cattle, sheep, or other animals. The only hormone that could be given to the children at first, therefore, was a tiny supply obtained from human bodies after death. In 1966, however, Li worked out the chemical composition of human growth hormone, and in 1970 his team synthesized it, greatly increasing the amount available for medical use. Growth hormone was the largest protein molecule made synthetically up to that time.

Li discovered several other pituitary hormones as well, including follicle-stimulating hormone, which is involved in female reproduction; melanocyte-stimulating hormone, which is somewhat similar to ACTH; and insulin-like growth factor I. He also learned new facts about how the hormones function, for instance, finding that growth hormone stimulates a female's mammary glands (breasts) to make milk after she has given birth. In 1962, Li won the Albert Lasker Medical Research Award. He also received the Ciba Award in Endocrinology in 1947, the Francis Emory Septennial Prize of the American Academy of Arts and Sciences in 1955, and a gold medal from the Minister of Education of the Republic of China in 1958. Li married Sheng-hwai (Annie) Lu, a fellow Chinese-born student, in 1938, and they had three children. He became an American citizen in 1955 and died on November 28, 1987.

Further Reading

"Li, C(hoh) H(ao)." *Current Biography Yearbook 1963*. New York: H. W. Wilson, 1963.

Lind, James
(1716–1794)
British
Surgeon

Scottish surgeon James Lind rediscovered a way to prevent the vitamin-deficiency disease scurvy and proved the method's effectiveness through the first controlled human tests of a medical treatment. He was born in 1716 in Edinburgh and trained at the College of Surgeons in that city. He began working for the British Navy in 1739.

Scurvy caused bleeding gums, swollen joints, weakness, and often death. Beginning in the late 1400s, when sea voyages lasting several months became possible, it struck sailors so frequently that it was called "the plague of the sea." Prisoners or others on very limited diets also sometimes suffered from it. Some captains had noted that adding fresh fruits and vegetables, especially citrus fruits, to sailors' diets appeared to prevent or cure scurvy, but no one had tested this idea or applied it systematically to stop the disease.

In 1747, after six weeks at sea, 80 of the 350 sailors aboard HMS *Salisbury*, on which Lind was ship's surgeon at the time, had developed scurvy, even though their food and water were still fresh. Lind decided to make a systematic test of all the treatments for scurvy that he had heard of: cider, seawater, vinegar, oranges and lemons, dilute sulfuric acid, and a mixture of garlic, mustard, and horseradish. He chose 12 sick sailors, divided them into pairs, and gave each pair one of the six treatments. The two sailors who received the oranges and lemons were ready for duty again in six days, but all the other men were still sick after two weeks.

Lind published an account of his test, along with a recommendation that citrus fruits be routinely added to sailors' rations, in *A Treatise on the Scurvy* in 1753. Unfortunately, although individual captains, such as explorer James Cook, followed his advice, the British Admiralty did not make it a part of naval regulations until

1795. When the order finally did go into effect, scurvy all but disappeared from navy ships, and the overall rate of sickness among British sailors dropped by 50 percent. The navy later switched from lemons, the original preferred fruit, to limes, earning English sailors the nickname of "limeys." Scientists later discovered that scurvy was caused, not by a poison generated in the body by damp air as Lind had thought (he believed that the citrus fruits provided an antidote to this poison), but by a lack of vitamin C (ascorbic acid), which the fruits supply.

Three years after *A Treatise on the Scurvy* was published, Lind wrote a second book, *An Essay on the Most Effectual Means of Preserving the Health of Seamen in the Royal Navy*, in which he expanded his concern about seamen's poor diet to cover all the terrible conditions under which navy sailors lived, which often cost them their health or their lives. "The number of seamen in time of war who died of shipwreck, capture, famine, fire or sword," he wrote, "are but inconsiderable in respect of [comparison to] such as are destroyed by the ship diseases, and by the usual maladies of intemperate [tropical] climates."

Lind became the physician at the Haslar Naval Hospital in Gosport, England, in 1758. While there, he experimented with distilling saltwater to produce fresh water, supplies of which often ran out or became foul during long sea voyages. Distillation had been tried before, but the resulting water had a burnt taste that sailors disliked. Potentially harmful materials, such as soap or chalk, were often added to the water to disguise this taste, but Lind found that it went away on its own if the water was exposed to air for a short time. He worked out a simple distillation process that could be used on any ship.

Lind later wrote two other books about diseases and their prevention. *Two Papers on Fevers and Infections* (1763) described typhus and other illnesses he had observed at sea. He recommended preventing typhus by fumigating ships with smoke from wood and gunpowder. (This probably worked because it killed the lice that, later scientists would learn, spread this microbe-caused disease.) *An Essay on Diseases Incidental to Europeans in Hot Climates* (1768) described illnesses that took the lives of many soldiers and settlers in tropical British colonies. James Lind died in 1794.

Further Reading

"James Lind 1716–1794." BBC Online History: Science and Discovery. Available online. URL: www.bbc.co.uk/history/historic_figures/lind_james.shtml/. Accessed 2001.

Linnaeus, Carolus (Carl von Linné)
(1707–1778)
Swedish
Botanist, Taxonomist

Carl von Linné, who preferred to be known by the Latin form of his name, Carolus Linnaeus, created the system of naming plants and animals that biologists still use. He was born Carl Linné in South Rashult, a village near Lund in the province of Smaland in southern Sweden, on May 23, 1707. His father, Nils Linné, a Lutheran minister, loved plants and passed this enthusiasm on to his son. As a child, Carl was known as "the little botanist."

Carl's parents hoped that he, too, would become a minister, but instead he began studying medicine at the University of Lund in 1727. A year later, he transferred to the University of Uppsala, where he also continued to study botany. In time, he became a lecturer in botany at the university and won a small grant from the Uppsala Academy of Sciences to visit Lapland, a primitive area in the far northern part of Sweden. During this expedition, made in 1732, Linné traveled almost 5,000 miles, became a great admirer of the area's reindeer-herding Sami natives, and interviewed their healers to learn about their use of medicinal plants. He found

100 plant species previously unknown to the rest of Europe.

The University of Uppsala was not equipped to grant medical degrees, so Linné finished his training at Holland's University of Harderwijk in 1735. Linné had become engaged to a young woman named Sara Moraeus, and her physician father helped to pay for the young man's travels. While doing further studies in Leiden in that same year, Linné obtained the aid of another patron and published the first edition of what would become his most famous book, the *Systema Naturae* (*System of Nature*), which set out the essentials of his plant classification system. In a second book, *Genera Plantarum* (1737), Linné recommended classifying plants according to the number, shape, and arrangement of their stamens (male reproductive organs) and pistils (female organs). He returned to Sweden in 1738 and married Sara the following year; they eventually had six children.

In 1742, Linnaeus, as he was now known, became professor of botany at the University of Uppsala, a position he kept for the rest of his life. His lectures proved extremely popular, and he inspired a generation of students, such as Daniel Solander, the naturalist on Captain James Cook's first round-the-world voyage, to travel to distant lands and collect plants. An ardent Swedish nationalist, Linnaeus also tried to improve the country's economy and make it independent of foreign trade. He attempted to grow valuable tropical plants such as coffee in Sweden or to find native plants that could fulfill the same functions, but most of these efforts were unsuccessful. He practiced medicine as well, eventually becoming the private physician of Sweden's royal family.

Before Linnaeus's time, botanists had used a variety of systems for naming plants. Most plants had several names, including common names in different European languages as well as unwieldy Latin names, sometimes consisting of seven or eight words, that described the plants' characteristics. This multiplicity of terms made it difficult to bring together all the information about a particular plant. Animal names were in the same confusing state.

Linnaeus decided to create an orderly system of naming that would reflect what he saw as the orderly way in which God had arranged nature. In his naming system, each kind (species) of plant or animal has a Latin name with two parts. The first part, which begins with a capital letter, names the genus, a small group with many features in common to which the individual type belongs. The second part, beginning with a lowercase letter, designates the species. Domestic cats, for instance, are classified as *Felis* (the genus of cats, which includes lions, bobcats, and so on as well as domestic cats) *domesticus* (the species of domestic cats). Linnaeus was not the first to propose a binomial (two-name) system, but he was the first to develop it in detail and use it consistently.

Linnaeus described his system most fully in *Species Plantarum* (*Species of Plants*), published in 1753, and in the 10th edition of *Systema Naturae*, published in 1758, the latter of which applies the system to animals as well as plants. Although names in this system mention only species and genus, Linnaeus carried his classification scheme further, arranging living things in a hierarchy of nested categories. He grouped genera (plural of genus) into orders, orders into classes, and classes into kingdoms. Later biologists added two further levels: families, placed between genera and orders, and phyla, groups larger than classes. Some use the terms *kingdoms* and *domains* for still larger groups.

Linnaeus's new system caused controversy at first. Some scientists did not want to give up the old names, and others pointed out that his classification scheme sometimes grouped together types of plants or animals that did not really have much in common. Nonetheless, by the time of Linnaeus's death (probably from a stroke) in Uppsala on January 10, 1778, the system he invented had come to be almost universally accepted because, as evolutionary biologist and science historian ERNST MAYR has written, it "brought consensus and simplicity back into taxonomy [biological classification] and nomenclature [scientific naming] where there had been a threat of total chaos." Having a single, logical naming system that everyone could agree on had become a necessity because travel to distant countries was increasing and hundreds of previously unknown species were being discovered, creating an onslaught of new information that could be handled in no other way.

The acceptance of Linnaeus's system, in turn, changed the way biologists thought about nature. Most earlier thinkers had organized living things in a single "Great Chain of Being" that stretched from the simplest to the most complex types (with humans, of course, at the top), or else had classified them according to their usefulness to humans. Linnaeus's system, however, produced an arrangement more like that of a tree, with large limbs dividing into smaller and smaller branches and finally into twigs. Whether Linnaeus chose it by foresight or luck, this tree image allowed his system to survive a later revolution in biological thinking for which it was never designed—the acceptance of evolution. For most of his life, Linnaeus himself refused to believe that species could have changed over time, but his treelike structure fitted well with the theory developed a century later by CHARLES ROBERT DARWIN, in which a single ancestral species slowly subdivides into multiple new types.

In his later years, Linnaeus, called "the prince of botanists," was widely honored for his naming system and his attempts to help his country. In 1761, for instance, he was made a noble, with the right to use "von" before his name. His memory is still honored today, although the details of his system have been modified many times. Some biologists feel that a different system is now needed to reflect new information about evolution and genetics and properly incorporate the vast number of species that have been discovered since Linnaeus's time, but others think that, perhaps with further alterations, the Linnaean system will continue to serve science as well as it has for 250 years.

Further Reading

Blunt, Wilfrid, and William T. Stearn. *Linnaeus: The Compleat Naturalist*. Princeton, N.J.: Princeton University Press, 2002.

Koerner, Lisbet. *Linnaeus: Nature and Nation*. Cambridge, Mass.: Harvard University Press, 1999.

Waggoner, Ben. "Carl Linnaeus (1707–1778)." University of California, Berkeley, Museum of Paleontology. Available online. URL: www.ucmp.berkeley.edu/history/linnaeus.html. Accessed 2003.

Lister, Joseph

(1827–1912)
British
Surgeon

Before Joseph Lister recognized bacteria as the cause of wound infections and developed a way to fight them, "the man laid on the operating table in one of our surgical hospitals" was, as one mid-19th-century physician put it, "exposed to more chances of death than was the English soldier on the field of Waterloo," the famous battle at which Napoleon was defeated. Lister was born on April 5, 1827, in Upton, England, to Joseph Jackson Lister, a well-to-do Quaker wine merchant, and his wife, Isabella.

Joseph Jackson Lister's hobby was making microscopes, and his son undoubtedly looked through them, but Joseph Lister's scientific interest turned in a different direction. While still a child, he decided to become a surgeon. He trained at University College, part of the University of London, and earned his M.D. with honors in 1852. His professors then recommended that he extend his experience by visiting other universities, such as the renowned one in Edinburgh, Scotland. Lister went to Edinburgh in 1853 and remained there for seven years, working under surgeon James Syme at the city's Royal Infirmary (Hospital). He married Syme's oldest daughter, Agnes, in 1856.

Lister became Regius Professor of Surgery at the University of Glasgow, another large Scottish city, and a surgeon at Glasgow's Royal Infirmary in 1859. There, he watched almost half his patients die of "wound diseases," in which their wounds and surgical incisions turned red and oozed pus and surrounding tissue rotted away. The same thing happened at most hospitals, and other surgeons simply accepted it. Lister, however, looked for a way to fight it.

Some surgeons thought that wound diseases were caused by chemical reactions that took place when wounds were exposed to air. Lister suspected that dust was the true cause, but he had no idea why it should have this effect until 1865, when a coworker showed him a translated article by French chemist LOUIS PASTEUR. Pasteur described microscopic living things, which he called "microbes" or "germs," that could be found in such materials as soured wine—and on dust particles. Pasteur wrote that some kinds of germs produced putrefaction, or rotting, of meat, and Lister recognized that infected wounds went through a process much like putrefaction. He therefore guessed that wound diseases might be caused by microbes that entered wounds from the air.

If this was the case, Lister reasoned, he might prevent wound infections by adding a microbe-killing chemical to bandages that covered the wounds. He remembered reading a report from the Scottish city of Carlisle that described the spreading of phenol, or carbolic acid, on sewage dumped in fields near the town. The report said that this chemical had destroyed the sewage's bad smell and also prevented a cattle disease associated with the sewage. Lister suspected that carbolic acid had had these effects because it killed microbes in the sewage.

Pure carbolic acid burned the skin, but diluted solutions of it could be applied fairly safely. Later in 1865, therefore, Lister began soaking bandages in diluted carbolic acid and applying them to compound fractures, in which broken bones protruded through the skin. He reported on 11 such cases in "On a New Method of Treating Compound Fractures," an article published in the British medical journal *Lancet* in 1867. He wrote that the wounds of nine patients had healed without any sign of disease. He called his treatment antisepsis, meaning "against infection," because its purpose was to prevent wound infection.

British surgeon Joseph Lister found a way to prevent wound infections by killing microorganisms in wounds with diluted carbolic acid; he published an account of this "antiseptic" treatment in 1867. *(National Library of Medicine)*

Later in 1867, after successfully extending his technique to surgical incisions, Lister expanded his description of antisepsis into a short book called *On the Antiseptic Principle in the Practice of Surgery*. In it, he described using carbolic acid not only for treating bandages but also for cleaning surgical instruments and everything else in an operating room that might touch a patient. In the nine months since he had begun applying his antiseptic techniques on a large scale, he wrote, there had been no wound infections in his surgical wards.

Most British surgeons had simply ignored Lister's *Lancet* report, but they met his book with outright hostility. Few had heard of what Lister called "the beautiful researches of Pasteur," and fewer still accepted the idea, which Pasteur himself was also beginning to propound, that organisms too small to see could cause disease. The surgeons had no desire to wash their hands in a harsh chemical or change the coats, spattered with blood and pus, in which they performed one operation after another. Nurses, similarly, balked at carrying out Lister's time-consuming cleaning procedures or working in the cloud of foul-smelling carbolic acid mist with which he recommended spraying operating rooms.

American surgeons were no more impressed with Lister's ideas than their British colleagues. Many surgeons in Germany and France, on the other hand, recognized his techniques for the breakthrough they were and began using them almost at once. As a result, antisepsis saved many lives during the Franco-Prussian War in 1870. Lister's methods also became popular in Edinburgh, to which he returned in 1869 to take over James Syme's position as professor of clinical surgery at the University of Edinburgh when Syme retired.

In 1877, perhaps because of his determination to persuade doubters that antisepsis really worked, Lister left friendly Edinburgh to become professor of clinical surgery at King's College, London, a stronghold of "anti-Listerian" skeptics. His patient efforts, combined with the evidence of reduced death rates among patients of surgeons who carefully followed his methods and German bacteriologist ROBERT KOCH's linkage of six different kinds of bacteria to six kinds of wound infections, convinced most surgeons by the 1880s. By then, some of Lister's followers had developed better methods of killing germs in wounds, and other surgeons were keeping microbes out of wounds entirely by, for instance, boiling surgical instruments and cloths to kill microbes on them before they were used. This approach, called asepsis ("no infection"), eventually replaced antisepsis. Antisepsis and asepsis greatly expanded the range of surgery; for instance, they made abdominal surgery practical for the first time.

Even most people who had regarded Lister's ideas with disdain respected and liked him personally. Once his methods were accepted, he was almost idolized. The British government made him a baronet in 1883 and a baron in 1897 and awarded him the Order of Merit in 1902. In 1891, he became chairman of the new British Institute of Preventive Medicine, later called the Lister Institute, and he was president of the Royal Society, Britain's chief science organization, from 1895 to 1900. The American ambassador to England once said to him, "My lord, it is not a profession, it is not a nation, it is humanity itself which, with uncovered head, salutes you." Lister retired in 1893 and died in Walmer, Kent, on February 10, 1912.

Further Reading

Fisher, Joseph B. *Joseph Lister, 1827–1912*. London: Macdonald and Jane's, 1977.

McTavish, Douglas. *Joseph Lister*. New York: Franklin Watts, 1992.

Pasteur, Louis, and Joseph Lister. *Germ Theory and Its Applications to Medicine and on the Antiseptic Principle in the Practice of Surgery*. Reprint, Amherst, N.Y.: Prometheus Books, 1996.

Lorenz, Konrad

(1903–1989)
Austrian
Zoologist, Ethologist

Konrad Zacharias Lorenz, along with KARL VON FRISCH and NIKO TINBERGEN, established the new zoological field of ethology, the comparative study of animal behavior in the wild. For this pioneering work, they shared the Nobel Prize in physiology or medicine in 1973.

Lorenz was born to Adolf Lorenz, an orthopedic surgeon, and Emma (Lecher) Lorenz, also a physician, on November 7, 1903, in Vienna, Austria. He grew up in Vienna and at his family's large country home in Altenberg, where he kept pets ranging from salamanders to monkeys. At age 10, after reading about CHARLES ROBERT DARWIN's theory of evolution, he thought he wanted to study animal descent further by becoming a paleontologist. However, he was at least equally fascinated by the behavior of the living animals he raised, especially geese, ducks, and other waterbirds. "What I did for fun as a child, I now do as research," he wrote later.

Lorenz wanted to study zoology, but his father insisted that he train for medicine and sent him to Columbia University Medical School in New York in 1922. Lorenz stayed there two years, then returned to Austria. At first, he studied medicine at the University of Vienna, earning his degree in 1928, but one of his teachers interested him in comparative anatomy, which let him reconnect with zoology. A year before his graduation, he married Margarethe Gebhardt, a fellow physician, and they later had two daughters and a son. Remaining at the university as an assistant in its Anatomical Institute after obtaining his medical degree, Lorenz began studying zoology formally, and he earned a Ph.D. in the subject in 1933.

While studying comparative anatomy, Lorenz wrote later, he realized that the comparative method could be applied to behavior as well as to body structure. He also read the leading books on animal psychology, most of which focused on experiments done in laboratories, and concluded that "none of these people knew animals." He therefore decided that a new branch of science dealing with animal behavior in natural surroundings needed to be founded and that this task "was my responsibility." In 1936, he met Tinbergen, who proved to share his views, and most of the basic concepts of ethology grew out of their discussions.

By that time, Lorenz had already made one major discovery about animal behavior. It stemmed from the observation, which he had first made as a child, that a newborn duckling would follow the first large, moving, quacking thing it saw—even if that thing was a human like himself—and apparently remain convinced for the rest of its life that this creature was its mother. Lorenz first wrote about this behavior, which he called imprinting, in 1935. When he studied it systematically during the late 1930s in a colony of greylag geese, he found that goslings possessed the ability to imprint only during a short period soon after birth. He showed that imprinting occurred in many species of birds and concluded that this behavior was instinctive, or genetically programmed, rather than learned.

Indeed, Lorenz, like his fellow ethologists, believed that most behaviors were genetic, products of evolution that had developed over many generations in response to environmental pressures. This view caused controversy because behaviorism, the most popular theory of animal psychology at the time, held that most, if not all, behaviors were learned—conditioned, or programmed, by stimuli in an individual animal's environment. (Later scientists would conclude that many behaviors have both instinctive and learned components.)

Perhaps Lorenz's chief specific contribution to ethological theory, which he developed in the late 1930s, was the proposal that, although the urge to carry out many behaviors is inborn, elements of the nervous system keep the behaviors from occurring until appropriate environmental stimuli trigger or "release" them. He believed that if an animal does not encounter appropriate stimuli for a behavior, a desire to carry out the behavior will build up, like water behind a dam, until eventually the behavior is released even by inappropriate stimuli. Thus, for instance, if goslings are kept away from the appropriate stimulus for imprinting—their mother—during a crucial early stage in their lives, they will imprint on any other stimulus object that has certain basic characteristics.

Beginning in 1937, Lorenz taught comparative anatomy and animal psychology at the University of Vienna, meanwhile carrying on his research. He became head of the psychology department at Albertus University in Königsburg, Germany, in 1940. A year later, he was drafted into the German medical corps. The Russians captured him in June 1944, and he remained a prisoner of war, working in hospitals in Soviet Armenia, until February 1948, when he was allowed to return to Austria.

Lorenz initially went back to his animal studies at Altenberg, where the Austrian Academy of Sciences established a small research station for him that it called the Institute of Comparative Ethology. In 1951, Germany's Max Planck Institute offered him a larger research station in the West German town of Buldern. During his stay there, he began to concentrate on aggressive behavior in animals and humans. He moved in 1955 to Seeweisen, Bavaria, where the Max Planck Institute established the Institute for Behavioral Physiology for him and two other zoologists. After the other two died, Lorenz became sole director of the institute in 1961. He kept this position until he retired in 1973 and returned to the Institute of Comparative Ethology at Altenberg, where he founded and directed an animal sociology department.

During the postwar phase of his career, Lorenz did more writing and lecturing than research. His works included several books for the general public as well as scientific books and papers. One of the most popular was *King Solomon's Ring* (first published in English in 1952), originally intended as a children's book, in which he described how he became interested in animals and some of his experiences with them. Praised for its charm and humor, it also introduced basic ideas of ethology to the general public.

A more controversial book was *On Aggression* (first English edition 1966), in which Lorenz claimed that human aggression is an inborn behavior. He thought that such aggression had become especially destructive because humans had learned how to use artificial weapons before they developed the instinctive inhibitions against killing their own kind that powerful animals such as wolves had evolved. Many of his later books, such as *Civilized Man's Eight Deadly Sins* (1974), dealt primarily with human behavior and ethical issues such as destruction of the environment.

Lorenz's human-oriented books, especially *On Aggression*, aroused even more widespread debate than his earlier ones because many anthropologists and behavioral psychologists questioned the validity of comparing animal and human behavior. Even those who accepted such comparisons sometimes disagreed with the specific conclusions Lorenz reached. For instance, some critics felt that by calling aggression instinctive, Lorenz was defending war. Lorenz was also criticized for having written a paper in 1940 that seemed to support the Nazi ideology, although he later said he regretted this work.

In spite of these disputes, Lorenz won many awards for his research. In addition to the Nobel Prize, he was given, for example, the City of Vienna Prize (1959), the Austrian Distinction for Science and Art (1964), and UNESCO's Kalinga Prize (1970). Lorenz died on February 27, 1989.

Further Reading

Evans, Richard Isidore. *Konrad Lorenz: The Man and His Ideas*. New York: Harcourt Brace, 1975.

Lorenz, Konrad. *King Solomon's Ring*. 1952. Reprint, New York: Plume, 1997.

———. "Konrad Lorenz–Autobiography." *Les Prix Nobel 1973*. Available online. URL: www.nobel.se/medicine/laureates/1973/lorenz-autobio.html. Last updated 2001.

———. *On Aggression*. Translated by Marjorie K. Wilson. New York: Harvest Books, 1974.

"Lorenz, Konrad (Zacharias)." *Current Biography Yearbook 1977*. New York: H. W. Wilson, 1977.

Lovelock, James

(1919–)
British
Chemist, Ecologist

In addition to discoveries in disciplines ranging from chemistry to medicine and instrument science, not to mention some 60 patented inventions, James Lovelock developed one of the most controversial and influential theories of the late 20th century: the idea that Earth is a "superorganism" that constantly regulates itself to keep its environment favorable to life. Following a suggestion by author William Golding, Lovelock termed this the "Gaia theory," after the ancient Greek goddess of the Earth.

James Ephraim Lovelock was born on July 26, 1919, in Letchworth, Hertfordshire, England. Thomas A. Lovelock, his father, was an art dealer, and Nellie A. E. (March) Lovelock, his mother, was a local official. Lovelock, an only child, became attracted to nature during long walks with his father in the countryside. His father also interested him in invention by giving him an assortment of wires, batteries, a bell, and "other oddments" as a Christmas present when Lovelock was four years old.

Lovelock earned a B.Sc. degree in chemistry from Manchester University in 1941. He later obtained a Ph.D. in medicine from the London School of Hygiene and Tropical Medicine (1948) and a D.Sc. in biophysics from the University of London (1959). He married Helen Hysop in 1942, and they had four children. Hysop died in 1989, and Lovelock remarried, to Sandra J. Orchard, in 1991.

In 1941, during World War II, Lovelock began working for Britain's Medical Research Council. He remained with the council for 20 years, chiefly at the National Institute for Medical Research in London. His projects ranged from attempts to understand and prevent the spread of the common cold to studies of what happens to living tissues and organisms during freezing and thawing. "I must have gone through every single division of the institute: chemistry, biophysics, experimental biology, virology, physiology, you name it!" he said later.

When Lovelock lacked devices to identify or measure phenomena he was studying, he often invented what he needed, and these inventions, in turn, proved valuable to other scientists. The electron capture detector, which he created in 1957, was probably the most widely adopted. This tool, which picks up trace gases in the atmosphere, has what Lovelock calls "the uncanny capacity selectively to respond to substances which are important environmentally or socially." In the early 1960s, for instance, it was used to prove that residues from pesticides had spread worldwide, an important piece of data that RACHEL LOUISE CARSON cited to support her claim that pesticides were a major threat to the environment in her influential book *Silent Spring* (1962).

Lovelock's inventiveness led the U.S. National Aeronautics and Space Administration (NASA) to hire him in 1961 as a consultant on developing lunar and planetary landers. He resigned from his National Institute job and moved with his family to Texas, where he became a professor of chemistry at Baylor College of Medicine in Houston as well as working with a NASA team at the Jet Propulsion Laboratory in Pasadena, California. Lovelock remained in the United States only until 1964, but he continued consulting with NASA for several years after that. NASA used several of his inventions in its planetary exploration programs.

NASA asked Lovelock how the agency might find out whether there was life on Mars, and Lovelock concluded around 1965 that the best way would be to examine the composition of the planet's atmosphere. With Dian Hitchcock at the Jet Propulsion Laboratory, he compared the atmospheres of Mars and Venus with that of Earth, the one planet on which life was known to exist. As he had predicted, they found that the atmospheres of Mars and Venus were unchanging, in chemical equilibrium, whereas Earth's atmosphere changed constantly in small ways and yet, overall, also remained remarkably stable.

While he was doing this research, Lovelock wrote later, "the personal revelation of Gaia came quite suddenly . . . like a flash of enlightenment." He concluded that the changes in Earth's atmosphere occurred because the metabolism of living things constantly added and subtracted tiny amounts of gases and that the effect of these changes was to keep the climate and chemical composition of the atmosphere hospitable to life. In short, life on Earth "interacts so closely with the atmosphere that the atmosphere itself might be considered as an extension of life." In this self-regulation, he felt, Earth acted as if it were a single living organism.

Lovelock first presented this idea to other scientists in 1968 and 1969. Both then and later, with few exceptions (the most notable being microbiologist LYNN ALEXANDER MARGULIS, who became a strong supporter of Lovelock's theory and helped him develop it further), they were either uninterested or highly skeptical. Evolutionary biologists such as RICHARD DAWKINS were the most critical, claiming that Lovelock's ideas went against CHARLES ROBERT DARWIN's theory of evolution by natural selection. The Earth, they said, could not evolve like a living thing because it could not reproduce and therefore had neither ancestors nor descendants. It could not respond to natural selection because it had no competitors.

In contrast, when Lovelock described his theory to the public in a book called *Gaia: A New Look at Life on Earth*, in 1979, the book became a best-seller. It was especially popular among environmentalists, who liked the idea of Earth as a living thing or even a modern goddess. They also hailed Lovelock's focus on symbiosis, or cooperation and interdependence among species, as opposed to the seemingly brutal competition that Darwin's theory stresses. Lovelock has complained, however, that some of his more extreme supporters have misinterpreted his writings just as badly as his critics. He never intended to ascribe conscious planning or purpose to "Gaia"'s activities, he says, nor to personify Earth as an intelligent being, let alone a religious figure.

Since returning to Britain, Lovelock has worked independently from his country home, although he has also maintained academic connections by being a visiting professor, most recently at Green College in Oxford University. He has devoted most of his time to speaking and writing about the Gaia theory, modifying it as new tests of it are performed. His popular books on the subject include *The Ages of Gaia* (1988), *Healing Gaia: Practical Medicine for the Planet* (1991), and his autobiography, *Homage to Gaia* (2001). He considers his chief scientific field to be "geophysiology," which he defines as the systems science of the Earth.

A number of predictions that Lovelock and others have made on the basis of the Gaia theory, such as the suggestion that living things, rather than nonliving mechanisms such as volcanoes, would prove to be the chief recyclers of the element sulfur from the sea to the land, have been tested and found to be correct. The theory as a whole remains controversial, however. Some scientists think that some of "Gaia"'s activities can be explained by the mathematical theory of complexity, which explains how orderly and complicated actions can be generated by, say, a computer program built on only a few simple rules. Lovelock insists, and many of his critics agree, that even if his theory proves to be wrong in whole or in part, it is useful to science because it provides a fertile new way of looking at natural phenomena.

Awards given to Lovelock reflect the wide range of his contributions to science and his success at being what he has called an "interdisciplinary wanderer." They include the World Meteorological Association's Norbert Gerbier Prize (1988), Japan's Blue Planet Prize (1997), and a Lifetime of Discovery award by the Discovery Channel and Britain's Royal Geographic Society (2002). Queen Elizabeth made Lovelock a Commander of the British Empire in 1990.

Further Reading

"Detailed Biography of James Lovelock." Available online. URL: www.ecolo.org/lovelock/lovedeten.htm. Posted 2001.

Lovelock, James. *Gaia: A New Look at Life on Earth*. New York: Oxford University Press, 1979.

———. *Homage to Gaia: The Life of an Independent Scientist*. New York: Oxford University Press, 2001.

"Lovelock, James." *Current Biography Yearbook 1992*. New York: H. W. Wilson, 1992.

Morton, Oliver. "Is the Earth Alive?" *Discover*, October 1999.

Ludwig, Karl Friedrich Wilhelm

(1816–1895)
Hessian/German
Physiologist

Karl Ludwig developed devices for recording events in the body and discovered important facts about the way several organs work. He is considered a founder of modern physiology. He was born in Witzenhausen, Hesse (later part of Germany), on December 29, 1816, and obtained his medical degree from the University of Marburg in 1840. He taught at the Universities of Marburg (1841–49) and Zurich (1849–55), the Austrian military medical academy in Vienna (1855–65), and, finally, the University of Leipzig, where he helped to establish a new Institute of Physiology and headed it until his death.

Ludwig is best known for his inventions, which were some of the first specific laboratory tools available to the science of physiology. One was the kymograph, a rotating paper-covered drum on which an automatic stylus constantly recorded blood pressure and respiration. Invented in 1846, it is the ancestor of today's digital hospital monitors. Other Ludwig inventions include the mercurial blood pump (first described in 1859), which allowed scientists to separate oxygen and other gases from the blood, and a device to measure the rate of blood flow in veins (1867).

Ludwig also made several advances in physiology. In 1844, while at Marburg, he discovered that the membranes of tubules in the kidney act as filters to remove harmful materials from the blood. He used measurements of nitrogen in the urine to determine the rate at which the body breaks down and uses proteins, of which nitrogen is a component. He also identified two different kinds of nerves that connect to the heart and produce different effects on it. He demonstrated that mechanical forces move blood through the body and stressed that every action in living bodies can be explained by principles of physics and chemistry.

In 1865, Ludwig showed that a frog's heart could be kept alive in a laboratory dish by pumping blood or a salt solution through its blood vessels. This process, called perfusion, became the basic technique for maintaining tissues and organs outside living bodies. It was adapted in the late twentieth century to preserve organs for transplantation.

Finally, Ludwig was renowned as a teacher. His *Textbook of Physiology*, first published as two volumes in 1852 and 1856, is considered the first modern physiology textbook. Ludwig died on April 23, 1895, in Leipzig.

Further Reading

"Ludwig, Karl Friedrich Wilhelm." In *The Biographical Dictionary of Scientists: Biologists*, edited by David Abbott. New York: Peter Bedrick Books, 1983.

Luria, Salvador

(1912–1991)
Italian/American
Molecular Biologist, Microbiologist

With MAX DELBRÜCK and ALFRED DAY HERSHEY, Salvador Edward Luria "set the solid foundation on which modern molecular biology rests" (as the Nobel Prize committee put it) through his studies of the genetics of bacteria and the viruses that infect them. The three men shared the prize in physiology or medicine in 1969.

Luria was born on August 13, 1912, in Turin, Italy, to David Luria, the manager and accountant of a small printing firm, and Esther (Sacerdote) Luria. He studied medicine at Turin University, graduating with highest honors in 1935. Like RITA LEVI-MONTALCINI, who became his friend, he trained under histologist Giuseppe Levi. After finishing his medical studies, he served in the Italian army for two years and then studied radiology and physics at the University of Rome.

Also like Levi-Montalcini, Luria was Jewish and therefore lost all chance of advancement when Italy's Fascist leader, Benito Mussolini, banned Jews from academic posts in 1938. Levi-Montalcini remained with her family in Italy, but Luria emigrated to France, only to flee that country in turn when the Germans occupied it in 1940. He moved to the United States, where he remained the rest of his life, becoming a naturalized citizen in 1947.

In the United States, Luria taught and researched first at Columbia University in New York (1940–42), then at Indiana University, Bloomington (1943–50); the University of Illinois, Urbana-Champaign (1950–59); and, finally, the Massachusetts Institute of Technology (MIT), where he headed the microbiology department. At MIT, he was named the Sedgwick Professor of Biology in 1964 and made an Institute Professor in 1970. He became director of the university's new Center for Cancer Research in 1974 and held this post until 1985. He married Zella Hurwitz, a psychologist, in 1945, and they had one son.

The title of Luria's autobiography (1984), *A Slot Machine, A Broken Test Tube*, stresses the important role that he saw chance and serendipity playing in his life. For instance, a casual conversation with the passenger next to him on an Italian trolley car in 1938 introduced him to bacteriophages ("bacteria eaters"), a group of viruses that infect bacteria. The other man proved to be a scientist who was studying these organisms, and after talking to him, Luria also decided to study bacteriophages, even though very little was known about viruses of any kind at the time.

Luria continued his investigation of bacteriophages and the bacteria that they infected after he moved to the United States. He met Delbrück in 1942 at Vanderbilt University in Tennessee, where he was studying briefly on a Guggenheim Fellowship, and the two collaborated on some bacteriophage experiments and agreed to keep in touch. They and Hershey, whom they met shortly afterward, formed the nucleus of what Delbrück called the Phage Group, an informal organization of scientists working on bacteriophages that Delbrülck organized.

At the time, biologists were not sure that bacteria had genes. Luria, however, believed that they did and that, as with higher organisms, random mutations in those genes could account for differences that developed among bacterial strains. Specifically, he believed that such mutations, rather than differences among bacteriophage viruses, accounted for some bacteria's ability to resist infection by bacteriophages. He was unsure how to prove this idea, however, until another chance occurrence, in 1943—a friend of his won the jackpot on a slot machine at a University of Indiana faculty party—gave him the idea of applying the theory of probability to determine the statistical likelihood of resistance developing in bacterial colonies.

Luria reasoned that if resistance occurred in groups of bacterial colonies that grew near one another, that would support the mutation theory, because such clusters would probably be made up of the descendants of the same mutated bacteria. If resistance was distributed randomly throughout a whole culture of bacteria, however, the resistance would probably be due to a weakness in the bacteriophage rather than to changes in the bacteria. Luria found his clusters, and Delbrück supplied mathematical analysis that supported his hypothesis about them. On the basis of these experiments, the two developed a "fluctuation test" that could predict the frequency of spontaneous mutations.

The other chance event referred to in the title of Luria's autobiography, the broken test tube, occurred in 1952 at the University of Illinois. He had been studying mutant bacteria that could be infected by bacteriophages but did not permit the viruses to reproduce, as normal bacteria did. When a test tube containing some of these bacteria broke, Luria had to study a different strain instead. Comparing the two led other scientists to discover that the mutant bacteria crippled the viruses' reproduction by cutting apart their DNA with substances called restriction enzymes. These enzymes became essential tools in genetic engineering experiments.

In addition to the Nobel Prize, Luria won the Lenghi Prize from the Italian National Academy of Science (1965) and the Louisa Gross Horwitz Prize from Columbia University (1969). He wrote a respected textbook, *General Virology* (1953), and a book on genetics for a lay audience, *Life: The Unfinished Experiment* (1973), the latter of which won the National Book Award in science. Outside the laboratory, he was known for his support of socialist and pacifist causes, such as opposition to the United States war in Vietnam; he donated most of his Nobel Prize money to antiwar groups. Luria died of a heart attack in Lexington, Massachusetts, on February 6, 1991.

Further Reading

Luria, Salvador Edward. *A Slot Machine, A Broken Test Tube: An Autobiography*. New York: HarperCollins, 1984.

"Luria, Salvador (Edward)." *Current Biography Yearbook 1970*. New York: H. W. Wilson, 1970.

Lyell, Charles

(1797–1875)
British
Geologist

Although he himself refused to believe in CHARLES ROBERT DARWIN's theory of evolution at first, Charles Lyell's influential view of geology laid the groundwork on which Darwin built. Lyell was born on his well-to-do family's estate in Kinnordy, Forfarshire, Scotland, on November 14, 1797, the oldest of his parents' 10 children. The family moved to Hampshire, England, when Lyell was still a baby, and he grew up there. Walks in the countryside with his father, a lawyer and botanist also named Charles, interested him in nature.

Lyell entered Exeter College, part of Oxford University, in 1816 and studied classics, in which he earned a bachelor's degree in 1819 and an M.A. in 1821. He also took classes in geology and mathematics. He was granted the right to practice law in 1822 and began doing so in 1825, but by then geology was his chief interest. He gave up law in 1828 and became a professor of geology at King's College, London, in

Charles Lyell's *Principles of Geology,* first published in 1830, stressed that changes in the Earth had been slow and gradual; his book inspired Charles Robert Darwin and helped to lay the groundwork for Darwin's theory of evolution. *(National Library of Medicine)*

1831. He married Mary Horner, the daughter of another geologist, a year later, and she helped him with his geological research during the rest of his career.

Many scientists in the 1820s believed that the Earth had been shaped chiefly by cataclysmic, worldwide natural disasters such as floods or volcanic eruptions, which had been much more violent in ancient times than they were at present. Famed French naturalist GEORGES CUVIER was the leading supporter of this theory, which was known as catastrophism. After studying rock formations all around Europe, however, Lyell concluded that, throughout its existence, the planet had been altered only by the same physical forces, operating with the same intensity, as were active during historical times. Changes such as the rise and fall of sea levels and the creation and destruction of mountains, he said, had taken place slowly, gradually, and cyclically rather than suddenly.

Lyell described these ideas, which came to be called uniformitarianism, in a three-volume text, *Principles of Geology: An Attempt to Explain the Former Changes of the Earth's Surface by Reference to Causes Now in Operation*, first published between 1830 and 1833. Paleontologist and science historian STEPHEN JAY GOULD called this work "perhaps the most important scientific textbook ever written." Lyell kept revising it all his life, incorporating new geological discoveries; it had gone through 11 editions by the time of his death. He also wrote other books, including *The Elements of Geology* (1838), which related geology to the study of fossils, and two books about the geology of the United States (1845, 1849). Lyell's writings showed his interest in biology as well as geology; for instance, he used differences in fossil shells to trace past changes in the Earth.

Lyell did not originate uniformitarianism—another British geologist, James Hutton, had reached similar conclusions about 50 years earlier, for instance—but he presented the theory and organized the evidence that supported it better than anyone else had done. His books, although controversial, proved extremely popular and influential, inspiring other geologists to gather additional evidence that supported uniformitarianism and leading to the theory's wide acceptance. (Scientists today think that both the uniformitarians and the catastrophists were partly right.)

Charles Darwin took *Principles of Geology* on his famous voyage on the HMS *Beagle*, during which he made the observations that led to his theory of evolution by natural selection. He was impressed by Lyell's statements that the Earth was very old and probably had not been repeatedly swept clean of life by worldwide catastrophes, which meant that species of plants and animals would have had time and opportunity to alter slowly. According to evolutionary biologist and science historian ERNST MAYR, Lyell also influenced Darwin by focusing attention on the questions of how species went extinct and were replaced by others.

Darwin and Lyell eventually met and became friends. For a long time, however, Lyell refused to believe that species could have changed, as Darwin proposed, rather than simply disappearing and being replaced. Like most people of his time, Lyell thought that throughout their existence, species had remained just as God first made them. By the early 1860s, however, Lyell came to think that Darwin might be right. In 1863, he published a book on the ancestry and archaeology of ancient humans, *The Geological Evidence of the Antiquity of Man with Remarks on Theories of the Origin of Species by Variation*, which expressed both a belief in the antiquity of human beings and an acceptance of Darwin's theory.

Lyell was greatly honored for his geological writing and research. Britain's top science organization, the Royal Society, awarded him both its Copley Medal and its Royal Medal. The British government knighted him in 1848 and made him a baronet in 1864. Lyell died in London on February 22, 1875.

Further Reading

Lyell, Charles. *Life, Letters and Journals of Sir Charles Lyell*. Reprint, New York: AMS Press, 1983.

———. *Principles of Geology*. 1833. Edited by James A. Secord. Reprint (abridged), New York: Penguin, 1998.

Wool, David. "Charles Lyell, the 'Father of Geology,' as a Forerunner of Modern Ecology." *Oikos*, September 2001.

MacArthur, Robert Helmer
(1930–1972)
Canadian/American
Ecologist

Robert MacArthur played a major role in changing the field of ecology from simple descriptive data collecting into a quantitative science. His frequent collaborator, EDWARD O. WILSON, called him "the most important ecologist of his generation."

MacArthur was born in Toronto, Canada, on April 7, 1930, but he moved to the United States when he was 17 years old. He first studied mathematics at Marlboro College in Vermont, obtaining a bachelor's degree in 1951, then transferred to Brown University in Providence, Rhode Island, from which he earned a master's degree in 1953. At this point, he became interested in ecology and went to Yale to study under famed ecologist G. Evelyn Hutchinson.

For his Ph.D. project, which won the Mercer Award for the best ecology thesis of the year in 1958, MacArthur studied five species of warblers that lived together in the New England woods. A rule of evolutionary biology, the competitive exclusion principle, stated that two animal species of the same general type could not occupy the same niche, or precise role in an ecological community, in the same place at the same time. These birds (now known as MacArthur's warblers) had seemed to break the rule, but MacArthur showed that, even though they occupied the same trees, the different species lived in different parts of the trees and had somewhat different ways of hunting for food. They therefore followed the rule after all.

MacArthur taught at the University of Pennsylvania from 1958 to 1964, first as an assistant professor (1958–61) and then as an associate professor of zoology (1961–64). He specialized in population biology, which contains elements of biogeography, genetics, evolutionary biology, taxonomy, and ecology. Unlike most ecologists of the time, he wanted to use mathematics to discover patterns and extract basic principles from ecological data. He then planned to apply these patterns and principles to the creation of models that could make testable predictions about the ways relationships among species would change under various environmental conditions.

MacArthur met Wilson, an expert on ants, in 1959, and the two "hit it off immediately," Wilson has said. Wilson shared MacArthur's enthusiasm for developing ecological theories and models, and he was impressed with MacArthur's strong mathematical background, which he himself lacked. He interested

MacArthur in his own specialty, biogeography, and they came to focus on the biogeography of islands, of which Wilson had visited many during his research.

After looking over many observations from Wilson and others, MacArthur and Wilson concluded that, over time, the rate at which new species immigrate to an island tends to drop and the rate at which species already on the island become extinct (in that location) tends to rise until these changes occur at equal rates, creating a balance, or equilibrium. At that point, the total number of species on the island stays about the same, even though the list of particular species constantly changes. The number of species on an island at equilibrium is affected by the size of the island and its distance from a mainland or much larger island: Larger islands, and those closer to mainlands, have more species than smaller or more remote islands.

In late 1962, MacArthur developed a mathematical model that could predict how fast a new island of a particular size and distance from a mainland would reach equilibrium and how many species it would contain when it did so. He and Wilson tested the model by examining data from periodic surveys of the Indonesian island of Krakatau (now Rakata), where the cataclysmic eruption of a volcano in 1883 had wiped out all local life, essentially creating a "new" island that was slowly repopulated. The Krakatau data matched the model's predictions nicely. MacArthur and Wilson published their findings in a book, *The Theory of Island Biogeography*, in 1967. Among other things, it has helped ecologists understand what happens to life in wilderness areas that, as human populations spread, become "islands" surrounded by seas of human settlement.

MacArthur, who became a professor of biology at Princeton University in 1965, developed several other mathematical models and theories that have proved useful to ecologists. The index of vegetational complexity, or foliage height diversity, which he devised in 1961, allows the number of bird species in an area to be predicted when the structure of their habitat is known. Another fruitful concept was MacArthur's division of animals into r species and K species. He said that r species produce many offspring, develop quickly, and have short life spans and high mortality rates, whereas K species are larger, produce smaller numbers of offspring but invest more energy in helping them develop and survive, and have more stable populations. Rabbits are an example of an r species and lions of a K species.

Unfortunately, MacArthur's career was cut short by kidney cancer, of which he died in Princeton on November 1, 1972, when he was just 42 years old. *Geographical Ecology: Patterns in the Distribution of Species*, a textbook he had written in his last year, was published shortly after his death. It applies the theory of equilibrium island biogeography to predicting diversity in other kinds of habitats and contains his thoughts on the relationships among species diversity, competition, and the organization and structure of animal communities.

Further Reading

MacArthur, Robert H. *Geographical Ecology: Patterns in the Distribution of Species*. 1972. Reprint, Princeton, N.J.: Princeton University Press, 1984.

———, and Edward O. Wilson. *The Theory of Island Biogeography*. 1967. Reprint, Princeton, N.J.: Princeton University Press, 2001.

Quammen, David. "Life in Equilibrium." *Discover*, March 1996.

Malpighi, Marcello

(1628–1694)
Italian
Naturalist, Histologist

Like ANTONI VAN LEEUWENHOEK, who was working at the same time, Italian physician

Marcello Malpighi used homemade microscopes to discover the inner structure of living things. The son of Marc-Antonio Malpighi, he was born into a wealthy family on March 10, 1628, in Crevalcore, Italy.

Malpighi earned a degree in medicine and philosophy from the University of Bologna, near his home, in 1653. He taught logic at the university for three years, then moved to the University of Pisa to teach theoretical medicine. He returned to Bologna in 1659 and stayed at that university as a professor of medicine for most of the rest of his life, except for four years (1662–66) during which he taught medicine at the University of Messina. He married Francesca Massari in 1667. In addition to teaching, he worked as a physician throughout his career, and he moved to Rome in 1691 to become the personal physician of Pope Innocent XII. Malpighi died of a stroke in Rome in late 1694.

Seventeenth-century Italian microscopist Marcello Malpighi observed the tiny blood vessels called capillaries, which connect arteries with veins, in the lungs of a frog in 1661; he also discovered other features of microscopic anatomy. *(National Library of Medicine)*

Like Leeuwenhoek, Malpighi seems to have made his own single-lens microscopes, which were essentially very powerful magnifying glasses. He described one of his first—as well as most important—observations through them in two letters to a friend, published in 1661. Studying the thin membranes of a frog's lung, he found that they consisted of air sacs intertwined with tiny blood vessels, which he called capillaries. He watched blood flow from small arteries in the lungs into the capillaries and from there into small veins.

The capillaries were the one element that had been missing in the groundbreaking description of the circulation of the blood that British physician WILLIAM HARVEY had published in the year of Malpighi's birth. Harvey had shown that the heart pumped blood into the body and lungs through the arteries and that the blood returned to the heart through the veins. He had theorized that blood made repeated circles through the body, but, lacking a microscope, he had been unable to discover how blood traveled from arteries to veins.

Also like Leeuwenhoek, Malpighi described most of his microscopic discoveries in letters to the Royal Society of London, one of the first and best known organizations of scientists. The society invited him to write in 1667 and, after he did so in the following year, made him a foreign member. Malpighi kept up the correspondence for the rest of his life. As the society did for Leeuwenhoek, it published Malpighi's letters in its journal, the *Philosophical Transactions*. Some of his letters were later collected in books.

The capillaries were not Malpighi's only important discovery about vertebrate anatomy. He observed taste buds in the tongue and theorized that they were the endings of specialized nerves. He studied the brain and spinal cord and showed that the spinal cord was a twisted rope of nerve fibers connected to the brain. He found tubules in the kidney, still called Malpighian tubules, and showed that they were involved in the production of urine. He also made detailed studies of the development of the chick embryo in the egg, published in the 1670s, which constitute some of the first major research in embryology.

Malpighi broke ground with his studies of invertebrate anatomy as well. His 1669 book on the silkworm (the larva, or young form, of a kind of moth) was probably the most precise description of an insect's anatomy and life cycle published up to that time. Among other things, he found that the silkworm breathes through branching tubes that open onto the outside of its abdomen. He called these structures tracheae.

Finally, in the 1670s, Malpighi turned his attention to plants. He detected tubes in them that he thought, incorrectly, served the same function as insects' tracheae. He also identified stomata, the pores through which leaves absorb gases from the air, although he did not recognize their function. He described plants' outer structure and circulatory systems. With plants, as with animals, Malpighi's guesses about the meaning of what he saw were often inaccurate, but in his actual observations this early microscopist was virtually without peer.

Further Reading

Adelmann, Howard W., ed. *The Correspondence of Marcello Malpighi*. Ithaca, N.Y.: Cornell University Press, 1975.

Meli, Domenico Bertoloni, ed. *Marcello Malpighi: Anatomist and Physician*. Florence, Italy: Leo S. Olschke, 1997.

Piccolino, Marco. "Marcello Malpighi and the Difficult Birth of Modern Life Sciences." *Endeavour*, December 1999.

Malthus, Thomas Robert

(1766–1834)
British
Economist

Although Thomas Malthus was not a biologist, his writings had a significant influence on CHARLES ROBERT DARWIN's development of the theory of evolution by natural selection. Malthus was the second son among eight children in the well-to-do family of Daniel Malthus and his wife, Henrietta (Graham). He was born in Dorking, Surrey, England, on February 16, 1766.

At Jesus College in Cambridge University, Malthus studied mathematics and trained for the ministry. He became an ordained Anglican minister in 1788 and the curate at Albury in 1796. Many of his parishioners were poor, and his observation of their unhappy lives influenced his thinking about economics.

Probably Malthus's most influential writing, and the one that affected Darwin, was *Essay on the Principle of Population*, which Malthus published anonymously in 1798. In it, he wrote that human populations tend to increase geometrically (in the ratio 2, 4, 8, 16, 32 . . .), whereas food supplies increase only arithmetically (in the ratio 2, 4, 6, 8, 10 . . .). The result is that population, if left unchecked, grows much faster than food supply. Eventually, food runs out, and starvation and disease force a shrinkage in the population.

Malthus wrote his essay in protest against the common belief that population growth was desirable for a nation. Specifically, he wanted to rebut two political theorists who had proposed that human society could eventually perfect itself to the point of eliminating poverty. Malthus claimed that, even if such a utopian condition came to pass, it would ultimately backfire disastrously because of the increase in population it would produce. He was probably wise to keep his name off the essay at first (he admitted to authorship in all of its later editions), because it aroused a storm of controversy that has continued to the present day. Many commentators saw his criticism of laws and programs designed to help the poor as heartless and cruel, but his theories won praise from such respected thinkers as philosopher John Stuart Mill.

Apparently taking his own advice to marry late, Malthus wed a cousin, Harriet Eckersal, in 1804, and they had three children. A year later, he became professor of history and political economy—the first such post in England—at the East India College at Haileybury, Hertfordshire, founded by Britain's East India Company to teach its employees. He was elected to Britain's top science organization, the Royal Society, in 1819. In addition to revisions of his famous essay, which went through a total of seven editions, Malthus published a major book, *Principles of Political Economy*, in 1820, as well as numerous other writings. He was the first economist to study population trends seriously and is considered the founder of modern demography—the statistical study of human populations. Malthus died of heart failure in Bath, Somerset, on December 23, 1834.

Malthus was primarily concerned with human society, but it was most likely his mention of competition among plants and animals that drew Darwin's attention when he read Malthus's essay in 1838. Darwin had already been thinking about how and why species changed over long periods of time, and he had come to suspect that competition between species accounted for some of these changes. According to evolutionary biologist and science historian ERNST MAYR, the Malthus essay, combined with other readings and conversations with plant and animal breeders, made Darwin begin to think about competition within species as well.

Darwin's reading of Malthus led him to believe that the unavoidable competition for food among members of a species would make variations within the species very important. Changes in a species began, Darwin hypothesized, when a member of the species happened to be born with a difference that gave it an advantage in competition, increasing its chances of surviving long enough to bear offspring. Some of those offspring would be likely to inherit the difference, and, if that characteristic continued to convey an advantage in the animal's environment, the number of animals with that characteristic would slowly increase until they became a new species or replaced the old one. Darwin wrote later that his views about the role of competition in evolution were "the doctrine of Malthus applied . . . to the whole animal and vegetable kingdoms." Mathus similarly influenced ALFRED RUSSEL WALLACE, who independently developed a theory of evolution almost identical to Darwin's.

Further Reading

James, Patricia D. *Population Malthus, His Life and Times*. Boston: Routledge and Kegan Paul, 1979.

Malthus, Thomas Robert. *An Essay on the Principle of Population*. Edited by Geoffrey Gilbert. Reprint, New York: Oxford University Press, 1999.

Petersen, William. *Malthus: Founder of Modern Demography*. Reprint, New Brunswick, N.J.: Transaction Publishers, 1998.

Rickard, Suzanne. "Conversations with Malthus." *History Today*, December 1999.

Riley, Mark T. "Thomas Robert Malthus," in *Great Thinkers of the Western World Annual 1999*. New York: HarperCollins, 1999.

Margulis, Lynn Alexander

(1938–)

American

Geneticist, Evolutionary Biologist, Microbiologist

Lynn Margulis has proposed or supported several ideas that, although rejected at first, eventually produced major changes in biologists' thinking. A colleague has called her "one of the most outspoken people in biology."

Margulis, as she is known today, was born Lynn Alexander on March 5, 1938, in Chicago, the oldest of four daughters of Morris and Leona Alexander. Her father headed a company that made marker stripes for roads, and he was also a lawyer and politician. Her mother was, she says, a "glamorous housewife." In an autobiographical essay, Margulis described her child self as "passionate, hungry for knowledge, grabby of the leading roles, . . . and nature loving."

Alexander entered the University of Chicago when she was only 15. Inspired by an innovative program there, she decided to become a scientist. She also met Carl Sagan, a physics graduate student who later became famous for popular books and television shows about science. They married in 1957, just after Alexander earned her B.A., and had two sons, Dorion and Jeremy. Lynn Sagan earned an M.A. in zoology and genetics at the University of Wisconsin, Madison, in 1960 and a Ph.D. at the University of California, Berkeley, in 1963.

Scientists at the time believed that in most living things, all genetic information was carried in the nucleus of cells; the only exceptions were thought to be bacteria, which lack a nucleus, and viruses, which are not cells. While at Berkeley, however, Lynn Sagan learned that some geneticists in the early part of the century had proposed that certain other bodies within the cell, called organelles, might also contain genetic material. These scientists had suggested that the organelles were once free-living bacteria that had come to reside inside other bacteria early in evolution. Eventually, the bacteria formed a mutually beneficial relationship, or symbiosis, that became so close that they could not survive without each other; indeed, they became a single organism. This microorganism was the ancestor of cells with nuclei.

Most geneticists thought this "serial endosymbiosis theory" was ridiculous, but Sagan gathered a wealth of evidence for it from her own research and that of others. For instance, in the early 1960s, she and others found that chloroplasts, the organelles that make food in green plants, and mitochondria, the organelles that help cells use energy, both contain DNA. This DNA was more like the DNA in bacteria than like that in cell nuclei, just as the endosymbiosis theory predicted. Furthermore, both chloroplasts and mitochondria proved to resemble certain types of free-living microorganisms.

Sagan assembled her ideas into a long paper and began submitting it to scientific journals. Fifteen rejected or lost it before the *Journal of Theoretical Biology* finally printed it in 1966. By then, her "turbulent" marriage to Carl Sagan had ended in divorce, and she had just begun teaching and researching at Boston University. She married Thomas N. Margulis, a crystallographer, in 1967 and with him had two more children, Zachary and Jennifer. Although she has described this marriage as "healthier and happier" than the one with Sagan, it, too, ended in divorce, in 1980.

Margulis expanded her 1966 paper into her first book, *Origin of Eukaryotic Cells* [cells with nuclei], which was published in 1970 in spite of a letter from the National Science Foundation saying that the book's ideas were "totally unacceptable to important molecular biologists." Eleven years later, when the book was issued in a revised edition as *Symbiosis in Cell Evolution*, the ideas in it had become widely accepted. William Culbertson, professor of botany at Duke University, has said, "The reason that the symbiotic theory [of cell development] is taken seriously is Margulis." Even evolutionary biologist RICHARD DAWKINS, who has been highly critical of some of Margulis's ideas, calls her almost single-handed establishment of the endosymbiosis theory "one of the great achievements of twentieth-century evolutionary biology."

Margulis has continued to expand the endosymbiosis theory, for instance, by maintaining that microorganisms called spirochetes were the ancestors of cell organelles that provide movement and even of sensory and nerve cells. Although she has provided evidence for it, this idea is considered more controversial than the claim that mitochondria and chloroplasts once led an independent existence.

Lynn Margulis helped to guide other unpopular ideas to acceptance as well. For instance, she supported a novel classification scheme first proposed by the late Robert H. Whittaker of Cornell University. Instead of dividing living things into the traditional two kingdoms of plants and animals, Whittaker listed five kingdoms: animals, plants, fungi, protists (organisms with cell nuclei that do not belong to the first three groups), and monera (bacteria and other microorganisms without nuclei). Thanks partly to Margulis, this classification system is widely used today.

Perhaps the most fiercely debated of Margulis's stands is her support of JAMES LOVELOCK's Gaia theory. Lovelock, a British chemist, first proposed in the late 1960s that, as Margulis puts it, "life does not randomly 'adapt' to an inert environment; rather, the nonliving environment of the Earth is actively made, modulated and altered by the . . . sum of the life on the surface of the planet." For instance, processes in the bodies of living things keep the planet's surface temperature within the narrow limits that can sustain life. Some have interpreted this theory, named after the ancient Greek goddess of the Earth, to mean that the planet is, in essence, a single living organism. Margulis continues to investigate aspects of Lovelock's theory, particularly the roles that microorganisms play in maintaining conditions that support life.

Controversy did not discourage Margulis in the past, and it does so even less today, when she has the position and honors to make others take her views seriously. She was elected to the National Academy of Sciences in 1983 and joined the University of Massachusetts, Amherst, as a Distinguished University Professor, the highest faculty rank, in 1988. (She taught originally in the university's botany department but transferred "with great delight" into the department of geosciences in 1997.) She won the university's Chancellor's Medal for Distinguished Faculty in 1992 and the Distinguished Service Award of the American Institute of Biological Sciences in 1998.

Margulis's writings, both scientific and popular, about evolution and microorganisms (some coauthored with her oldest son, Dorion Sagan) have been prolific, and she has also developed science education materials for young people. Margulis and Sagan's most recent book, *Acquiring Genomes: A Theory of the Origins of Species* (2002), is as controversial as ever. It maintains that what the authors call symbiogenesis, or the inheritance of genomes acquired through symbiosis, began in ancient organisms and lies at the root of evolution and the creation of new species. Because she has constantly espoused new and interesting ideas, paleontologist Niles Eldredge has called Margulis "one of the most original and creative biologists of our time."

Further Reading

"Dr. Lynn Margulis: Microbiological Collaboration of the Gaia Hypothesis." 1996. Mountain Man Graphics. Available online. URL: www.magna.com.au/~prfbrown/gaia_lyn.html. Accessed 2003.

Mann, Charles. "Lynn Margulis: Science's Unruly Earth Mother." *Science*, April 19, 1991.

Margulis, Lynn, and Dorion Sagan. *Acquiring Genomes: A Theory of the Origins of Species*. New York: Basic Books, 2002.

"Margulis, Lynn." *Current Biography Yearbook 1992*. New York: H. W. Wilson, 1992.

Royte, Elizabeth. "Attack of the Microbiologists." *New York Times Magazine*, January 14, 1996.

Mayr, Ernst

(1904–)
German/American
Evolutionary Biologist, Taxonomist, Philosopher of Science

Ernst Walter Mayr helped to form the version of CHARLES ROBERT DARWIN's theory of evolution by natural selection that most biologists accept today. He also developed new concepts of what species are and how they arise. He is considered to be one of the 20th century's leading evolutionary biologists as well as a major historian of biology.

Mayr was born on July 5, 1904, in Kempten, Germany. Although his father, Otto Mayr, was a judge, there had been physicians in the family for four generations, and Mayr originally intended to become a medical doctor, too. From childhood, however, he had been an avid birdwatcher, and in 1923, while studying medicine at the University of Greifswald, he spotted two red-crested pochards, a rare type of duck that had not been seen in central Germany since 1846. A report of this sighting, his first scientific paper, introduced him to Erwin Stresemann, curator of birds at the University of Berlin's Natural History Museum. Stresemann persuaded Mayr to keep up his interest in zoology by working at the museum during the summers. In 1925, Mayr abandoned his medical training and switched to zoology, in which he received a Ph.D. from the University of Berlin the following year, at the age of 21. Immediately afterward he became an assistant curator at the museum, a post he continued to hold until 1932.

One of the things that had drawn Mayr to zoology was the hope of "following in the footsteps of Darwin and other great explorers of the tropics." He found his chance in 1927, when British zoologist Walter Rothschild hired him to study the birds of New Guinea, a large island off northern Australia then controlled by the Dutch. Mayr spent much of 1928 and 1929 in this remote area. He then explored the Solomon Islands, small islands in the southwestern Pacific, as part of an expedition sponsored by the American Museum of Natural History in 1930.

When the Solomons trip was over in 1931, Mayr moved to New York City to catalog the expedition's bird collection at the museum. His employment, originally intended to be temporary, lasted 20 years, during which he became a naturalized citizen of the United States. He married Margarete Simon in 1935, and they had two daughters. While at the Museum of Natural History he turned his New Guinea and Solomon Islands research into an exhaustive work, *List of New Guinea Birds* (1941). He wrote two more books on birds of the Pacific in the early 1940s.

Mayr's bird research made him begin thinking about what species are. Scientists of the time disagreed about whether species should be defined according to features of anatomy (as CAROLUS LINNAEUS had done in his classic naming system), genetics, or some other quality. Many felt that the whole concept of a species was simply a convenience of classification rather than a true reflection of nature. Mayr, however, believed that species were real, and while preparing his bird books, he worked out a new definition for the term. He wrote in a 1940 paper that "species are groups of actually or potentially interbreeding natural populations which are reproductively isolated from other such groups." Most biologists now accept this definition.

At the same time, Mayr developed a theory about speciation, or the way new species form. Unlike many geneticists, who thought that massive genetic changes could create a new species within a single generation, Mayr believed that new species evolve slowly. His Pacific island research also convinced him that they normally develop from populations separated from others of their kind by a geographical barrier, such as a newly formed lake or mountain range. Such separation forces the isolated population, which is often quite small, to breed only within itself, amplifying whatever genetic differences it possesses. If the population survives, it eventually accumulates so many differences that it will not interbreed with members of the original species even if the geographical barrier is removed. The divergent population is then a new species.

Mayr was not the first to suggest this idea of "geographic speciation," but he gathered convincing evidence for it and made it popular for the first time. He described his theory in perhaps the most important of his many books, *Systematics and the Origin of Species*, published in 1942. "That geographic speciation is the prevailing process of speciation, at least in animals, was no longer questioned after this date," he wrote later in his usual emphatic style.

Systematics and the Origin of Species marked Mayr as a major thinker in evolutionary biology. His theories, along with those of Theodosius Dobzhansky, GEORGE GAYLORD SIMPSON, and G. L. Stebbins, integrated modern genetics with classical Darwinian views and changed the field's focus from individuals to populations. Mayr showed that natural selection operated on individual genes and biochemical molecules as well as on whole organisms.

Interested in encouraging others to do research on evolutionary biology, Mayr founded the Society for the Study of Evolution in 1946. In 1953, he left the American Museum of Natural History to become the Alexander Agassiz Professor of Zoology at Harvard University. He headed Harvard's Museum of Comparative Zoology from 1961 to 1970. His teaching and writing inspired a generation of students, including STEPHEN JAY GOULD, who in the late 1990s called Mayr "the greatest living evolutionary biologist."

At Harvard, Mayr further refined his theory of speciation. He had noted during his Pacific trips that individuals living at the edge of a species's range (the area in which the species is normally found) are often somewhat different from others of their kind. He came to feel that these border areas are the most probable places for new species to form, both because populations there are more likely to become geographically isolated and because the environment is likely to be somewhat different from the one the species commonly experiences. Environmental stress is therefore added to inbreeding, further increasing differences in the isolated population. Mayr regards this theory, which he calls peripatric speciation, as his greatest achievement. He described it in *Animal Species and Evolution*, published in 1963. He and many other evolutionary biologists now believe that most new species arise under these conditions.

In his late fifties, a time when many people are starting to think about retirement, Mayr began a new career as a historian and philosopher of science. He did retire, becoming a professor emeritus, in 1975, but he continued writing full time. His exhaustive and highly acclaimed 1982 book, *The Growth of Biological Thought*, traces important controversies through the history of biology, with an emphasis on the development of evolutionary theory. Other Mayr books on the history and philosophy of biology include *Toward a New Philosophy of Biology* (1988) and *One Long Argument: Charles Darwin and the Genesis of Modern Evolutionary Thought* (1991). *What Evolution Is* (2001) explains this subject to a general audience, stressing the importance of individuals and populations rather than genes.

Mayr's many awards testify to his preeminence in 20th-century biology. They include the Balzan Prize (1984); the International Prize for Biology, also called the Japan Prize (1994); and the Crafoord Prize, from the Royal Swedish Academy of Sciences (1999). A Public Broadcasting System sketch of Mayr called these three awards "biology's triple crown." Mayr has also won the National Medal of Science (1970), the Distinguished Service Award of the American Institute of Biological Sciences (1999), and the Sarton Medal, the highest honor for science history.

Ernst Mayr told an interviewer in 1998, "I always wanted to know everything, read everything, and that included not just science but literature, the arts. I'm still very active." In 2002, at age 98, he was still hard at work.

Further Reading

Dreifus, Claudia. "An Insatiably Curious Observer Looks Back on a Life in Evolution." *New York Times*, April 16, 2002.

Mayr, Ernst. *The Growth of Biological Thought: Diversity, Evolution, and Inheritance*. 1982. Reprint, Cambridge, Mass.: Belknap Press/Harvard University Press, 1985.

———. *What Evolution Is*. New York: Basic Books, 2001.

"Mayr, Ernst." *Current Biography Yearbook 1984*. New York: H. W. Wilson, 1984.

Young, Robert. "Ernst Mayr: An Informal Chronology." Available online. URL: http://library.mcz.harvard.edu/history/ernstmayr.html. Updated 2001.

McClintock, Barbara
(1902–1992)
American
Geneticist

Working in her cornfields while other geneticists investigated molecules, Barbara McClintock was ignored for decades because her discoveries ran counter to the mainstream of genetics as her methods. In the end, though, she proved that, contrary to what almost everyone had thought, genes could move and control other genes. Organisms thus could partly shape their own evolution.

McClintock preferred her own company almost from her birth in Hartford, Connecticut, on June 16, 1902. She grew up in Brooklyn, then a somewhat rural suburb of New York City, to which her father, Thomas, a physician for Standard Oil, moved the family when she was six. She became determined to go to college, even though, as she said later, her mother, Sara, feared that a college education would make her "a strange person." Her father took her side, and in 1919 she enrolled in Cornell University in Ithaca, New York.

By the time she graduated in 1923, McClintock had decided to make the relatively new science of genetics her career. As a graduate student in the university's botany department, she studied maize, or corn. She amazed fellow geneticists by devising a way to tell a maize cell's 10 pairs of chromosomes apart, which had not been possible before. She earned her M.A. in botany in 1925 and her Ph.D. in 1927, after which Cornell hired her as an instructor.

In 1931, McClintock and another woman scientist, Harriet Creighton, performed an experiment that clearly linked physical changes in chromosomes to changes in genetic makeup, ending some scientists' lingering doubt that genes were part of chromosomes. In spite of what was now a national reputation in genetics, however, she gained no promotion at Cornell, so she left the university shortly after the paper describing this landmark research was published. For the next several years, she led an unstable academic life, surviving on grants and dividing her research time among three universities in different parts of the country.

The University of Missouri at Columbia, one of the institutions at which McClintock had done part-time research, gave her a full-time position as an assistant professor in 1936. While there, she studied changes in chromosomes and inherited characteristics made by X rays, which damage genetic material and greatly increase the number of mutations that occur in it. But this university, too, refused to treat her with the respect she felt she deserved, and she resigned in 1941. Fortunately, she then learned about the genetics laboratory at Cold Spring Harbor, on Long Island, New York. This facility, part of the Carnegie Institution of Washington, had been the first genetics laboratory established in the United States. McClintock moved to Cold Spring Harbor in 1942 and remained there for the rest of her life. She was a Distinguished Service Member of the institution from 1967 on.

The discovery in the 1940s that DNA is the carrier of most genetic information and the working out of DNA's chemical structure and method of reproduction in 1953 revolutionized genetics, turning attention away from whole organisms or even cells and toward molecules.

Geneticists saw genes, now shown to be parts of DNA molecules, as unalterable except by chance or the sort of damage that X rays produced. FRANCIS CRICK, the codiscoverer of DNA's structure, expressed what he called the "central dogma" of the new genetics by saying, "Once 'information' has passed into protein *it cannot get out again*."

McClintock meanwhile went her own way, working with her unfashionable corn and "letting the material tell" her what was happening in its genes. Beginning in 1944, contrary to Crick's later proposed central dogma, she found genes that apparently could change both their own position on a chromosome and that of certain other genes, even moving from one chromosome to another. This movement, which she called transposition, appeared to be a controlled rather than a random process. Furthermore, if a transposed gene landed next to another gene, it could turn that gene on (make it active, or capable of making the protein for which it carries the coded instructions) if it had been off, or vice versa. Genes that could control their own activity and that of other genes had not been recognized before. McClintock suspected that such genes and their movement played a vital part in organisms' development before birth.

Even more remarkable, some controlling genes appeared able to increase the rate at which mutations occur in the cell. McClintock theorized that these genes might become active when an organism finds itself in a stressful environment. Increasing the mutation rate increases the chances of a mutation that might help the organism's offspring survive. If a gene that increases the mutation rate could be turned on by something in the environment, then organisms and their environment could affect their own evolution, something no one had thought possible.

McClintock attempted to explain her findings at genetics meetings in the early 1950s, but other scientists met her presentations with blank stares or even laughter. She offered ample evidence for her claims, but her conclusions were too different from the prevailing view to be accepted. After a while, she stopped trying to communicate her research, and most other geneticists forgot, or never learned, who she was; one called her "just an old bag who'd been hanging around Cold Spring Harbor for years." She did not let rejection stop her work, however. "If you know you're right, you don't care," she said later.

McClintock's work began to be recognized in the late 1960s. For instance, in 1967, the same year in which she officially retired (in reality, her work schedule continued unchanged), she received the Kimber Genetics Award from the National Academy of Sciences. She was awarded the National Medal of Science in 1970. Only in the late 1970s, however, did other geneticists' work begin to support hers in a major way. Researchers found transposable elements, or "jumping genes" as they became popularly known, in fruit flies and other organisms, including humans. The idea that some genes could control others was also proved.

The trickle of honors became a flood in the late 1970s and early 1980s. McClintock won eight awards in 1981 alone, the three most important of which—the MacArthur Laureate Award, the Lasker Award, and Israel's Wolf Prize—came in a single week. Then, in 1983, when she was 81 years old, she won the greatest scientific award of all, the Nobel Prize. She was the first woman to earn an unshared Nobel in physiology or medicine and only the second woman (after Marie Curie) to win an unshared Nobel in any category.

These honors and their attendant publicity irritated McClintock more than they pleased her. She complained, "At my age I should be allowed to . . . have my fun," which meant doing her research in peace. McClintock continued to have her scientific "fun" among the corn plants almost until her death on September 2, 1992, just a few months after her 90th birthday.

Further Reading

Comfort, Nathaniel C. *The Tangled Field: Barbara McClintock's Search for the Patterns of Genetic Control*. Cambridge, Mass.: Harvard University Press, 2001.

Keller, Evelyn Fox. *A Feeling for the Organism: The Life and Work of Barbara McClintock*. San Francisco: W. H. Freeman, 1983.

Kittredge, Mary. *Barbara McClintock*. New York: Chelsea House, 1991.

"McClintock, Barbara." *Current Biography Yearbook 1984*. New York: H. W. Wilson, 1984.

Mechnikov, Ilya Ilyich
(Elie Metchnikoff)
(1845–1916)
Russian
Immunologist, Zoologist, Embryologist

Russian biologist Ilya Mechnikov discovered one of the main methods by which the body protects itself against disease. He was born in Ivanovka, the Ukraine, on May 16, 1845, to landowner Ilya Mechnikov, who was also an officer of the Imperial Guard, and his wife, Emilia Nevahovna, daughter of a well-known Jewish writer. He was interested in natural history from childhood, and his first scientific paper was published when he was only 18. He attended the University of Kharkov, finishing its four-year zoology program in two years. After his graduation in 1864, he studied further at the Universities of Giessen, Göttingen, and Munich in what is now Germany.

In 1865, while at Giessen, Mechnikov discovered that certain cells in a flatworm could crawl independently through the worm's body. He watched these cells, which resembled microorganisms called amoebas, surround food particles that the worm swallowed. They appeared to digest the food, then distribute its nutrients to the rest of the animal's body. He would later find similar cells in a variety of living things.

Mechnikov returned to Russia in 1867 and began teaching zoology, first at the University of Odessa and then at the University of St. Petersburg. He also did extensive research in comparative embryology, especially of insects and marine invertebrates. This work won the von Baer Prize and helped Mechnikov earn his master's degree in 1867 and his doctorate in zoology from St. Petersburg in 1868.

The success of Mechnikov's early career soon gave way to tragedy. In 1870, he married Ludmilla Feodorovna, who was so ill with tuberculosis that she had to be carried to her wedding. Soon afterward, the couple returned to Odessa, where Mechnikov became a professor of zoology and comparative anatomy at the university. In spite of Mechnikov's constant care, Ludmilla died in 1873. Depressed over her death, conflict at the university, and his own failing health and eyesight, Mechnikov tried to kill himself with an overdose of opium, but his attempt failed.

Mechnikov married again in 1875, to a 15-year-old girl named Olga Belokopitova. When she caught typhoid fever, a serious infectious disease, in 1880, he attempted suicide once more. This time, he tried to make his death useful to science by injecting himself with blood from a person with relapsing fever, another serious microbe-caused illness, in order to find out whether the disease could be transmitted through blood. He did catch the fever, but he recovered, as did Olga.

In 1882, Mechnikov resigned from the Odessa faculty and traveled with Olga and her younger brothers and sisters to Messina, on the Italian island of Sicily. There, he set up his own laboratory and continued his studies of marine animals, including the larvae of starfish. These tiny creatures were as clear as glass, so he could see everything that happened inside them. He found that they, like his flatworms, contained unusual cells that wandered through their bodies. He watched these cells devour the dead and dying cells that were produced when the larvae changed into adult starfish.

While Mechnikov was studying the larvae one day in December 1882, he wrote later, "a new thought suddenly flashed across my brain"—the idea that the moving cells might "serve in the defense of the organism against intruders." To test this theory, he picked thorns from a rose bush in his garden and stuck them into the larvae. "I was too excited to sleep that night," he recalled. The next day, he examined the larvae under his microscope and saw that, as he had expected, the free-moving cells were clustered around the wounds.

The crawling cells in the starfish larvae reminded Mechnikov of whitish cells called leukocytes ("milky cells") that he and others had seen in the blood of humans and other vertebrates. These cells, too, congregated around wounds, especially if the wounds became infected with bacteria. Mechnikov now suspected that the cells defended the body by devouring bacteria. He explained his idea to Carl Claus, an Austrian zoology professor, and Claus suggested calling the cells phagocytes, Greek for "eater cells."

Mechnikov presented a paper describing his theory about phagocytes in Odessa in 1883. Most of his audience was skeptical, so he began to look for other organisms in which he could demonstrate a cellular defense system. He found that tiny shrimplike creatures called Daphnia, or water fleas, possessed phagocytes and were sometimes infected with a fungus. When he exposed Daphnia (which, like the starfish larvae, were transparent) to the fungus, he saw their phagocytes attack and apparently digest the invader's threadlike filaments. Mechnikov described Daphnia as a model of infection and defense in a journal article published in 1884.

Meanwhile, renowned French chemist LOUIS PASTEUR had developed vaccines to protect people and animals against several infectious diseases, and the Russian government set up a bacteriological institute in Odessa in 1886 to manufacture similar vaccines. Mechnikov was asked to run the institute, but when its products failed to live up to people's expectations, critics blamed him because he lacked a medical degree. He went to Paris in 1888 to ask Pasteur for advice about making the vaccines, and the aging scientist not only welcomed him but offered him his own laboratory at Pasteur's new research facility in Paris, the Pasteur Institute. Mechnikov was delighted to accept, and he stayed at the institute for the rest of his life, becoming its director in 1895.

During his years at the Pasteur Institute, Mechnikov, increasingly known by the French spelling of his name, Elie Metchnikoff, studied subjects ranging from the venereal disease syphilis (which he and French researcher Émile Roux succeeded in giving to apes in 1903, thereby providing an animal model for research on the disease) to the control of agricultural insect pests. His chief interest, however, continued to be the immune system and, especially, phagocytes. He confirmed that phagocytes in vertebrate blood devoured invading microorganisms, and he became convinced that these cells, even though they were only one of several types of leukocytes, were the body's chief form of defense. He described this theory in *L'Immunité dans les Malades Infectieuses* (*Immunity in Infectious Diseases*), published in 1901.

Mechnikov's emphasis on phagocytes was controversial because researchers such as the Germans EMIL VON BEHRING and PAUL EHRLICH had found that the body also defends itself by making chemicals called antisera or antitoxins, which destroy poisons that disease-causing microbes make. These scientists and their supporters thought that such liquid substances (made, later scientists learned, by other white cells) rather than phagocytes were the most important defense against disease. The two camps disputed their theories vigorously; for instance, Roux described the bushy-bearded

Mechnikov at a scientific meeting in 1891, "arguing with [his] adversaries, [his] face red, [his] eyes burning, [his] hair dishevelled. . . . appear[ing] to be the Demon of Science." Both groups in fact were partly right, which the Swedish Royal Academy of Science apparently recognized when they awarded the 1908 Nobel Prize in physiology or medicine jointly to Mechnikov and Ehrlich. Mechnikov's research also won the Copley Medal from Britain's top science organization, the Royal Society.

Mechnikov explored the role of phagocytes in inflammation, a process in which tissue becomes swollen, red, and sometimes filled with pus. He showed that inflamed tissue, which can appear in numerous diseases, is a battlefield in which phagocytes (and, other scientists later learned, many other immune system cells and biochemicals) are attacking microbes or other materials that they identify as not belonging to the body. Pus is made up of cells that have been killed in the fight. Mechnikov suggested that harmful inflammation sometimes occurs when phagocytes mistakenly attack the body's own tissues, a theory later found to explain diseases such as rheumatoid arthritis.

Mechnikov also believed that phagocytes play a part in aging, the subject that occupied the last 15 years of his life. Indeed, he is considered to have founded the study of aging, a field called gerontology. He concluded that certain kinds of bacteria in the digestive system produce toxins that cause aging and that eating foods such as yogurt, which contain other bacteria that make lactic acid, would hamper the first kind of bacteria and thereby prolong life. The treatment failed to extend his own life, however. After a series of heart attacks, augmented by grief over World War I, he died in Paris on July 16, 1916, at the age of 71. Although later research showed that Mechnikov had considerably overestimated the importance of phagocytes, his studies of these "eater cells" encouraged other scientists to focus on the immune system and helped to create the field of immunology.

Further Reading

Bibel, Debra Jan. "Centennial of the Rise of Cellular Immunology: Metchnikoff's Discovery at Messina." *ASM News*, December 1982. Available online. URL: www.asmusa.org/mbrsrc/archive/pdfs/481282p558.pdf.

de Kruif, Paul. *Microbe Hunters*. Reprint, San Diego, Calif.: Harvest Books, 1996.

"Ilya Ilyich Mechnikov—Biography." *Nobel Lectures, Physiology or Medicine 1901–1921*. Available online. URL: www.nobel.se/medicine/laureates/1908/mechnikov-bio.html. Last updated 2001.

Metchnikoff, Olga. *Life of Elie Metchnikoff*. Reprint, Manchester, N.H.: Ayer Co., 1972.

Tauber, Alfred, et al. *Metchnikoff and the Origins of Immunology*. New York: Oxford University Press, 1991.

Medawar, Peter Brian

(1915–1987)
British
Immunologist, Zoologist

Peter Brian Medawar proved Australian immunologist FRANK MACFARLANE BURNET's theories about acquired immunological tolerance, helping to pave the way for successful organ and tissue transplants. Medawar and Burnet shared the Nobel Prize in physiology or medicine in 1960.

Medawar, the son of Lebanese-born businessman Nicholas Medawar and his wife, Edith (Dowling) Medawar, was born on February 28, 1915, in Rio de Janeiro, Brazil. Shortly afterward, his parents returned to Britain, their home country, and Peter grew up there. He attended Marlborough College and then Magdalen College, the latter of which is part of Oxford University. He taught and did research at Oxford (Magdalen College and St. John's College) from 1935 to 1947, earning the Edward Chapman

Research Prize in 1938. He obtained a degree in zoology in 1939. He married Jean Taylor in 1937, and they later had four children.

During World War II, when he worked for Britain's Medical Research Council in addition to teaching at Oxford, Medawar developed a way to use fibrinogen, a natural substance that takes part in blood clotting, to make a glue that helped to keep skin grafts in place and reunite the ends of cut nerves. This innovation solved some of the surgical problems involved in transplanting tissues, but Medawar nonetheless noticed that grafts from unrelated donors, unlike grafts taken from a different part of a patient's own body, almost always failed to remain attached and grow. He was one of the first researchers to suggest that the reason for this lay in the immune system, which attacks any cells that do not belong to the body. Usually, such attacks aid survival, but if the "foreign" cells belong to transplanted tissue that an injured body needs, the immune system's reaction becomes harmful.

Medawar moved to the University of Birmingham as Mason Professor of Zoology in 1947 and continued to study immunology. Scientists knew by then that the immune system bases its reactions on antigens, substances that appear on the surface of cells. A person's immune system will attack any cells carrying antigens not found on cells in the person's own body. Researchers disagreed, however, about when and how the immune system "learns" which antigens to classify as belonging to the body. Some thought this information was inherited, but Burnet—drawing on some of Medawar's earlier experiments—proposed that the determination of which antigens to tolerate and which to reject occurred some time during development before birth. He suggested that if an animal were given cells from another animal before this determination occurred, the first animal's immune system would accept grafts from the second animal when both were adults.

Burnet's ideas echoed conclusions that Medawar had drawn from his own research on cattle at Birmingham. At University College, part of the University of London, where he had become Jodrell Professor of Zoology and Comparative Anatomy in 1951, Medawar tested Burnet's prediction, using two unrelated strains of mice. He injected cells from strain B mice into strain A mice before the latter's birth. After the mice were born, he tested them and found no sign of immune response to strain B antigens. He then grafted tissue from strain B mice onto strain A mice, some of which had received the earlier cell injections and some of which had not. The grafts grew on the mice that had received the injections, but the immune systems of the untreated mice destroyed the grafts.

Medawar's experiments, which he described in 1953, showed that Burnet was correct: Immunological tolerance, which includes the ability to accept grafts, is not inborn but rather is acquired before birth. This discovery did not in itself show a way to make transplants practical, but it led to a better understanding of the way in which the immune system reacts to transplanted tissue. By showing that the system could be influenced during an organism's life, it also gave hope that some future alteration might make transplants possible. (Certain drugs are used for this purpose today.) Medawar's research indicated, furthermore, that the antigens of transplant donors and recipients should be matched as closely as possible.

Medawar became head of the National Institute for Medical Research in London in 1962. He retired from this position in 1975, then became professor of experimental medicine at the Royal Institution in 1977. He remained active in spite of a series of severe strokes that began when he was only 54 years old. In addition to doing further research on the immune system, he wrote seven popular books on ethics, human society, and the philosophy of science, including *The Future of Man* (1960), *Advice to a*

Young Scientist (1979), and an autobiography, *Memoir of a Thinking Radish* (1986).

Medawar received many honors in addition to the Nobel Prize, including the Royal Medal of Britain's chief science organization, the Royal Society, in 1959. He was made a Commander of the British Empire in 1958 and was knighted in 1965. Additional strokes severely incapacitated him in the 1980s, and one finally killed him on October 2, 1987.

Further Reading

Medawar, Jean S. *A Very Decided Preference: Life with Peter Medawar.* New York: W. W. Norton, 1990.

Medawar, Peter Brian. *Memoir of a Thinking Radish: An Autobiography.* New York: Oxford University Press, 1986.

"Medawar, Peter Brian." *Current Biography Yearbook 1961.* New York: H. W. Wilson, 1961.

Gregor Mendel, an Augustinian monk in what is now the Czech Republic, discovered the basic laws of heredity in the early 1860s by growing thousands of pea plants in the monastery garden. *(National Library of Medicine)*

Mendel, Gregor

(1822–1884)
Austro-Hungarian
Botanist, Geneticist

Johann Mendel, who took the name Gregor when he became an Augustinian monk, performed experiments in his monastery garden that led to the founding of genetics. Mendel was born to Anton and Rosine Mendel, peasant farmers and gardeners, on July 22, 1884. His birthplace, the village of Heinzendorf (now Hyncice), was then part of the Moravia region of the Austro-Hungarian Empire; it now belongs to the Czech Republic.

Mendel learned about plant breeding from his father and other relatives. His obvious intelligence led his family and the local priest to obtain secondary schooling for him, and for a while, with financial help from his sister, he attended Olmütz University. Money and health problems forced him to leave, however. Finally, in 1843, when he was 21 years old, he joined the Augustinians, an order of monks whose main work was teaching, in order to obtain an education and free himself from financial pressures, even though he felt no religious vocation. He became a member of the order's St. Thomas Monastery at Brünn (now Brno).

Working as a substitute science teacher in the local secondary school, Mendel proved popular with his students, but he failed the examination required for a regular teaching position. In 1851, therefore, the Augustinians sent him to the University of Vienna for two years. Although he learned such subjects as chemistry, botany, and plant physiology there, he failed the teaching examination again in 1856. After that, although he continued teaching as a substitute, the abbot, or head of the monastery, increasingly let him follow his interests by working in the monastery garden.

Hoping to find ways to improve crop plants, Mendel began an eight-year breeding study of common (edible) garden peas in 1856. Many naturalists and breeders of plants and animals had tried to discover how individual characteristics, or traits, are passed from parents to offspring, but Mendel's approach differed from those of most of the others. Instead of studying his plants as a whole, he focused on seven specific, easily observable characteristics, such as plant height and the shape and color of seeds (peas). He also observed far more plants than other researchers had—about 28,000—and he followed some family lines through seven generations. He carefully tabulated how many offspring from each mating showed each form of the traits he studied, and he used mathematics to discover patterns in the data he collected. He was one of the first to use statistics, the analysis of numerical information, in science.

Each of Mendel's chosen characteristics existed in two definite forms. Some plants had smooth seeds, for instance, while others had wrinkled ones. He began by mating a plant with one form of a characteristic and a plant with the other form. He found that all their offspring, called hybrids, showed only one form of the trait; none showed the other form or a form intermediate between those of the two parents. When he created another generation by self-fertilization (using male and female sex cells from the same plant), however, some of the offspring showed the "missing" form. Over many breedings, he found that, on average, three out of every four second-generation plants showed the form that had appeared in the first-generation hybrids, while the fourth plant showed the form that had existed in one of the parents but had seemingly vanished in the hybrids.

After analyzing the results of his experiments, Mendel concluded that traits were controlled by "factors" somehow passed from parents to offspring. Each offspring inherited one factor for a particular trait from its mother and one from its father. Mendel made no guesses about what physical form these factors might take.

Some factors had more powerful effects than others, Mendel thought. If an offspring inherited factors for the same form of a trait from both parents, the offspring would show and pass on that form of the trait. If an offspring inherited a factor for a different form of a trait from each parent, however—say, a factor for smooth seeds from its father and a factor for wrinkled seeds from its mother—it would show the form of the trait specified by the factor that was stronger, or dominant. Nonetheless, it would still contain both factors. It would pass on the dominant factor to half of its own offspring and the weaker, or recessive, factor, to the other half. If hybrid plants (those that contained factors for different forms of a characteristic) were then self-fertilized, an average of three out of four of their offspring—one containing two dominant factors and two containing one dominant and one recessive factor—would show the dominant form of the trait. The fourth plant, having inherited two recessive factors, would show the recessive form of the trait.

Mendel worked out four laws that seemed to govern the breeding of his plants: The law of segregation stated that each offspring inherits only one factor for a characteristic from each parent, even though each parent possesses two such factors; the factors somehow separate during the formation of the plant's sex cells. The law of independent assortment said that each characteristic is inherited separately, unaffected by others. The other laws described the behavior of the dominant and recessive factors and stated that differences in factors are responsible for differences in traits.

Mendel presented his ideas in lectures to the Brünn Natural Sciences Society in February and March 1865 and published them in a paper called "Experiments in Plant Hybridization," which appeared in the society's journal in 1866. Most biologists either did not read the paper or paid it no attention. Mendel was elected abbot of his monastery two years later, and after that he had little time for gardening.

Mendel died in 1884, his work still unknown. In 1900, however, scientists from three different countries—Dutchman HUGO DE VRIES, German Carl Correns, and Austrian Erich Tschermak von Saysenegg—independently rediscovered Mendel's laws through their own breeding experiments and, when examining the scientific literature before reporting their results, found the monk's original paper as well. Their published acknowledgments, along with the support of British zoologist WILLIAM BATESON, finally brought Mendel's discoveries to the attention of the scientific world. The new field of genetics was the result.

Although Mendel had designed his experiments primarily with plant breeding in mind, he had foreseen that they might have wider application. He knew about CHARLES ROBERT DARWIN's theory of evolution by natural selection and wrote in his paper that his research might offer "the solution of a question the importance of which cannot be overestimated in connection with the history of the evolution of organic forms [living things]." The scientists who rediscovered his work realized that he had filled in an important gap in Darwin's theory by providing a mechanism by which traits favored by natural selection could be passed on to future generations. Although later work showed that Mendel's laws did not always hold—traits are not always inherited independently of each other, for instance—his work nonetheless became an essential part of biologists' understanding of how living things develop and change.

Further Reading

Blumberg, Roger B. "Mendelweb." Available online. URL: www.netspace.org/MendelWeb/. Accessed 2001.

Edelson, Edward. *Gregor Mendel and the Roots of Genetics*. New York: Oxford University Press, 1999.

Henig, Robin Marantz. *The Monk in the Garden: The Lost and Found Genius of Gregor Mendel, the Father of Genetics*. Boston: Houghton Mifflin, 2000.

Orel, Vitezslav. *Gregor Mendel, the First Geneticist*. New York: Oxford University Press, 1996.

Meyerhof, Otto Fritz

(1884–1951)
German
Biochemist

In a biographical article on Otto Meyerhof, David M. States claims that the research done by Meyerhof and his coworkers in Germany in the 1920s and 1930s "pioneered our understanding of how energy is biochemically transformed, stored and released for work in the cell." Meyerhof played a major part in working out the chemical processes that take place in muscle and helped to found the field of bioenergetics. He was one of the first researchers to combine physics with chemistry and cell physiology.

Meyerhof was born to Felix Meyerhof, a Jewish merchant, and Bettina May Meyerhof in Hannover, Germany, on April 12, 1884. His parents moved to Berlin when he was a child, and he grew up there. He trained as a physician at the Universities of Freiburg, Berlin, Strasbourg, and Heidelberg, obtaining his medical degree from the University of Heidelberg in 1909. His first interest was psychology, but while working as an assistant at the Heidelberg Clinic between 1909 and 1912 he met biochemist Otto Warburg, who persuaded him to specialize in that field. During his career in Germany, he first researched and taught in the physiology department of the University of Kiel (1913–24), becoming a professor in 1918. He was the director of physiology at the Kaiser Wilhelm Institute for Biology in Berlin from 1924 to 1929, then directed the new Kaiser Wilhelm Institute for

Medical Research in Heidelberg from 1929 to 1938. He married Hedwig Schallenberg, a painter, in 1914, and they had three children.

Meyerhof's first research was on how cells use oxygen. He studied the oxidation process, or respiration, in plantlike single-celled organisms called yeasts and compared it with another major chemical process, fermentation, which yeasts also carry out. He discovered certain substances, called coenzymes, that take part in both respiration and fermentation. Both of these essential processes involve the breakdown of substances to release energy used for cellular activities, and Meyerhof went on to show that they have many features in common.

Meyerhof studied respiration in frog muscles as well as in yeasts, and this research led him to examine other energy changes in muscle. At Kiel in 1919, he and his coworkers showed a relationship between frog muscles' breakdown of glycogen, a carbohydrate, and their production of a chemical called lactic acid. Lactic acid, which accumulates in muscle after repeated contractions, was thought to be a mere waste product, but Meyerhof showed that its production by glycogen breakdown in the absence of oxygen, a process he called glycolysis, was actually part of a very important cycle. In this cycle, about a quarter of the lactic acid produced is oxidized to release energy, and this energy, in turn, is used to turn the rest of the lactic acid back into glycogen to be used again.

Meyerhof shared the Nobel Prize in physiology or medicine in 1922 with British physiologist Archibald Vivian Hill, who had also studied energy metabolism and heat production in muscles. Meyerhof and Hill, who had worked together at times, both demonstrated that energy in cells is transformed through cyclical processes.

Chemical Dynamics of Life Phenomena, considered to be Meyerhof's most important book, was published in 1924 and described his discoveries about cells' use of energy up to that point. He went on to isolate the enzymes involved in glycolysis in 1925 and shortly afterward recreated the whole complex process in a test tube, without living cells, which made it much easier to study. Glycolysis, which Meyerhof and a competing German researcher, Gustav Embden, described in detail in the late 1920s and early 1930s, came to be known as the Embden-Meyerhof pathway.

Research in Meyerhof's laboratory took a new direction in 1929, when Kurt Lohman, a member of the laboratory, discovered a molecule called adenosine triphosphate (ATP). Meyerhof's group and others eventually learned that energy produced by the breakdown of carbohydrates is stored in ATP, which States calls the "universal energy donor." The energy, released when the phosphate bonds in the ATP molecule are broken, is used to power processes such as muscle contraction and transmission of messages through the nervous system. In the early 1930s, Meyerhof came to realize that ATP and other phosphorus-containing compounds were more important than lactic acid, on which he had previously focused. Indeed, his laboratory showed that the chief function of the lactic acid cycle is to take part in the formation of ATP.

Meyerhof, as a Jew, found himself in increasing danger after the National Socialists seized control of Germany in 1933. He was reluctant to leave his research, but in 1938 he and his family finally fled to France. They had to leave secretly, abandoning all of Meyerhof's books and papers. For two years, Meyerhof was director of research at the Institute of Physicochemical Biology in Paris, but he was forced out of his home again when the Nazis occupied France in 1940. The Meyerhofs crossed the Pyrenees into Spain, then went to Portugal and, finally, the United States.

Meyerhof joined the University of Pennsylvania School of Medicine, which had created a physiological chemistry department just for him. He remained there for the rest of his life, becoming a naturalized citizen in 1946. At the university, working with Hill and others, he further refined his studies of the glycogen cycle. He died of a heart attack in Philadelphia on October 6, 1951.

Further Reading

States, David M. "Otto Meyerhof and the Physiology Department: The Birth of Modern Biochemistry." Max Planck Institute for Medical Research History. Available online. URL: http://sun0.mpimf-heidelberg.mpg.de/History/Meyerhof.html. Accessed 2001.

Miller, Stanley Lloyd

(1930–)
American
Chemist

While still a graduate student, Stanley Miller became famous for an experiment in which he duplicated conditions believed to have existed on the primitive Earth. His experiment produced amino acids, the building blocks of proteins. Miller is still a leader in exobiology, which considers how life might have originated on Earth and whether it might exist on other planets.

Miller was born to Nathan Miller, a lawyer, and Edith (Levy) Miller on March 7, 1930, in Oakland, California. He attended the University of California, Berkeley, graduating in 1951. He then went to the University of Chicago, where he became interested in the origins of life after hearing a speech by Nobel chemistry laureate Harold C. Urey later that same year. Building on proposals made in the 1920s by British biochemist J. B. S. HALDANE and Russian biochemist Aleksander Oparin, Urey proposed that the environment in which life first arose on Earth included a "reducing atmosphere" consisting of methane, ammonia, hydrogen, and water. He said he hoped that someone would try to find out whether molecules found in the bodies of living things would form spontaneously under the conditions he described.

Miller was so excited by Urey's ideas that he offered to abandon his existing Ph.D. project and take on the experiments Urey had described instead. Urey tried to discourage him at first, fearing that Miller would delay his progress toward a degree without producing any useful results. The two eventually agreed that Miller would devote a year to the research, then abandon it if he failed to obtain anything he could use in a thesis. In fact, however, Miller achieved a breakthrough within weeks.

Miller prepared an apparatus containing a miniature "ocean" of liquid water, which he boiled at the start of the experiment; an "atmosphere" consisting of gaseous methane, hydrogen, and ammonia, plus water vapor from the ocean (oxygen, a vital component of the planet's atmosphere today, was not thought to have been present in significant amounts before the formation of life); and "lightning" in the form of a continuous electric discharge. He would have been happy to find even traces of amino acids accumulating in the water, but in fact, after several days of exposure to the discharge, he discovered that about 4 percent of what Urey called the "primordial soup" consisted of these compounds.

Miller's success not only earned his Ph.D. but resulted in an article, "A Production of Amino Acids Under Possible Primitive Earth Conditions," which made the cover of *Science* magazine in 1953 and attracted considerable attention from the popular press. (Some stories claimed that Miller had created life in a test tube.) It also created controversy. Some scientists believed that bacteria had contaminated Miller's apparatus and produced the amino acids, but, during a year of postdoctoral research at the California Institute of Technology in Pasadena, Miller repeated his experiment under sterile conditions and obtained the same results. Other scientists also confirmed his findings. According to fellow exobiologist Gustaf Arrhenius, Miller's experiment "had a tremendously important role in making chemists aware that the whole question of origin of life could be approached by lab experiments," whether or not they agreed with his conclusions.

Miller taught and researched at the College of Physicians and Surgeons at Columbia University in New York from 1955 to 1960, then joined the University of California, San Diego, as a professor of chemistry. He has devoted his career primarily to expanding on his famous experiment and further exploring the conditions under which the first living things might have arisen. In the mid-1990s, for example, he suggested that life probably began along shorelines and in drying lagoons, where chemicals were likely to be more concentrated than they would be in the open sea. He has shown that precursors of nucleic acids and another important biochemical called pantetheine can form under conditions like those he believes existed on the primitive Earth.

Miller's work earned the Oparin Medal from the International Society for the Study of Life in 1983, and his and Urey's theory of the origin of life has received some outside confirmation. A meteorite that landed in Australia in 1969, for instance, proved to contain many of the same amino acids, in the same proportions, that had appeared in Miller's experiment.

Not all exobiologists accept the Miller-Urey theory, however. Some question the pair's assumption that primitive Earth had a reducing atmosphere. They claim that too much water was present to allow the amount of free hydrogen that Miller and Urey proposed. They also say that ultraviolet light from the sun would have destroyed the methane and ammonia that provided source material for Miller's amino acids.

Other scientists think that the sun was less luminous in Earth's early days than it is now. If that was the case, temperatures on Earth would have been too low to support Miller's chemical reactions. Miller, however, insists that his scenario is as likely as any. "There's no evidence" about what the primitive Earth and its atmosphere were really like, he says.

In 1992, the National Aeronautics and Space Administration (NASA) made Miller, four other scholars at different universities, and their students part of a NASA Specialized Center of Research and Training (NSCORT) for studies in exobiology. Miller felt that NASA's support was logical as well as welcome. "If you're going to search for life on other planets, understanding how it started on Earth is essential," he said in 1995. Miller retired and became a professor emeritus at about this time, but he continued to be involved in research on compounds and conditions that might have contributed to the origin of life.

Further Reading

Cohen, Jon. "Novel Center Seeks to Add Spark to Origins of Life." *Science*, December 22, 1995.

Henahan, Sean. "From Primordial Soup to the Prebiotic Beach: An Interview with Exobiology Pioneer, Dr. Stanley L. Miller, University of California, San Diego." 1996. Access Excellence. Available online. URL: www.accessexcellence.org/WN/NM/miller.html. Accessed 2003.

Milstein, César

(1927–2002)
Argentinean/British
Molecular Biologist

César Milstein, working with Georges Köhler and others at Britain's Cambridge University, developed a way to make large numbers of cells that all manufacture precisely the same kind of antibody, a substance made by the immune system. Their discovery, which has proved useful in immunology and in the diagnosis and treatment of certain illnesses, earned both men the Nobel Prize in physiology or medicine in 1984.

Milstein was born in Bahía Blanca, Argentina, to a Jewish immigrant family on October 8, 1927. He earned a bachelor's degree in chemistry from the Colegio Nacional in Bahía

Blanca in 1944. He then went to the University of Buenos Aires, from which he graduated in 1952. He met his future wife, Celia Prilleltensky, through political activities on campus. They married in 1953 and hitchhiked around Europe on their honeymoon.

Milstein completed his doctorate in biochemistry at Buenos Aires in 1957. Between 1958 and 1961, he studied enzymes in the laboratory of FREDERICK SANGER at Cambridge University in England, earning a second doctorate in 1960. He then headed the division of molecular biology at the National Institute of Microbiology in Buenos Aires, but he resigned after two years as a protest against the Argentinian military government's political persecution of other professors. He went back to England and joined the Medical Research Council Laboratory of Molecular Biology, where he became codirector, with Sanger, of the protein and nucleic acid chemistry division. He became the division's sole head in 1983.

On his return, at Sanger's suggestion, Milstein changed his research focus to immunoglobin, the chemical of which antibodies are made. Instructions for making the complex immunoglobulin molecule are carried on several different genes, and Milstein helped to work out the sequence of bases in the gene that specifies the part of the molecule called the light chain.

Milstein also wanted to learn how the immune system, when exposed to proteins called antigens on the surfaces of invading cells such as bacteria, is able to produce antibodies exactly tailored to fit those antigens. (Antibodies attach to cells bearing the antigen they fit and thereby signal the immune system to destroy those cells.) Research on cells that make antibodies was difficult, however, because most such cells would not grow in laboratory cultures. The only exceptions Milstein knew of were cells from a mouse blood cancer called myeloma, but these cells could not be programmed to make specific types of antibody by exposing them to particular antigens, as other antibody-forming cells could, so they were not suitable for his experiments.

"Since we could not get a known cell line [type of cell that could reproduce itself dependably in culture] to do what we wanted, we were forced to construct such a cell line," Milstein wrote later. He and others in his laboratory, which since 1974 had included the German-born Köhler, decided to try to fuse, or hybridize, myeloma cells with cells from the spleen (an abdominal organ in which antibody-producing cells are manufactured), which are programmable but do not grow in culture. After many failures, they finally created what they called a hybridoma.

With their new technique, Milstein and Köhler could produce cells with the best features of both parents. They could expose a mouse to an antigen to get its immune system to make antibodies to that antigen, remove antibody-producing cells from its spleen, and fuse them with myeloma cells. This produced many small hybridomas, each of which was a colony of genetically identical, or clone, cells descended from a single fused spleen-myeloma cell. Different hybridomas made different antibodies, but all the cells in a single hybridoma made the same antibody. The scientists could use tests to identify the hybridomas that produced the kind of antibody they wanted to study, then allow those cells to multiply indefinitely, creating relatively large amounts of pure antibody.

Milstein and Köhler's description of what they called monoclonal (single-clone) antibodies was published in 1975. They and other scientists quickly realized that the power to make large amounts of a single kind of antibody could be useful in medicine as well as in research. Cancer cells, for instance, carry antigens different from those on normal cells. Researchers hoped that monoclonal antibodies that reacted with particular cancer antigens either could destroy tumors themselves or, more probably, could be attached to molecules of poisonous or radioactive substances that would do so. Because the antibodies would attach only to cancer cells, the theory went, they would be "smart missiles" that carried deadly substances directly to tumors, both increasing the poisons' effectiveness and decreasing the damage to normal cells that standard cancer chemotherapy causes.

Although early efforts to use monoclonal antibodies as medical treatments were disappointing, more than 50 different kinds had been approved for use by 2002 to treat cancer, heart disease, rheumatoid arthritis, and other conditions, and many more were in development. The antibodies are also used to diagnose medical conditions, for instance, in home pregnancy test kits, and in biotechnology and biomedical research. They can help molecular biologists identify particular proteins on the surface of cells, for example.

For several years after his breakthrough discovery, Milstein concentrated on improving the technology of making monoclonal antibodies, developing new uses for them, and persuading other scientists to adopt them. He also continued his studies of the genetics of the immune system and the question of how the system is able to produce so many different kinds of antibodies. Later, he studied mutations that occur in antibody-producing cells in response to vaccines. He also worked to advance science throughout the world, especially in Latin American countries. In addition to the Nobel Prize, his honors included such prestigious awards as Israel's Wolf Prize (1980), the Royal Medal of Britain's Royal Society (1982), and the Albert Lasker Basic Medical Research Award (1984). The British government made him a Companion of Honour for his services to molecular biology. Milstein died on March 24, 2002.

Further Reading

Milstein, César. "César Milstein—Autobiography." *Les Prix Nobel 1984*. Available online. URL: www.nobel.se/medicine/laureates/1984/milstein-autobio.html. Last updated 2002.

———. "Monoclonal Antibodies." *Scientific American*, October 1980.

Mitchell, Peter Dennis

(1920–1992)
British
Biochemist

Peter Mitchell described a key process involved in the body's storage and release of energy. He was born in Mitcham, Surrey, England, on September 29, 1920, the son of Christopher and Kate (Taplin) Mitchell. His father, a civil servant, earned an Order of the British Empire. His mother, Mitchell said, taught him the importance of rationality and of taking responsibility for his life.

Mitchell attended Queens College in Taunton and then Jesus College, part of Cambridge University. He earned his bachelor's degree in 1943 and his doctorate in 1951, both from Jesus College, and worked in the Cambridge biochemistry department from 1943 to 1955. He then was chosen to head a new chemical biology unit in the department of zoology at Edinburgh University, where he became a reader, the equivalent of a professor, in 1962.

Ill health forced Mitchell to leave Edinburgh in 1963. Restoring old houses was one of his hobbies, and almost on a whim he bought Glynn House, a burned-out manor house near Bodmin, Cornwall, and began rebuilding it. Glynn House became not only a home for Mitchell and his wife, Helen, but the site of a private institute, Glynn Research Laboratories, which Mitchell opened around 1965. He and a small number of colleagues worked there until his death in 1992.

When Mitchell started his most important research, on the way cells generate energy, in the early 1960s, biochemists knew that the cell's energy is stored in the phosphate bonds of a chemical called adenosine triphosphate, or ATP. The energy is released when one of these bonds is broken, turning ATP into adenosine diphosphate (ADP). ADP is built up into ATP again by a process called phosphorylation, which requires energy. The energy for phosphorylation comes from electrons being passed through a sort of bucket brigade of proteins in the double membranes of certain cell organelles—mitochondria in animal cells and chloroplasts in plant cells. Even though these basic facts were understood, many details of the processes remained unclear.

Most biochemists of the time thought that electron transport was connected to phosphorylation by means of a string of chemical reactions controlled by enzymes. Mitchell, however, theorized that, at the same time the electrons are being passed along, protons, subatomic particles with an electrical charge opposite to that of electrons, are pushed off the inner mitochondrial membrane and diffuse through the outer membrane. This makes the electric charge on the outside of the membrane different from that on the inside. Mitchell proposed that energy needed to make ATP from ADP is stored in the form of this electrical potential difference and that the chemiosmotic movement set up by differences in the concentration of chemicals on the two sides of the membrane drives the phosphorylation process directly rather than through chemical reactions.

Many scientists rejected this chemiosmotic theory, as it was called, when Mitchell first proposed it in 1961. Over the years, however, he and his coworkers gathered more and more evidence to support it, and after about a decade it came to be accepted. Later researchers found that a similar mechanism is involved in other cell processes that require energy.

In addition to revealing an important aspect of energy metabolism, Mitchell's work showed that membranes in the cell are not merely walls but rather are active participants in cellular processes. His chemiosmotic theory is now considered a basic principle in the field of bioenergetics. Indeed, in an article written for the University of Illinois at Urbana-Champaign, Antony Crofts says that the theory "started a revolution which has echoed beyond bioenergetics to all biology and shaped our understanding of the fundamental mechanisms of . . . every aspect of life."

Mitchell won the Nobel Prize in chemistry in 1978, as well as many other awards, including the CIBA Medal and Prize of the British Biochemical Society (1973), the Freedman Foundation Award of the New York Academy of Sciences (1974), and the Copley Medal of Britain's Royal Society (1981). He was interested in whole human beings as well as cells, especially in the ways people communicate and attempt to get along with one another, and he made studies of interactions among scientists and in political and economic systems.

Further Reading

Crofts, Antony. "Peter Mitchell (1920–1992) and the Chemiosmotic Hypothesis." University of Illinois, Urbana-Champaign. Available online. URL: www.life.uiuc.edu/crofts/bioph354/mitchell.html. Posted 1996.

Prebble, John, and Bruce Webster. *Wandering in the Gardens of the Mind: A Biography of Peter Mitchell and Glynn*. New York: Oxford University Press, 2002.

Monod, Jacques
(1910–1976)
French
Molecular Biologist

With François Jacob and others, Jacques-Lucien Monod showed how the operation of genes is controlled and helped to reveal the way genes make proteins. For this work he, Jacob, and André Lwoff, their mentor at France's Pasteur Institute, shared the Nobel Prize in physiology or medicine in 1965.

Monod was born in Paris on February 9, 1910, but he grew up in the south of France. His father, painter Lucien Monod, introduced him to the work of CHARLES ROBERT DARWIN and interested him in biology. His mother was Charlotte MacGregor, an American of Scottish descent. He earned a bachelor's degree from the Lycée de Cannes in 1928 and advanced degrees from the University of Paris in 1931 (licensate in science) and 1941 (doctor of science). He was an assistant professor of zoology at the university from 1934 to 1944. When the Germans occupied France during World War II, Monod was in charge of a military unit in the Resistance, the underground movement to fight the occupiers. For this highly risky work, he earned the croix de guerre, the Legion of Honor, and the American Bronze Star.

Monod joined the Pasteur Institute in 1945. He headed the microbic physiology laboratory until 1954 and the cellular biochemistry department from 1954 to 1971, when he was made the director of the entire institute. He also became a professor at the University of Paris in 1953, a professor of the chemistry of metabolism at the Sorbonne, and, in 1967, a professor of molecular biology at the Collège de France.

French molecular biologists Jacques Monod, shown here, and François Jacob proposed in the early 1960s that segments of DNA called operators and repressors control the operation of structural genes, the parts of the DNA molecule that direct the making of proteins. *(National Library of Medicine)*

In 1958, Monod and Jacob began investigating the way that genes of the common intestinal bacterium *Escherichia coli* make proteins. Scientists had known since the early 1940s that genes produce physical characteristics in living things by making proteins and that each gene normally carries the instructions for making one protein. In the 1950s, JAMES WATSON, FRANCIS CRICK, and others had determined that these instructions are encoded in the sequence of bases in the segment of DNA that makes up the gene. DNA normally remains in the nucleus of the cell, however, whereas proteins are made in the outer part of the cell (the cytoplasm). No one knew how genetic information moved from the nucleus to the cytoplasm.

Researchers had discovered a nucleic acid similar to DNA, called RNA, in the cytoplasm during the 1950s. Monod and Jacob proposed in 1961 that DNA copies its instructions into a short-lived form of this substance, which they called messenger RNA. They said that, as its name suggested, messenger RNA carries the DNA instructions into the cytoplasm. This theory was later shown to be correct.

Monod and Jacob also proposed an answer to a second puzzle about the way genes work. All cells in the body contain the same genes, yet a nerve cell, say, makes different proteins than a muscle cell. A single kind of cell may also make different proteins at different stages of its life. Clearly, most genes are inactive (do not produce proteins) most of the time. A gene is "turned on" only when a cell develops a need for that gene's particular protein. Scientists wondered what kept genes turned off and what allowed them to begin functioning when they were needed.

Monod and Jacob suggested that each protein is produced by a three-gene cluster that they called an operon. An operon, they proposed, consists of a structural gene, which carries the actual instructions for making the protein; an operator, which can activate the structural gene and allow it to produce messenger RNA; and a repressor, which binds to the operator and keeps it from permitting the structural gene to function. When the cell detects a need for the protein, chemicals signal the repressor to remove itself from the operator, and the operon's functional cycle begins. It was this theory that earned the Nobel Prize. WALTER GILBERT and others later confirmed the theory experimentally.

In addition to the Nobel Prize, Monod received the Montyon Physiology Prize (1955) and the Leopold Mayer Prize (1962) of the French Academy of Sciences and the British Louis Rapkine Medal (1958). He wrote *Chance and Necessity* (1971), a best-selling work of scientific philosophy that emphasized the roles of chance and natural selection (necessity) in evolution. He married Odette Bruhl, a museum curator and archaeologist, in 1938, and they had two sons. Monod died on May 31, 1976, in Cannes.

Further Reading

"Monod, Jacques." *Current Biography Yearbook 1971*. New York: H. W. Wilson, 1971.

Montagnier, Luc

(1932–)
French
Virologist

After a decade-long dispute with American researcher ROBERT GALLO, Luc Montagnier was recognized as the first person to isolate HIV, the virus that causes AIDS. Montagnier was born on August 18, 1932, in Chabris, France, near the city of Tours. His father, Antoine, was a certified public accountant and also an amateur scientist who did chemistry experiments in his garage. Imitating him, Luc developed a desire to "explain the world through science." He attended the College de Chatellerault and then the Universities of Poitiers and Paris. He obtained his preliminary degree from Poitiers in 1953, a licensate in science from Paris in 1955, and a medical degree from Paris in 1960. He married Dorothy Ackerman in 1961, and they have three children.

Montagnier did postdoctoral studies in Britain in the early 1960s, focusing on viruses and cancer. Researchers had learned that some viruses that cause cancer in animals belong to an unusual group later called retroviruses, which have a genome made of RNA rather than the more common DNA. In 1963, Montagnier helped to discover how these viruses reproduce, which he still considers "one of my major contributions." A year later, at the Institute of Virology in Glasgow, Scotland, he and coworker Ian MacPherson developed a new technique for growing cancer cells in the laboratory. Montagnier became laboratory director of the Radium Institute in Orsay, France, in 1965. In 1972, he won the Prix Rosen de Cancérologie and moved to the Pasteur Institute, the famous laboratory established in Paris by chemist and microbiologist LOUIS PASTEUR. He became a professor there in 1985.

At the Pasteur Institute, Montagnier continued his studies of retroviruses and cancer in a laboratory that he founded and directed. He also investigated interferon, a substance produced by the immune system that researchers of the time hoped would be an effective new treatment against cancer. His work on the genes that carry the instructions for making this compound helped to make relatively large amounts of it available for research. (Unfortunately, interferon proved to work only against a few rare tumors.)

When word began to spread in 1981 that an unusual disease was destroying the immune systems of homosexual men, Montagnier became interested, especially when evidence accumulated that the illness might be caused by a retrovirus similar to those he had been studying. At first, he recalled later, "I was told we should not touch a marginal disease in marginal people. This could do something bad for the reputation of the Pasteur Institute." At the end of 1982, however, the department of the institute that made medical products for sale became concerned that the still-unknown virus might contaminate blood products that it imported from the United States and asked him to hunt for the virus full time.

In early 1983, Montagnier and his coworkers succeeded in isolating a virus from immune system cells of a patient with symptoms associated with the new disease, which had come to be called AIDS. He named the virus LAV, for lymphadenopathy-associated virus, in a paper published in *Science* in May. (Lymphadenopathy is swelling of the lymph glands, a common symptom in people infected with the virus.) He did not directly claim that LAV caused AIDS, however, and his report caused little stir.

A number of other laboratories, including Gallo's at the U.S. National Institutes of Health, were also trying to find the cause of AIDS. In April 1984, Montagnier was startled and dismayed to hear that the secretary of Health and Human Services had announced with much fanfare that Gallo had isolated the virus that caused AIDS. Gallo called his virus HTLV-III because he believed that it was related to two cancer-causing retroviruses he had identified a few years before and named HTLV-I and HTLV-II. Montagnier had sent Gallo some cells containing LAV twice in 1983, however (researchers often trade samples in this way), and he suspected that the American scientist, accidentally or otherwise, had simply reisolated LAV.

Even the French, Montagnier complained later, gave Gallo credit for isolating the AIDS virus at first, but eventually officials at the Pasteur Institute came to share Montagnier's suspicion. After Gallo's group obtained a patent for a blood test to identify antibodies to the virus (a sign of infection), which promised to generate millions of dollars in royalties, the Pasteur Institute sued them in December 1985. The institute stated that it had filed for a patent on a similar test earlier than the Gallo group but had been ignored by the U.S. Patent Office, which it also sued. By this time, genetic studies had shown that HTLV and LAV were indeed the same virus, which was later renamed HIV (human immunodeficiency virus). The Pasteur Institute demanded both a share of the blood test royalties and an official acknowledgment that Montagnier had isolated the virus before Gallo.

An American court dismissed the Pasteur suit in July 1986, but bad feelings continued. In March 1987, U.S. president Ronald Reagan and French premier Jacques Chirac attempted to end the dispute by signing an unprecedented international agreement that divided the royalties equally between the two countries (80 percent of the money went to a foundation for AIDS research) and the credit between the two laboratories. Further genetic studies published in 1991 proved, however, that the virus Gallo isolated had definitely come from Montagnier's laboratory. Gallo was cleared of all charges of wrongdoing, but Montagnier became generally recognized as the first person to isolate HIV. Because Gallo had contributed substantially to the discovery by, for instance, tying the virus more clearly to AIDS than Montagnier had, however, both researchers agreed around 2001 to return to the view that credit for the overall discovery of HIV should be shared equally.

Meanwhile, leaving arguments and lawsuits to others, Montagnier continued his research on HIV. In 1986, his group found a second, somewhat weaker strain of the virus in West Africa and named it HIV-2. He became head of a new department of AIDS and retroviruses at the Pasteur Institute in 1990. Much of his research, then and since, has concentrated on attempts to develop a vaccine to keep people infected with HIV from developing AIDS, an approach he believes is more likely to succeed than the search for drugs to cure the disease. He has also studied the mechanisms by which HIV attacks cells and the possibility that other microorganisms, particularly a type of microbe called mycoplasma, may augment HIV's effects.

Awards that Montagnier has received for his work include the Prix Gallien (1985), the Albert Lasker Medical Research Award (shared with Gallo in 1986), and the Japan Prize (1988). France made him a Chevalier of the Legion of Honor in 1984 and a Commander of the National Order of Merit in 1986. In the late 1990s, he divided his time between the Pasteur Institute and Queens College, part of the City University of New York, where he was a distinguished professor and director of the college's Center for Molecular and Cellular Biology. In early 2002, however, Montagnier and former rival Gallo agreed to work together in attempts to develop an effective vaccine for AIDS as codirectors of the Program for International Viral Collaboration. UNESCO and the World Foundation for AIDS Research and Prevention, of which Luc Montagnier is president, sponsor the program.

Further Reading

Cohen, Jon. "Longtime Rivalry Ends in Collaboration." *Science*, February 22, 2002.

Montagnier, Luc. "A History of HIV Discovery." *Science*, November 29, 2002.

———. *Virus: The Co-Discoverer of HIV Tracks Its Rampage and Charts the Future*. Translated by Stephen Sartarelli. New York: W. W. Norton, 1999.

Morgagni, Giovanni Battista
(1682–1771)
Italian
Pathologist, Anatomist, Physician

Along with ANDREAS VESALIUS and WILLIAM HARVEY, Giovanni Morgagni helped to change medicine from a combination of unorganized observation and mysticism to a science firmly focused on physical features of the body. Morgagni's contribution was to tie diseases to damage in specific organs.

Morgagni was born on February 25, 1682, in Forli, a town near Bologna, Italy. A precocious student, he began studying philosophy and medicine at the University of Bologna when he was only 16. After obtaining his medical degree in 1701, he became an assistant to the university's professor of anatomy, Antonio Valsalva. He returned to Forli in 1709 and married Paola Verazeri, a local noblewoman. They raised 12 daughters and 3 sons. Morgagni was a popular teacher at the University of Padua from 1711 on; he was appointed professor of anatomy, the university's most highly regarded faculty position, in 1715.

In addition to being a professor, Morgagni was a respected working physician, and other doctors often consulted him. These consultations, along with his own lifetime of experience in both treating living patients and dissecting the bodies of patients who had died (performing autopsies), provided the storehouse of information that went into his great work, *De Sedibus et Causis Morborum per Anatomen Indagatis* (*The Seats and Causes of Disease Investigated by Anatomy*). This book was published in 1761, when Morgagni was 79 years old.

Before Morgagni, most doctors still followed the teachings of the ancient physicians HIPPOCRATES and GALEN, who had believed that disease was a result of an imbalance among four fluids called humors that were thought to flow throughout the body. Morgagni, however, saw disease as a process that began in particular locations in the body, not in the body as a whole. He further believed that the signs of illness seen in living patients could be tied to specific types of organ damage that were revealed in autopsies. Symptoms, he wrote, were "the cry of the suffering organs." Morgagni was by no means the first to connect symptoms with specific organs, but he was the first to tie bedside observations to autopsy findings in a systematic way.

Morgagni's book was divided into five parts, covering diseases of the head, thorax (chest), and abdomen, diseases of a general nature and those requiring surgery, and indexes of symptoms and of organ changes seen at autopsy. It presented some 700 case histories in the form of 70 letters to a young physician. Almost all the case histories contained both a description of a sick person and an account of an autopsy on that same person. Some also included surveys of previous authorities' opinions about the diseases being discussed or reports of experiments Morgagni had performed to establish physiological facts. The conditions Morgagni wrote about include ruptured appendix, atherosclerosis (hardening of the arteries), stomach ulcers, stroke, and bowel cancer. He described some of them for the first time in medical literature.

Physicians flocked to buy Morgagni's book. An American doctor wrote in 1764 that *De Sedibus* was held "in the highest estimation throughout all Europe, and all the copies of the last [third] edition [are] already bought up." Because of this book, Morgagni is considered to be the founder of pathological anatomy, the medical specialty that shows how the structure of body parts changes in disease. Eminent 19th-century German pathologist RUDOLF VIRCHOW wrote that "beginning with Morgagni and resulting from his work, the dogmatism of the old schools [of medical thinking] was completely shattered, and . . . with him the new medicine begins." Admired as much for his calm nature and kindness as for his learning, Morgagni died of a stroke in Padua on December 6, 1771, at the age of 89.

Further Reading

Morgagni, Giovanni. *Seats and Causes of Disease Investigated by Anatomy*. 1761. Reprint, Armonk, N.Y.: Futura Publishing Co., 1980.

Nuland, Sherwin B. *Doctors: The Biography of Medicine*. New York: Random House, 1988.

Morgan, Thomas Hunt
(1866–1945)
American
Geneticist, Embryologist

Thomas Hunt Morgan and his coworkers and students in Columbia University's famous "Fly Room" proved conclusively that genes are carried on chromosomes and worked out principles of genetic action that corrected and extended GREGOR MENDEL's basic laws of heredity. Morgan, the son of diplomat Charlton Hunt Morgan, belonged to a Kentucky family that had played a prominent role in the Civil War; his uncle, John Hunt Morgan, had been called "the Thunderbolt of the Confederacy." Morgan was born in Lexington on September 25, 1866, the same year Mendel published his famous paper.

Morgan was interested in nature from childhood, when he collected birds' eggs and fossils. He earned a bachelor's degree from the State College of Kentucky in 1886, then took a Ph.D. in zoology from Johns Hopkins University in Baltimore, Maryland, in 1890. His early research interest was experimental embryology, especially of invertebrate sea creatures. He became an associate professor of zoology at Bryn Mawr University, a women's college near Philadelphia, in 1891. One of his students was Lillian Sampson, who became his wife in 1904 and later helped him in his genetic work. They had four children.

Edmund B. Wilson, head of the zoology department at Columbia University in New York City, persuaded Morgan to become professor of experimental zoology there in 1904. He also interested Morgan in genetics, a field then in its infancy. Mendel's work had been rediscovered only four years earlier, and geneticists still had no idea what physical form Mendel's "factors," or units of inherited information, assumed. American geneticist Walter Sutton had suggested in 1902 that the factors were somehow connected with chromosomes, threadlike bodies in the cell nucleus, but Sutton's theory was by no means widely accepted. Morgan himself doubted it at first.

Morgan lacked the laboratory space to keep large colonies of mice or other animals commonly used in breeding experiments, but around 1909 a coworker told him about fruit flies or vinegar flies (*Drosophila melanogaster*), tiny insects that swarm around rotting fruit. Morgan found that the flies were easy to attract, cheap to raise, and able to live by the thousands in left-over milk bottles. Furthermore, they reproduced once every 10 days, producing hundreds of offspring each time, which made them ideal subjects for genetic studies. Morgan's laboratory at Columbia, filled with the smell of rotting bananas and the buzzing of thousands of fruit flies, soon became known as the Fly Room.

Morgan first exposed the insects to X rays and other treatments intended to produce genetic mutations, but nothing seemed to work. Then, early in 1910, he spotted a male fly that had white eyes instead of the usual red ones. He bred this precious mutant to one of its red-eyed sisters and found that all their offspring had red eyes. When these offspring were mated, however, about one in four of the second-generation flies had white eyes. This result exactly confirmed the proportions Mendel had found for what he called dominant and recessive forms of a characteristic. Morgan's mutants showed one pattern that Mendel had never seen, however: All the white-eyed flies were male.

Geneticists knew by this time that chromosomes normally occur in matched pairs. However, Morgan's mentor, Wilson, and one of Morgan's former students, NETTIE MARIA STEVENS, had discovered independently in 1905 that one pair of chromosomes does not always fit this description. This pair consists of two matching chromosomes in females, but in males, one chromosome in the pair is much smaller than the other. Wilson and Stevens called the larger chromosome X and the smaller one, found only in males, Y. Because each

The scientists in Thomas Hunt Morgan's famous "Fly Room" at Columbia University in New York produced basic discoveries about genetics in the early years of the 20th century by studying fruit flies; Morgan is shown here. *(National Library of Medicine)*

offspring receives only one chromosome of a pair from each of its parents, any offspring is sure to get an X from its mother but can receive either an X or a Y from its father. Wilson and Stevens concluded that offspring receiving two Xs become females, whereas those receiving an X and a Y become males.

Morgan's discovery about the white-eyed flies convinced him that the factor, or gene (as factors were called after 1909), for eye color in fruit flies must be carried on the X chromosome. Because the form of the gene that produces red eyes apparently was what Mendel had called dominant, a fruit fly would have red eyes if it received even one copy of the gene. A female thus would almost always have red eyes because she would be likely to receive the red-eye form of the gene on one or the other of her X chromosomes. If she was the product of a mating between a red-eyed female and a white-eyed male, however, she would also have a copy of the white-eye form of the gene and would pass it on to about half of her offspring. Males who received the white-eye gene on the X chromosome donated by their mothers would have white eyes because they would have no other copy of the gene to offset it.

Fruit fly eye color was the first specific characteristic associated with a particular chromosome. Linkage of the eye color gene with the chromosome that determines sex, which Morgan announced in a paper published in July 1910, greatly strengthened the evidence for a connection between chromosomes and genes and persuaded Morgan, as well as many other scientists, that Sutton's theory was correct.

This was just the first of many discoveries to emerge from the Fly Room. Mendel had maintained that characteristics were inherited independently of each other, but Morgan and his students found that this often was not the case. They concluded that traits that were usually inherited together, or linked, probably were determined by genes on the same chromosome.

Late in 1911, however, the Fly Room scientists also noticed that linked characteristics sometimes were not inherited together. They tied this fact to microscopists' observation that during the formation of sex cells (sperm and eggs), the two chromosomes in each pair wind around each other just before they separate to go into the different daughter cells. Morgan suspected that during this twining process the chromosomes sometimes break apart and reassemble, with parts of one chromosome sticking to the other. His group called this phenomenon crossing over. Morgan concluded that the more often two linked traits become separated by crossing over, the farther apart they probably are on a chromosome. One of Morgan's students, HERMANN JOSEPH MÜLLER, wrote that "Morgan's evidence for crossing over and his suggestion that genes further apart cross over more frequently was a thunderclap: hardly second to the discovery of Mendelism."

Crossing over suggested a possible method for beginning to determine where genes were on a chromosome. Another Morgan student, Alfred Sturtevant, developed this idea further in 1911, when he was still an undergraduate. Ignoring his assigned homework, he stayed up most of one night creating a map of five genes believed to be on the fruit fly's X chromosome, using information from breeding experiments to determine the genes' relative distance from each other. Morgan called this chromosome map "one of the most amazing developments in the history of biology." Chromosome mapping grew into a powerful tool for geneticists.

Morgan, Müller, Sturtevant, and another coworker, Edward Lewis, described these and other conclusions from their fruit fly experiments in *The Mechanism of Mendelian Heredity*, published in 1915. This groundbreaking book tied theories derived from breeding experiments to evidence seen under a microscope. It offered the best evidence yet that genes were part of chromosomes and theorized that they were arranged on the chromosomes like beads on a string. It both proved that Mendel's laws of heredity were basically correct and explained cases in which they did not apply. Neurobiologist Eric Kandel has written that Morgan's book showed that the gene "was at once the unit of Mendelian heredity, the driving force for Darwinian evolution, and the control switch for [embryonic] development."

The Fly Room became as famous for its unusually democratic atmosphere as for its genetic discoveries. Sturtevant wrote later, "There can have been few times and places in scientific laboratories with such an atmosphere of excitement and such a record of sustained enthusiasm . . . The group worked as a unit. Each carried out his own experiments but each knew exactly what the others were doing, and each new result was freely discussed." No doubt partly because of this open attitude, Morgan was exceptionally successful at attracting talented students, whom one colleague called his greatest discoveries. Several went on to win Nobel Prizes. Morgan himself won the prize in physiology or medicine in 1933. He also won the Darwin Medal (1924) and the Copley Medal (1939) of Britain's Royal Society.

In 1928, Morgan accepted an offer to head a new biology laboratory at the California Institute of Technology (Caltech) in Pasadena. He devoted his early Caltech years to further research in genetics, but he later returned to his first love, experimental embryology. He died on December 4, 1945, in Pasadena.

Further Reading

Allen, Garland E. *Thomas Hunt Morgan, the Man and His Science*. Princeton, N.J.: Princeton University Press, 1978.

Kandel, Eric R. "Thomas Hunt Morgan at Columbia University: Genes, Chromosomes, and the Origins of Modern Biology." Available online. URL: www.columbia.edu/cu/alumni/Magazine/Legacies/Morgan/. Accessed 2001.

Shine, Ian, and Sylvia Wrobel. *Thomas Hunt Morgan, Pioneer of Genetics*. Lexington: University Press of Kentucky, 1976.

Morton, William Thomas Green

(1819–1868)
American
Dentist

Succeeding where his onetime dental business partner, Horace Wells, had failed, William Morton popularized the use of anesthesia in surgery. Morton was born on August 9, 1819, in Charlton, Massachusetts, the son of a shopkeeper. Little is known of his early life. He may have studied dentistry at the Baltimore College of Dental Surgery.

After briefly working with Wells, Morton established his own dentistry practice in Boston in 1844. About a year later, he was surprised to receive a visit from his former partner. The excited Wells explained that he had been to an entertainment, popular at the time, at which some people paid a showman to let them breathe nitrous oxide, nicknamed "laughing gas" for the euphoria it caused, while others paid to watch the silly antics that the people who breathed the gas often performed. Wells saw a fellow townsman fall and gash his shin while under the influence of the gas, yet the man said later that he had felt no pain. Wells then arranged with the showman to try the gas privately himself and asked a fellow dentist to pull one of his teeth after he had breathed the nitrous oxide. He, too, felt nothing, and he hailed the gas as the key to "a new era in tooth-pulling."

Wells also thought that laughing gas might be used in general surgery, perhaps promising an end to the agony that even the simplest operation inflicted on patients. He asked Morton, who had contacts among surgeons in Boston, to help him set up a demonstration of the gas. Morton introduced Wells to John Collins Warren, the senior surgeon at Massachusetts General Hospital, and Warren agreed to let Wells give a patient nitrous oxide before an audience of medical students. When Wells gave the gas to a young man who had volunteered to have a sore tooth extracted later in 1845, however, he apparently used too little, for when Wells began to pull the tooth, the man let out a piercing scream. The watching students laughed and booed Wells from the room.

Morton witnessed Wells's failure, but he believed that his fellow dentist was on the right track. (Even before Wells's visit, he himself had been experimenting with ways to make dentistry painless.) After consulting with chemist Charles

T. Jackson, Morton turned from laughing gas to ether, another substance known mostly for its euphoric effects. Ether, a liquid, evaporated into a strong-smelling gas that some medical students inhaled during "ether frolics." These students, like the people who took laughing gas, had noticed that they sometimes hurt themselves without knowing it while under ether's influence. Indeed, unknown to Morton, Crawford Long, a Georgia physician who had taken part in ether frolics while in medical school, had already let a friend inhale ether vapor before removing a small tumor from the man's neck in March 1842. The experiment had been a success—the first time a gas had been used successfully in surgery to deaden pain—but Long had not told others of his work or published an account of it.

Although he was not the first to use anesthesia, ambitious American dentist William Thomas Green Morton was the one who made surgeons aware that there was a way to free people from pain during operations. *(National Library of Medicine)*

After trying ether on everything from worms to his family dog to himself, Morton worked out a dose that seemed to be effective. When a music teacher named Eben Frost visited him one evening in September 1846 and asked Morton to hypnotize him to deaden the pain before extracting a decayed tooth, Morton used ether on him instead. The extraction was painless, and the ambitious Morton immediately had Frost sign a notarized statement to that effect and took the statement and the story to a local newspaper. He also applied for a patent on his treatment, carefully not naming the substance he used because ether was already known and thus could not legally be patented.

Seeking publicity for his new procedure, Morton somehow persuaded surgeon Warren to allow another test. On the scheduled day—October 16, 1846—the hospital demonstration room filled with onlookers hoping for a show as amusing as the one Horace Wells had inadvertently put on. This time, however, the treatment worked, and the operation, in which Warren removed a tumor from a man's jaw, took place with scarcely a murmur from the patient. "Gentlemen, this is no humbug [fake]," Warren proclaimed when the operation was over.

Word of Morton's technique spread quickly. Physician and author Oliver Wendell Holmes suggested calling it anesthesia, which means "lack of feeling." A report of Warren's operation appeared in a Boston medical journal, and copies soon reached Europe. On December 19, English surgeon Robert Liston used ether in an amputation with as much success as Morton and Warren had had in their more minor surgeries. "This yankee dodge . . . beats mesmerism [hypnotism] all hollow," Liston told his own amazed audience. Surgeons everywhere soon began anesthetizing their patients with ether, nitrous oxide, or (starting about a year after Morton's demonstration) a third gas, chloroform.

Morton, however, failed to make the profit from his discovery that he had hoped for. He obtained his patent, but word that his mystery substance was ether soon leaked out, making the patent useless. Jackson, the chemist with whom he had discussed the ether treatment, claimed to have thought of it first. Wells and, eventually, Long also joined the dispute. Morton spent the rest of his life entangled in claims, counterclaims, and lawsuits and died, bitter and poverty-stricken, of a stroke in New York City on July 15, 1868, when he was only 49 years old.

Although Morton was not the first to use anesthesia, later historians have generally given him credit for successfully introducing it into surgery. He was elected posthumously to the American Hall of Fame in 1920 for this invaluable achievement.

Further Reading

Nuland, Sherwin B. *Doctors: The Biography of Medicine*. New York: Random House, 1988.

Wolfe, Richard J. *Tarnished Idol: William Thomas Green Morton and the Introduction of Surgical Anesthesia*. Novato, Calif.: Norman Publishing, 2001.

Müller, Hermann Joseph
(1890–1967)
American
Geneticist

Hermann Müller developed a way to use X rays to increase the number of mutations in genes, greatly aiding the study of genetic changes. He was the son of Hermann Joseph and Frances (Lyons) Müller, both first-generation Americans. Hermann Joseph Müller continued the fine-art metal business that his own father, a German immigrant, had established.

Hermann Müller was born in New York City on December 21, 1890, and grew up there. His father, who had stirred his interest in both science and humanitarian pursuits, died when he was nine years old. Müller began to focus on genetics at about age 10 after seeing models of horses' evolution. He and two friends established what is thought to have been the first high school science club in the United States.

Müller won a scholarship to New York's Columbia University in 1907. He earned a bachelor's degree in zoology with honors in 1910 and a master's degree in 1911. He taught physiology for a year at Cornell University Medical School in Ithaca, then returned to Columbia to complete his Ph.D.

During his graduate years at Columbia, Müller worked in the famous Fly Room with geneticist THOMAS HUNT MORGAN. He played an important part in the Morgan laboratory's discovery of crossing over, in which chromosomes trade pieces during the formation of sex cells, and his thesis on crossing over earned his Ph.D. in 1916. He was also a coauthor, with Morgan and two other Morgan students, of *The Mechanics of Mendelian Heredity* (1915), a landmark book that linked GREGOR MENDEL's laws of heredity with discoveries about chromosomes, the bodies in the cell nucleus that Morgan's group had shown to contain genes.

Müller did postdoctoral work at Rice Institute in Houston, Texas (1916–17), and Columbia (1918–20), then returned to Texas. He taught genetics and evolution at the University of Texas, Austin, throughout the 1920s, becoming a full professor there in 1925.

Müller's most important discoveries occurred between about 1918 and 1926, when he was at Columbia and the University of Texas. While working with Morgan, Müller had become interested in mutations, but these genetic changes occurred in nature so seldom that they were hard to study systematically. Furthermore, Müller suspected that most mutations were recessive, which meant that they would produce physical changes only when two organisms carrying the mutated gene mated and passed on that gene to offspring—an event that might not occur for generations after the mutation actually occurred. The mutations might never become visible if, as Müller believed was often the case, they caused death during embryonic development.

Müller thought that mutations resulted from random damage to the still-unknown genetic material caused by chemical changes that, in turn, were induced by molecular motion. When looking for a way to increase the frequency with which mutations occurred, he first tried heat, which speeds up this motion, in 1919. He found in 1926, however, that X rays were far more effective. In a groundbreaking paper published in 1927, he announced that exposure to X rays increased the frequency of mutations in fruit flies to 100 times their natural level of 400 in 20,000,000. This was the first time that large numbers of mutations had been produced artificially. Müller also described techniques for detecting the occurrence of mutations without waiting for them to appear as physical changes.

Having more mutations to study offered the potential to advance genetics considerably. Geneticists could make statistical studies of mutation rates and correlate them with, for example, X-ray dosage. Müller's experiments also confirmed his belief that mutations were caused by chemical reactions, which suggested that chemical techniques could be used to investigate genetic changes. His artificial creation of mutations helped to lay the foundation for genetic engineering. Müller was awarded the Nobel Prize in physiology or medicine in 1946 for his X-ray work.

Work pressure, the failure of Müller's marriage to Jessie Jacobs (a mathematician whom he had married in 1923), and opposition to his socialist views brought on a nervous breakdown in 1932, and Müller decided to go overseas for a while. He traveled first to Germany, but in 1933, after the Nazis took control of that country's government, he moved to the Soviet Union. He was senior geneticist at the Soviet Academy of Sciences' Institute of Genetics in Leningrad and Moscow until 1937, when his criticism of the mistaken views of the institute's director, Trofim Lysenko, made him so unpopular with the Soviet government that he was forced to leave the country.

Müller joined the Institute of Animal Genetics at the University of Edinburgh, Scotland, later in 1937. While there he met and married Dorothea Kantorowicz, a German refugee physician, in 1939 and earned a doctorate of science in genetics in 1940. He returned to the United States in 1941, and from then until 1945 he taught at Amherst College, meanwhile studying the relationship between natural mutations and aging. He then transferred to the University of Indiana, Bloomington, where he stayed as a professor of zoology for the rest of his career. There, he continued his study of radiation-induced mutations, using them as a way of learning about both the mutation process and the effects of radiation.

Müller's discoveries about mutations combined with his political activism to make him an outspoken commentator on several controversial issues. In *Out of the Night: A Biologist's View of the Future* (1935), he recommended establishing banks to store the sperm of gifted men for use in voluntary breeding programs to offset harmful mutations and improve the human species. Later, he opposed aboveground nuclear bomb tests, fearing that radioactive fallout would increase mutations. He was also concerned about possible risks from peaceful uses of atomic power and from the use of X rays in diagnosis and medicine. Müller died at Bloomington on April 5, 1967.

Further Reading

Carlson, Elof Axel. *Genes, Radiation, and Society: The Life and Work of H. J. Müller*. Ithaca, N.Y.: Cornell University Press, 1981.

"Hermann Joseph Müller—Biography." *Nobel Lectures, Physiology or Medicine, 1942–1962*. Available online. URL: www.nobel.se/medicine/laureates/1946/muller-bio.html. Last updated 2001.

"Muller, H(ermann) J(oseph)." *Current Biography Yearbook 1947*. New York: H. W. Wilson, 1947.

Mullis, Kary B.
(1944–)
American
Biochemist

Kary Banks Mullis invented a process for reproducing small fragments of DNA quickly, cheaply, and in large quantities, thereby revolutionizing fields ranging from biotechnology to forensic science. He received the Nobel Prize in chemistry in 1993 for this achievement.

Mullis was born in Lenoir, North Carolina, on December 28, 1944, but he was raised in Columbia, South Carolina, after his parents, Cecil Banks Mullis, a salesman, and Bernice (Barker) Mullis, a real estate broker, separated. As a teenager, he invented devices ranging from an automatic dog door to homemade rockets.

Mullis obtained a bachelor's degree in chemistry from the Georgia Institute of Technology in Atlanta in 1966. While there, he met and married the first of his four wives, Richards Haley. (His later wives were Gail Hubbell, Cynthia Gibson, and Nancy Cosgrove. He had a daughter by Haley and two sons by Gibson.) He earned a Ph.D. in biochemistry from the University of California, Berkeley, in 1972 and did postdoctoral work at the University of Kansas Medical School in Kansas City (1973–77) and the University of California, San Francisco (1977–1979). During this period, he also tried his hand at writing fiction and managing a bakery.

In 1979, a friend, Thomas White, found Mullis a job at Cetus Corporation in Emeryville, California, one of the first biotechnology companies. Mullis's work there, synthesizing small DNA fragments for other scientists' research, presented no challenge to his inventive mind. He therefore began to look for biological problems to solve.

One Friday evening in April 1983, during the long commute from his office to his home, Mullis was trying to design improvements in techniques for discovering the sequence of bases in a DNA molecule, which determines the genetic information that the molecule carries. He suddenly realized that a process he had been considering offered not a better way of sequencing DNA but a better way of copying it. He stopped the car and began scribbling his ideas on a piece of paper. "It was difficult for me to sleep that night with deoxyribonuclear bombs exploding in my brain," he wrote later.

The process Mullis invented on that historic drive came to be known as the polymerase chain reaction, or PCR. It uses a bacterial enzyme called DNA polymerase to make copies of any fragment of DNA. PCR can be done over and over again, doubling the number of copies of the chosen fragment each time, to produce billions of copies in just a few hours. Soon after PCR's invention, a machine was developed to carry it out automatically, making it relatively inexpensive and reliable. It presented a great contrast to the only copying method previously available, which involved inserting target stretches of DNA into the genomes of bacteria and letting the bacteria multiply. That process took weeks and often incorporated errors made by the bacteria's genetic copying mechanism.

PCR can serve an almost infinite variety of purposes, making it, as *Current Biography Yearbook 1996* stated, "as fundamental to biological research as the screwdriver is to carpentry." Police investigators use it to copy minute amounts of DNA found at crime scenes so they can obtain amounts large enough to match against samples from suspects. Genetic counselors use it to test unborn babies for inherited diseases. Paleontologists use it to copy fragments of DNA from long-extinct organisms that have been preserved in amber, bones, or other remains. Indeed, the blockbuster 1993 movie *Jurassic Park* was based on the idea that whole dinosaurs could be created from DNA fragments in this way, although most scientists say that that would be impossible.

Mullis first described PCR to the scientific community in a paper published in late 1985 and a speech at a meeting in 1986. The speech's audience gave it a standing ovation. Cetus obtained a patent on the process in 1987 and later sold it to pharmaceutical giant Hoffman-LaRoche for $300 million—the largest amount paid for a single patent up to that time. Unhappy with the treatment he had received at Cetus, however, Mullis had left the company in 1986. He moved to San Diego, California, where he blended work with an easygoing lifestyle that came to include surfing, women, and writing. (*Dancing Naked in the Mind Field*, a book of his essays, was published in 1998.) Since then, he has been a consultant for several biotechnology companies; in 2002, he was vice president for molecular biology for Burstein Technologies in Irvine, California. He also explored whimsical projects, such as incorporating DNA from rock stars into jewelry, and espoused unpopular causes, most notably the claim that the virus called HIV has not been proved to cause AIDS.

Some observers thought that Mullis's controversial behavior and outspoken opinions would keep him from winning the Nobel Prize even for such an outstanding achievement as the invention of PCR, but that proved not to be the case. He also won many other awards, including the Gairdner Foundation International Award (1991), the Robert Koch Award (1992), and the Japan Prize (1993). He makes no apologies for his unconventional approach to science and life. "I think really good science doesn't come from hard work," he has said. "The striking advances come from people on the fringes, being playful." The creation of PCR certainly shows that this approach can sometimes be spectacularly successful.

Further Reading

"Autobiography of Kary B. Mullis." *Les Prix Nobel 1993*. Available online. URL: www.nobel.se/chemistry/laureates/1993/mullis-autobio.html. Last updated 2002.

Mullis, Kary B. *Dancing Naked in the Mind Field*. New York: Pantheon, 1998.

———. "The Unusual Origin of the Polymerase Chain Reaction." *Scientific American*, April 1990.

"Mullis, Kary B." *Current Biography Yearbook 1996*. New York: H. W. Wilson, 1996.

Wade, Nicholas. "After the 'Eureka,' a Nobelist Drops Out." In *The New York Times Scientists at Work*, edited by Laura Chang. New York: McGraw-Hill, 2000.

N

Nirenberg, Marshall

(1927–)
American
Biochemist, Molecular Biologist, Geneticist

In the early 1960s, Marshall Warren Nirenberg and his coworkers discovered how the cell uses the information carried in DNA to make proteins—in short, as the media of the time described it, they "cracked the genetic code." This groundbreaking achievement earned the 1968 Nobel Prize in physiology or medicine for Nirenberg and two other scientists who, working in other laboratories, also helped to decipher the code of life.

Nirenberg was born in New York City on April 10, 1927, to Harry and Minerva (Bykowsky) Nirenberg. Around age 10, he developed rheumatic fever, a bacterial disease that can damage the heart, and doctors recommended that the family move to a warmer climate to preserve his health. In 1941, therefore, the Nirenbergs went to Orlando, Florida, which at the time, Nirenberg has written, was "a natural paradise." He became fascinated by nature and "was happy exploring swamps and caves, and collecting spiders."

Nirenberg earned a B.S. in zoology and chemistry from the University of Florida, Gainesville, in 1948. He worked as a laboratory and teaching assistant while still an undergraduate. He obtained an M.S. in zoology from the university in 1952 with research on insects called caddisflies, but by this time biochemistry had become his chief interest. He earned a Ph.D. in that specialty from the University of Michigan, Ann Arbor, in 1957.

Nirenberg's Ph.D. project, which concerned the way certain cancer cells take up sugar, helped him obtain an American Cancer Society postdoctoral fellowship to the National Institutes of Health (NIH) in Bethesda, Maryland. He joined NIH permanently in 1960 and married a fellow NIH biochemist, Brazilian-born Perola Zaltzman, in 1961. Nirenberg became chief of the biochemical genetics section of the National Heart Institute in 1962 and head of the laboratory of biochemical genetics at the National Heart, Lung, and Blood Institute in 1966, a position he still holds.

Once settled at NIH, Nirenberg did something he later described as "very, very risky": He changed fields from biochemistry to genetics and molecular biology, in which he had no experience. These fields had been transformed just a few years before when FRANCIS CRICK and JAMES WATSON worked out the molecular structure of DNA, the carrier of genetic information in most organisms.

Scientists knew that the information in DNA was encoded in the sequence in which four types of smaller molecules called bases are joined together to make the long DNA molecules. This information specifies the order in which the cell must assemble 20 different kinds of molecules called amino acids to make particular proteins. Crick and others had suggested in the late 1950s that sets of three bases, termed codons, could represent particular amino acids, because three was the smallest number that allowed enough combinations (64, or $4 \times 4 \times 4$) to specify all the amino acids. This theory had not been proved, however, nor did anyone know which codons stood for which amino acids or how the information moved from DNA to protein.

Many molecular biologists suspected that the answer to the second question—and, indirectly, to the first one as well—lay in a chemical cousin of DNA called RNA. Unlike DNA, which stays in the nucleus of the cell, RNA can move from the nucleus into the cytoplasm, the substance that makes up the main part of the cell. Molecular biologists knew that DNA can copy itself into a molecule of RNA with bases in the same order, except that another base, uracil (U), substitutes for one of the DNA bases, thymine (T). Nirenberg thought that RNA might act as an intermediate between DNA and protein, providing a template, or pattern, in the sequence of its bases that chemicals in the cytoplasm could use to assemble amino acids in the correct order to make a certain protein.

Beginning in 1960, Nirenberg and a German postdoctoral researcher named J. Heinrich Matthaei created a "cell-free system" in which to test their theories. This system, made up of materials extracted from a common bacterium, contained everything thought to be necessary for the manufacture of proteins, including all 20 amino acids. Adapting a technique invented by Spanish-born American biochemist SEVERO OCHOA, Nirenberg and Matthaei then synthesized a piece of RNA containing just one base, uracil (U). When they added this synthetic RNA to their mixture, the system produced a protein made up of molecules of a single amino acid, phenylalanine. This landmark experiment simultaneously proved that RNA—specifically, the single-stranded form later called messenger RNA—could order the making of a protein and that the RNA code "word" for phenylalanine was U-U-U (T-T-T in DNA). "This was the first time anybody really showed that messenger RNA existed biochemically," Nirenberg said in a 1998 interview.

When Nirenberg presented his first "code translation" at the fifth International Congress of Biochemistry in Moscow in August 1961, his name was not well known, and few people came to hear his talk. DNA pioneer Crick was there, however, and he immediately understood the importance of what the younger man was saying. Crick arranged for Nirenberg to present his paper again at a session that nearly everyone at the conference attended. This time, many audience members were, as some reported later, "electrified." Nirenberg's work soon captured the attention of the media and the public as well. In a letter to Crick in January 1962, he commented wryly, "The American press has been saying that [my] work may result in (1) the cure of cancer and allied diseases, (2) the cause of cancer and the end of mankind, and (3) a better knowledge of the molecular structure of God. Well, it's all in a day's work."

Now that one letter of the genetic code had been deciphered, Nirenberg's and several other laboratories began a race to translate the rest. Nirenberg and coworker Philip Leder announced an improved technique for deciphering codons in 1964, and the laboratories of Nirenberg and his chief competitors, HAR GOBIND KHORANA, then at the University of Wisconsin, Madison, and Robert Holley, at Cornell University in New York, as well as Ochoa, worked out the meaning of all the triplets by 1966. They found that more than one triplet sometimes stands for the same amino acid and that some triplets do not represent amino acids at all but rather tell the cell when to start or stop protein synthesis or give other signals.

Khorana and Holley shared the 1968 Nobel Prize with Nirenberg for their parts in translating the code. Nirenberg also won other awards during the 1960s, including the National Medal of Science (1965), the Louisa Gross Horwitz Prize of Columbia University (1968), and the Albert Lasker Medical Research Award (1968).

With his work on the genetic code completed, Nirenberg changed the direction of his research once more and began studying the nervous system. In the late 1960s and early 1970s, he used laboratory cultures of brain cancer cells to examine the biochemistry of nerve cells. He studied neural receptors in the eyes of chicks from the mid-1970s to the late 1980s, then shifted his attention to genes active in embryonic development, especially "master" genes called homeobox genes that lay out the map of the developing embryo. He studied these genes chiefly in fruit flies (a common organism used in genetic experiments) and to a lesser extent in mice, discovering several new homeobox genes.

In the 1990s and beyond, Nirenberg has combined his two previous interests in a study of the genes that control the nervous system's development in the embryo. He is trying to learn how individual nerve cells find the right partners with which to establish connections, or synapses. He is also investigating how key genes and biochemicals affect each other's actions during development and in the daily metabolism of cells. Nirenberg has been active politically as well, warning of the dangers of environmental destruction and incautious alteration of genes, including human cloning.

Further Reading

Cohen, Richard. "Lasker Luminaries: An Interview with Marshall Nirenberg." Albert and Mary Lasker Foundation. Available online. URL: www.laskerfoundation.org/awards/library/lumin int_mn.html. Posted 1998.

National Library of Medicine. "Profiles in Science: The Marshall Nirenberg Papers." Available online. URL: http://profiles.nlm.nih.gov/JJ/Views/Exhibit/narrative/biographical.html. Downloaded 2002.

"Nirenberg, Marshall W(arren)." *Current Biography Yearbook 1965*. New York: H. W. Wilson, 1965.

Nüsslein-Volhard, Christiane
(1942–)
German
Geneticist, Embryologist

Christiane Nüsslein-Volhard helped to identify sets of genes that control the development of body structure in animals ranging from fruit flies to humans. For this work, she and two other researchers in the same field shared the Nobel Prize in physiology or medicine in 1995. Donald Brown, an embryologist at the Carnegie Institution of Washington, has called Nüsslein-Volhard "the most important developmental biologist of the second half of this century. . . . Perhaps of all time."

Nüsslein-Volhard was born Christiane ("Janni" to her friends) Volhard on October 20, 1942, in Magdeburg, Germany. She grew up in Frankfurt. As her interest in science developed ("I knew at the age of 12 at the latest that I wanted to be a biologist," she wrote in her Nobel Foundation autobiographical sketch), she found herself somewhat alone in a family primarily devoted to the arts. Her father, Rolf Volhard, was an architect, and her mother, the former Brigitte Haas, was the daughter of a painter. Her four brothers and sisters were also interested mostly in the arts. Her parents encouraged her to follow her own path, however, and she shared the family's fondness for music and art, so they remained close.

Nüsslein-Volhard (the first part of her last name is left over from an early, short marriage) studied biology, physics, and chemistry at the

Goethe University in Frankfurt, graduating in 1964. She earned a diploma in biochemistry from the University of Tübingen in 1969 and a Ph.D. in genetics from the same university in 1973. She did postdoctoral work in Basel, Switzerland, and Freiburg, Germany, in the mid-1970s.

While at Tübingen, Nüsslein-Volhard became interested in the way genes control the complex processes by which living things develop before birth. Like THOMAS HUNT MORGAN and many other geneticists, she chose fruit flies as her study organism. Most of these earlier geneticists had studied mutations that affect adult flies, but a few, including Morgan student Edward Lewis, one of the two scientists with whom Nüsslein-Volhard would later share the Nobel Prize, had focused on genes that affect development. Lewis had found a mutated fly that had two pairs of wings instead of the normal single pair and had shown that, in fact, the mutation had duplicated a whole segment of the flies' bodies.

Nüsslein-Volhard did her first fruit fly research in Basel. She also met Eric Wieschaus, who shared her interest and would later share her Nobel Prize as well. The two joined the European Molecular Biology Laboratory in Heidelberg in 1978 and became coheads of a small research team. During the next two years, they treated thousands of flies with chemicals that produced massive mutations, then studied the flies' offspring under microscopes to observe the mutations' effects. They worked out new techniques to identify mutations that affect development at early stages and to determine which patterns of development the genes altered. Their announcement of the identification of 15 developmental genes in 1980 demonstrated for the first time that the number of "master" developmental genes is fairly small and that the genes could be systematically identified.

Nüsslein-Volhard returned to the Max Planck Institutes in Tübingen, where she had done some of her diploma research, in 1981. She was a group leader at the Friedrich Miescher Laboratory there until 1985. From 1986 to 1990, she was one of five directors of the group's Institute of Developmental Biology. She then became director of the entire department of genetics for the Max Planck Institutes, a position she still holds.

Nüsslein-Volhard and her coworkers have identified some 120 "pattern genes" that control formation of body segments and the organs in them. These genes act partly by creating substances that flow through the embryo in gradients moving from high concentrations to low and turning on other genes as they go. The gradients lay out the map of embryonic development, determining, for instance, which part will be the embryo's head and which its tail. These master genes have been called homeotic genes. Other researchers showed in the 1990s that such genes exist in a wide variety of organisms, including humans. The fact that homeotic genes have been conserved so widely during evolution emphasizes their essential role in development.

Even before she won the Nobel Prize, Nüsslein-Volhard was internationally renowned for her work with fruit flies. For instance, she won the Albert Lasker Public Service Award in 1991 and the Gregor Mendel Medal from the Genetical Society of Great Britain and the General Motors Cancer Research Prize in 1992. Her coworkers were all the more surprised, therefore, when she announced in the early 1990s that she was changing the focus of her work to zebrafish, then almost unknown in genetic studies. She said that she wanted to learn more about homeotic genes in vertebrates and therefore needed to study a vertebrate. Furthermore, she pointed out, zebrafish eggs and the unborn fish inside them are transparent, which makes their development easier to observe than that of fruit flies or most other creatures.

Although Nüsslein-Volhard's switch to zebrafish shocked some others in her field at first ("This is a terrible thing for science," one wrote), these small, striped fish later became common tools of developmental research. She keeps hundreds of thousands of them in 7,000 tanks near her laboratory, and she and her coworkers have isolated some 1,200 valuable mutants among them. She published a massive atlas of zebrafish genes in 1996.

Nüsslein-Volhard's discoveries may shed light on spontaneous abortions (miscarriages) and birth defects, about 40 percent of which are thought to be due to malfunctioning of the developmental genes she studies. She compares her work to one of her favorite hobbies, making her own jigsaw puzzles. In genetic as well as physical puzzles, she says, "the most important thing is . . . finding enough pieces and enough connections between them to recognize the whole picture."

Further Reading

Ackerman, Jennifer. "Journey to the Center of the Egg." *New York Times Magazine*, October 12, 1997.

Angier, Natalie. "'Lady of the Flies' Dives into a New Pond." In *The New York Times Scientists at Work*, edited by Laura Chang. New York: McGraw-Hill, 2000.

Nüsslein-Volhard, Christiane. "Christiane Nüsslein-Volhard—Autobiography." *Les Prix Nobel 1995*. Available online. URL: www.nobel.se/medicine/laureates/1995/nusslein-volhard-autobio.html. Last updated 2002.

Ochoa, Severo

(1905–1993)
Spanish/American
Biochemist

Severo Ochoa revealed important facts about biochemical substances and processes, including the way cells use energy and the way nucleic acids reproduce. He was born on September 24, 1905, in Luarca, Spain. His father, also named Severo Ochoa, was a lawyer. His mother was the former Carmen Albornoz.

Ochoa obtained his bachelor's degree from the College of Málaga in 1921 and his medical degree with honors from the University of Madrid in 1929. He did postdoctoral work in the laboratories of OTTO FRITZ MEYERHOF at the Kaiser Wilhelm Institute in Germany and HENRY HALLETT DALE at the National Institute for Medical Research in England. He returned to Spain in 1931 and became a lecturer in biochemistry and physiology at the University of Madrid's medical school. He also married Carmen Garcia Cobian in that year. In 1935, he became head of the physiology division at the university's new Institute for Medical Research, but he left the country a year later because of the Spanish civil war. He spent another year in Meyerhof's laboratory, then went to England, where he worked at Oxford University from 1938 to 1941.

Ochoa came to the United States in 1941 and joined Washington University in St. Louis, where he worked in the laboratory of CARL FERDINAND CORI and GERTY THERESA RADNITZ CORI and taught pharmacology for a year. In 1942, he moved to the New York University College of Medicine, becoming a professor of pharmacology and head of the pharmacology department in 1946, a professor of biochemistry and head of the biochemistry department in 1954, and a naturalized U.S. citizen in 1956. He retired from New York University in 1974 and joined the Roche Institute of Molecular Biology in New Jersey. In 1985, he returned to Spain and became a professor of biology at the University Autonoma in Madrid. He died in Madrid on November 1, 1993.

Much of Ochoa's early research involved enzymes, which make many biochemical processes possible by speeding up the chemical reactions involved in them. Inspired by Meyerhof, he studied the ways cells, especially in muscle, store and use energy. He investigated the biological function of vitamin B1 (thiamine) while at Oxford in the late 1930s. With the Coris, he studied enzymes involved in the processes by which the carbohydrate glycogen is broken down to release energy. Around 1947, he isolated two enzymes that take part in the Krebs or citric acid cycle, named after HANS ADOLF KREBS, which is part of the glycogen breakdown process. Then, in the early 1950s, he isolated triphosphopyridine nucleotide (TPN), a key substance in photosynthesis, the process by which plants use energy from sunlight to change water and carbon dioxide into sugars and other foods. TPN is also found in animal cells, where it takes part in processes exactly opposite to photosynthesis, converting sugars back into carbon dioxide and water and releasing energy for the cells to use.

Ochoa's best known scientific contribution was announced in 1955. He used polynucleotide phosphorylase, an enzyme he had discovered in sewage bacteria, to join subunits (nucleotides) together into a chainlike molecule of ribonucleic acid (RNA) in the laboratory. RNA is an essential link between the genetic information carried in DNA and the proteins that the cell manufactures according to DNA's instructions. Although Ochoa's synthetic version was not exactly like natural RNA, the processes he used were similar to those that go on in the cell, and his creation of artificial RNA proved essential for discovering how cells make proteins.

ARTHUR KORNBERG of Stanford University, who had studied under Ochoa in 1946, adapted Ochoa's methods to synthesize DNA in 1957. Ochoa and Kornberg shared the Nobel Prize in physiology or medicine in 1959 for their achievements. In the early 1960s, Ochoa also helped to decipher the "genetic code," through which the order of smaller molecules called bases within DNA and RNA molecules specifies the order in which the cell must assemble smaller molecules called amino acids to make proteins. In addition to the Nobel Prize, Ochoa received the Neuberg Medal in Biochemistry (1951), the Charles Leopold Mayer Prize (1955), an award from the Societé de Chimie Biologique (1955), and the U.S. National Medal of Science (1980).

Further Reading

"Ochoa, Severo." *Current Biography Yearbook 1962*. New York: H. W. Wilson, 1962.

Ohta, Tomoko

(1933–)
Japanese
Geneticist, Evolutionary Biologist

Tomoko Ohta has helped to develop the relatively new scientific branch of population genetics, which studies the mechanisms of evolutionary change at the molecular level. She was born in Aichi Prefecture, Japan, in a village called Miyoshi-cho, on September 7, 1933. She grew up during the difficult days during and after World War II when many goods, even

Japanese geneticist Tomoko Ohta developed the "nearly neutral" hypothesis about evolution at the molecular level in the early 1970s. *(Tomoko Ohta)*

food, were often in short supply. On the other hand, she points out, the postwar period brought new educational opportunities for girls and women, including herself.

One of the first women to attend the University of Tokyo, Ohta studied horticulture (the science of growing garden plants) and obtained a bachelor's degree in agriculture in 1956. Lack of money delayed her graduate studies, but she finally earned a Ph.D. in population genetics from North Carolina State University in 1967. This field allowed her to combine an early interest in mathematics with a new interest in genetics. The University of Tokyo awarded her a doctorate in science in 1972.

After gaining her Ph.D., Ohta returned to Japan and joined the department of population genetics at the National Institute of Genetics in Mishima in 1967. There, she worked for MOTOO KIMURA, helping him find evidence to support his theory about the way proteins evolved. Kimura believed that most evolution at the molecular level is caused by genetic drift (inheritance of random changes in genes, whether they benefit the organism or not) rather than by the mechanism of natural selection described by CHARLES ROBERT DARWIN. This "neutral" theory of mutations, which Kimura first proposed in 1968, was and to some extent is still controversial, but geneticists have found it valuable because it provides testable predictions about the rate of molecular evolution. Ohta was a coauthor of Kimura's influential book *Theoretical Aspects of Population Genetics* (1971).

Beginning in the early 1970s, Ohta developed her own theory, the "nearly neutral" hypothesis, to explain some aspects of molecular evolution. In contrast to both Kimura and classical evolutionists, she believes that both random forces and natural selection affect the preservation of many mutations. She and Kimura argued about these ideas often, but, unlike traditional laboratory heads in Japan, she says, he encouraged her and other younger researchers to think for themselves. Some geneticists, including Kimura, felt that her theory was unnecessarily complex, but she showed in a 1995 paper that new techniques of DNA analysis provided evidence to support it.

Frustrated because there was no evidence at the time to prove any of the competing theories about molecular evolution, Ohta turned in the late 1970s and 1980s to a related problem, the way gene "families"—groups of closely related genes with similar functions—evolve. She wrote a book on this subject, *Evolution and Variation of Multigene Families*, which was published in 1980. Variation among these groups of seemingly redundant genes, she believes, helps to generate genetic diversity and plays an important role in evolution.

Ohta headed the first laboratory of population genetics at the National Institute of Genetics from 1977 to 1984 and was a professor at the institute from 1984 until 1997, when she retired to become a professor emerita and adjunct professor. She was a vice director of the institute in 1990–91. In 1981, she became the first winner of the Saruhashi Prize, a prize for women scientists established by the eminent Japanese female geochemist Katsuko Saruhashi and the organization Saruhashi founded, the Association for the Bright Future of Women Scientists. Ohta has also won the Japan Academy Prize (1985), the Avon Special Prize for Women (1986), and the Weldon Memorial Prize from Britain's Oxford University (1986). In 2002, she was elected a foreign member of the U.S. Academy of Sciences, the first Japanese woman to be so honored.

Further Reading

Kozai, Yoshihide, et al., eds. *My Life: Twenty Japanese Women Scientists*. Tokyo: Uchida Rokakuho Publishing Co., Ltd., 2001.

Myers, Fred. "Women Scientists: It's Lonely at the Top." *Science*, October 23, 1992.

P

⊠ Paracelsus (Philippus Aureolus Theophrastus Bombastus von Hohenheim)

(1493–1541)
Swiss
Physician, Chemist

A contentious man obsessed with mystical doctrines, the wandering Swiss physician who called himself Paracelsus nonetheless advanced medicine by introducing chemical concepts and urging physicians to rely on their own observations rather than the words of ancient authorities. His rebellious attitude led some contemporaries to compare him to the religious reformer Martin Luther, although he did not share Luther's beliefs.

Paracelsus was born Philippus Aureolus Theophrastus Bombastus von Hohenheim in Einsiedeln, Switzerland, around November 10, 1493. His father, Wilhelm Bombast von Hohenheim, the illegitimate son of a German noble family, was a practicing physician. Philippus's mother, the former Elsa Ochsner, was a local woman who worked for a large abbey in the town. She drowned herself when the boy was nine years old. Philippus and his father then moved to Villach, Austria, in 1502.

From his father, Philippus learned medicine and also probably alchemy, a forerunner of chemistry that combined a study of physical substances with complex religious and magical symbolism. Instructors in Villach, whose usual job was preparing men for work in the local mines, taught him about minerals. In 1507, when he was just 14 years old, he left home to become a wandering scholar and physician. He attended universities throughout Europe, but no one knows whether he earned a degree from any of them. The chief thing he seems to have acquired was a dislike of the entire academic approach.

Renaissance scholars had recently rediscovered and retranslated many ancient Greek and Roman manuscripts lost or distorted during the Middle Ages, and they held these books in the highest regard. In medicine, the chief authority was the Roman physician GALEN, who (following even older thinkers such as the Greek HIPPOCRATES) had taught that disease was due to an imbalance among four humors, or fluids, that were thought to flow through the body. Von Hohenheim showed his opinion of these ancient authorities—as well as himself—in the name he began using during this period: *Paracelsus* means "greater than Celsus," another Roman physician whose works were very popular. "One hair on my neck knows more than all you authors, and my shoe-buckles contain more wisdom than both Galen and Avicenna [a highly respected medieval Arabic physician]," Paracelsus once wrote.

Paracelsus insisted that, instead of following ancient authorities, physicians should learn from him or from those who had direct experience with treating the sick, even if such people belonged to groups that academics disregarded. "From time to time [the physician] must consult old women, gypsies, magicians, wayfarers and all manner of peasant folk and random people; for these have more knowledge about [medicine] than all the high colleges," he wrote. Most of all, he said, physicians should learn "by our own observation of nature, confirmed by extensive practice and long experience." Ironically, Galen, whom Paracelsus so outspokenly scorned, had said much the same thing.

Paracelsus questioned the four humor theory of disease, maintaining that illness is usually caused by forces outside the body rather than an imbalance inside. His approach to treatment was also new, relying chiefly on chemical compounds, especially those made from metals, instead of plant-based remedies or nonspecific treatments such as bleeding. He was the first to give most of these compounds internally. His stress on chemicals was tied to his continuing belief in alchemy. ("The object of alchemy is not to make gold but to prepare medicines," he once wrote.) His introduction of the concepts of chemistry, which were contained within alchemy, into medicine helped to lay the groundwork for pharmacology, the systematic study and development of drugs.

Many medicines that Paracelsus used, such as salts of mercury, arsenic, and antimony, were potentially poisonous. Nonetheless, because he paid careful attention to the way these compounds were prepared and the amounts, or dosages, he gave (something that few other doctors of the time bothered to do), he often cured patients whom no other physician was able to help. These successes, combined with his wide experience—he had been a surgeon in several armies as well as a wandering physician—brought him a considerable reputation as a healer.

Because of this reputation, Johannes Frobenius, a renowned humanist scholar and publisher in Basel, asked Paracelsus to come from the nearby city of Strasbourg, where the physician was then living, and treat Frobenius's infected leg in 1527. When the treatment succeeded, the grateful Frobenius persuaded the Basel city council to make Paracelsus both the city's physician and a professor of medicine at the local university. The other professors resented this move, and Paracelsus seemingly went out of his way to make the situation worse. For instance, he burned the works of Galen and Avicenna in a public bonfire and insisted on teaching his classes in the "vulgar" Germanic language rather than in the Latin that everyone else used.

After Frobenius died in 1528, Paracelsus's opponents lost no time in making him feel unwelcome. After losing a disastrous lawsuit, he left Basel and resumed his wandering through Europe. By 1541, when he was invited to become the personal physician of Duke Ernst of Bavaria in Salzburg, he was already in poor health, although he was only 48 years old. He died on September 24 of that year.

Paracelsus had his greatest influence after his death. He had written down many of his ideas, but only one book, *Grosse Wundartzney* (*Great Medical Book on Wounds*, 1536), had been published during his lifetime. After he died, however, he began to gain the following he had never had in life. Disciples recovered manuscripts that he had left in various cities and published them, often along with commentaries. These books included a volume on the diseases of miners that was the first book about illnesses related to a specific occupation. Paracelsus's writings contained many new and accurate observations about disease and recommendations for treatment. For example, he wrote that mental illnesses had a physical cause, rather than being due to possession by demons as most people thought, and should be treated gently.

By the mid-17th century, Paracelsus's incorporation of chemistry into medicine had become widely accepted in Europe. It encouraged advances in both sciences and opened the way for people such as ANDREAS VESALIUS and WILLIAM HARVEY to develop a new kind of medicine.

Further Reading

Debus, Allan G. "Paracelsus and the Medical Revolution of the Renaissance: A 500th-Anniversary Celebration." National Library of Medicine. Available online. URL: www.nlm.nih.gov/exhibition/paracelsus/paracelsus_2.html. Updated 1998.

Jacobi, Jolande, ed. *Paracelsus*. Translated by Norbert Guterman. Princeton, N.J.: Princeton University Press, 1995.

Hartman, Francis, and R. I. Robb, eds., *The Life and Teachings of Paracelsus*. Reprint, San Diego, Calif.: Wizards Bookshelf, 1997.

Louis Pasteur helped to establish that microorganisms could cause disease, and he developed ways to weaken microbes in the laboratory so they could be used in vaccines that would protect people and animals against microbe-caused illnesses. *(National Library of Medicine)*

Pasteur, Louis

(1822–1895)
French
Chemist, Microbiologist

Although trained as a chemist rather than a physician, Louis Pasteur caused a revolution in medicine by proposing and helping to prove that microorganisms could cause deadly diseases. He also developed vaccines to prevent several such diseases, saved millions of francs for French industries, and became one of the most revered scientists of all time.

Pasteur was born on December 27, 1822, in Dôle, a village in eastern France near the Jura Mountains, to Jean-Joseph Pasteur, a tanner, and his wife, Jeanne (Roqui). He grew up in the neighboring town of Arbois. As a youth he was more interested in art (for which he showed considerable talent) than in schoolwork, but his father urged him to train to be a teacher. After earning a bachelor of arts degree in 1840 and a bachelor of science degree in 1842 from the Royal College in Besançon, therefore, Pasteur entered the Ecole Normale Supérieure, a huge government-run teacher training college in Paris, in 1843. He qualified as a physical science teacher three years later and earned a doctorate of science in 1847.

By that time, however, Pasteur had become more interested in chemistry than in teaching. Guided by professors at the Sorbonne and other Paris universities, he did such impressive research on crystals that he was allowed to present his work to the French Academy of Sciences, the country's top science organization, in 1848. This was a tremendous honor for a 25-year-old man.

A year after his triumph at the academy, Pasteur became a professor of chemistry at the University of Strasbourg. There, he met Marie Laurent, the daughter of the college's minister, and they married later in 1849. Their long and happy marriage produced five children, though only two survived to adulthood. Pasteur was professor of chemistry and dean of the science faculty at the University of Lille from 1854 to 1857, when he returned to Paris to become the assistant director of scientific studies and general administration at the Ecole Normale Supérieure. Around 1867, he gave up these administrative posts, but he continued to do research at the school.

Solving an industrial problem for a factory owner in Lille interested Pasteur in the subject of fermentation around 1856. In fermentation, a widespread process in nature, sugar is broken down to produce carbon dioxide and other substances, such as alcohol. Most chemists of the time thought that fermentation was strictly a chemical process, but during the next 15 years or so, Pasteur proved that it was carried out by yeasts and other microorganisms. This research, showing that "fermentation is . . . correlative [always associated] with life," established the basic principle that microorganisms, or microbes, could produce major changes in living matter, which underlay all the rest of Pasteur's research. Pasteur believed that microorganisms also caused putrefaction, or decay, a breakdown process similar to fermentation that occurs in animal tissue.

Throughout his life, Pasteur combined basic science with practical applications. His fermentation research not only cast light on a basic natural process but yielded a way of controlling that process to meet human needs. In 1866, he showed that undesirable fermentations, such as types that spoiled wine, could be prevented by heating the liquids. This process, called pasteurization in his honor, was later used to kill disease-causing bacteria in milk and other drinks.

In 1858, while still studying fermentation, Pasteur also began research on spontaneous generation, the idea that living things could develop from nonliving matter. Many people still believed that this ancient theory held true at least for microorganisms, and some scientists claimed to have proved it. Pasteur, however, showed that microbes would not appear in nutrient solutions made sterile (free of microorganisms) to begin with and kept away from dust in the air, on which microbes could live. After reading about Pasteur's demonstration that microbes were associated with dust, which was published in 1862, British surgeon JOSEPH LISTER concluded that microbes in dust caused wound infections, and he developed a way to prevent such infections by destroying microorganisms in the wounds.

Two tragic events interrupted Pasteur's next research project, which, like the one on fermentation, involved a major threat to the French economy. In 1865, the French Ministry of Agriculture asked him to investigate a disease that was killing large numbers of silkworms and destroying the French silk industry. Silkworms are the caterpillars, or larvae, of a type of moth, and silk thread is made from the cocoons in which they change from larva to adult. Pasteur eventually identified not one but two silkworm diseases, one of which was associated with microorganisms (the other was probably caused by viruses, which were too small to see with microscopes of the day), and showed silkworm breeders how to prevent both. In 1868, however, he had to halt his studies for several months because he suffered a stroke that temporarily paralyzed his left side. Two years later, he published a two-volume work on his silkworm research, but 1870 was a bitter year for him because France lost the Franco-Prussian War.

Pasteur's silkworm studies, along with his earlier work on fermentation and putrefaction, convinced him that microbes, or germs, cause many diseases. "When we see beer and wine subjected to deep alterations because they have given refuge to microorganisms," he wrote around 1870, "it is impossible not to be pursued by the thought that similar facts may, *must*, take place in animals and in man." He devoted the rest of his life to research that grew out of this idea, which came to be known as the germ theory of disease.

During the 1870s, Pasteur investigated several diseases that affected birds, mammals, and humans, including anthrax, which killed sheep and cattle, and chicken cholera, which attacked fowls. An accident during the chicken cholera research, in which a laboratory dish containing the microorganisms that caused the disease was left behind during the researchers' summer vacation, led to the discovery that such aging weakened the microbes to the point where they no longer made chickens sick. Furthermore, chickens that had received injections of the weakened microbes remained healthy even when given injections of full-strength cholera germs.

Pasteur realized that he had accidentally rediscovered vaccination, a procedure that British physician EDWARD JENNER had developed almost 100 years before. Instead of relying on a naturally weak form of a microbe as Jenner had done, however, Pasteur had found a way to weaken microorganisms deliberately in the laboratory. He and his coworkers developed several other techniques to weaken bacteria and used them to make vaccines against several animal diseases, including anthrax. They demonstrated the anthrax vaccine successfully in a dramatic trial at a farm called Pouilly le Fort on June 2, 1882. After that day, as one of Pasteur's assistants wrote, "there were no longer any skeptics [about Pasteur's vaccine] but only admirers." The vaccine was quickly mass-produced and given to thousands of animals. Pasteur was awarded the Great Ribbon of the Legion of Honor for this work.

Pasteur's final triumph was a vaccine against rabies, a brain disease spread by the bite of sick animals. Although fairly rare, rabies was painful and always fatal, so this vaccine, Pasteur's only one against a human disease (anthrax can affect humans, but Pasteur's anthrax vaccine was given only to animals), "increased [his] . . . popularity . . . more than all his former works," according to an assistant.

Pasteur was still testing the rabies vaccine, made from the dried spinal cords of rabbits given the disease, on animals when nine-year-old Joseph Meister and his mother appeared on the scientist's doorstep on July 6, 1885. Meister's mother begged Pasteur to give the vaccine to the boy, who had been mauled by a rabid dog, and, after much hesitation, Pasteur did so. Fortunately, Meister remained healthy. After Pasteur reported the successful vaccination to the Academy of Sciences in October 1885, people from as far away as Russia and the United States came to Paris to receive the vaccine.

By this time, France saw the aging Pasteur as a national hero, almost a saint. In the late 1880s, the French government combined its money with donations from other governments and individuals all over Europe to build an institute that would carry on his work. The Pasteur Institute, still a world-famous research center, opened its doors on November 14, 1888. Pasteur attended this celebration and another, held at the Sorbonne in 1892 to honor his 70th birthday, but by then a second stroke had greatly weakened him. He died of kidney disease on September 28, 1895. After a magnificent state funeral, he was buried on the grounds of the Pasteur Institute.

Further Reading

Cohn, David V. "The Life and Times of Louis Pasteur." 1996. University of Louisville (Kentucky). Available online. URL: www.louisville.edu/library/ekstrom/special/pasteur/cohn.html. Accessed 2003.

Geison, Gerald L. *The Private Science of Louis Pasteur.* Princeton, N.J.: Princeton University Press, 1995.

Robbins, Louise. *Pasteur and the Hidden World of Microbes.* New York: Oxford University Press, 2001.

Trachtmann, Gerald. "Hero for Our Time." *Smithsonian,* January 2002.

Vallery-Radot, René, *The Life of Pasteur.* Translated by R. L. Devonshire. Garden City, N.Y.: Garden City Publishing Co., 1927.

Patrick, Ruth
(1907–)
American
Ecologist

Ruth Patrick made pioneering studies on the effects of pollution on freshwater ecology and invented a sensitive tool for evaluating water pollution. She was born on November 26, 1907, in Topeka, Kansas, but grew up in Kansas City, Missouri. Her father, Frank Patrick, was a lawyer, but his hobby was studying diatoms, one-celled algae (plantlike living things) that are the base of the food chain in freshwater. He took Ruth and her sister on weekly "expeditions" into the nearby woods, during which the girls gathered plants and other specimens. He also used a tin can suspended from a pole to scrape water plants and other organisms from rocks in a stream. The group then took their finds home and examined them under Frank Patrick's microscope. These experiences began Ruth Patrick's lifelong studies of freshwater ecosystems.

Ruth Patrick is credited with cofounding the field of limnology, or freshwater ecology; she also developed a device that aids in the detection of pollution in rivers and streams by measuring populations of microorganisms called diatoms. *(Academy of Natural Sciences, Philadelphia)*

Patrick attended Coker College, a women's college in Hartsville, South Carolina, graduating in 1929. She earned a master's degree in 1931 and a Ph.D. in botany in 1934, both from the University of Virginia. During the summers of her college years, she supplemented her science education by studying at such famous biological research institutions as Cold Spring Harbor Laboratory in New York. There, she met Charles Hodge IV, who was studying to be an entomologist (insect specialist). They married in 1931, but Patrick continued to use her maiden name professionally as, she has said, a tribute to her father. She and Hodge had one son. Hodge died in 1985, and Patrick married Lewis H. Van Dusen Jr. in 1995.

Patrick has spent most of her working life at the Academy of Natural Sciences in Philadelphia, which she joined in 1937. She was an unpaid assistant curator of microscopy until 1945, when she was finally put on the payroll. In 1947, she established a department of limnology (freshwater ecology) at the academy, now called the Patrick Center for Environmental Research, and directed it until 1973; she is still its curator. From 1973, she held the Francis Boyer Research Chair at the academy. She was the chairperson of the academy's board of trustees from 1973 to 1976—the first woman to hold this position—and is now its honorary chairperson. She has also taught at the University of Pennsylvania.

Patrick's first study subject was diatoms. With Charles Reimer, she produced a monumental two-part work on the subject, *Diatoms of the United States*, which was published in 1966. She then expanded her research to include general ecology and biodiversity in rivers, studying hundreds of streams, rivers, and lakes in North and South America. She is considered to be the cofounder of the field of limnology.

Patrick studied pollution's effect on streams long before RACHEL LOUISE CARSON made concern about pollution fashionable. Beginning in the 1940s, she showed that diatoms are sensitive indicators of pollution in freshwater, and she invented the diatometer, which measures the numbers and sizes of different species of diatoms in a water sample. She developed a mathematical model of a natural diatom community and showed how the degree of pollution could be determined by measuring changes in the community. She was the first to point out that scientists need to study whole ecological communities, not just single species, to determine the effects of pollution. Maintenance of diversity in such communities, she says, is a sign of ecological health.

Unlike Carson, Patrick has had a relatively cordial relationship with government and industry. She has often worked as a consultant for both. She has even been a director of the Pennsylvania Power and Light Company and of the chemical company du Pont—the first woman and the first environmentalist ever on the board of the latter. "We have to develop an atmosphere where the industrialist trusts the scientist and the scientist trusts the industrialist," she has said. She currently heads The Environmental Associates of the Academy of Natural Sciences, a group of corporate executive officers concerned about environmental effects of industrial activities.

Patrick was elected to the National Academy of Sciences in 1970. In 1975, she won the John and Alice Tyler Ecology Award, at the time the world's highest-paying award in science. She used the money to prepare a five-volume masterwork, *Rivers of the United States*, the first volume of which was published in 1994. In 1996, she won the American Society of Limnology and Oceanography's Lifetime Achievement Award as well as the National Medal of Science and was also inducted into the South Carolina Hall of Science and Technology. The University of South Carolina, Aiken, has named its science education center after her.

Further Reading

Holden, Constance. "Ruth Patrick: Hard Work Brings Its Own (and Tyler) Award." *Science*, June 6, 1975.

Pauling, Linus Carl
(1901–1994)
American
Chemist

Linus Pauling, the first person to win two unshared Nobel Prizes, made contributions to scientific fields ranging from quantum physics to cancer treatment. His discoveries in biology included the basic structure of protein molecules and the nature of the defect in the common inherited disease sickle-cell anemia.

Pauling was born in Portland, Oregon, on February 28, 1901, to Herman and Lucy (Darling) Pauling. He spent part of his childhood in tiny Condon, Oregon, where Herman Pauling ran a drugstore, and his interest in nature probably began in that rural environment. The family was again living in Portland, however, when Herman Pauling died around 1910. Linus had to take part-time jobs to contribute to the family income, but he still found time to collect insects and minerals and do chemistry experiments. He failed to graduate from high school because he insisted on substituting his own reading for the school's required civics class.

Although his mother urged him to end his schooling and go to work, Pauling entered Oregon Agricultural College, later Oregon State University, in Corvallis in 1917. Here, too, he took many part-time jobs, including working as a laboratory assistant and teaching chemistry to home economics majors. He fell in love with one of the students in that class, Ava Helen Miller, and they married in 1923. They had four children.

After graduating in 1922 with a B.Sc. in chemical engineering, Pauling went to California Institute of Technology (Caltech), in Pasadena. His advisor, Roscoe Dickinson, introduced him to the new technique of X-ray crystallography, which helped chemists determine the arrangement of atoms within molecules. Pauling became expert in applying this tool to complex organic, or carbon-containing, compounds. He earned a Ph.D. in chemistry with highest honors in 1925 and then did postdoctoral work in chemistry and physics in Europe, studying with such eminent scientists as Niels Bohr. He returned to Caltech in 1927 and remained there until 1963, becoming a full professor in 1931. From 1937 to 1958, he headed Caltech's department of chemistry and chemical engineering as well as its Gates and Crellin Chemical Laboratories.

Pauling's chief interest had always been the structural arrangement of atoms within molecules and the electrical forces, called bonds, that hold the atoms together. In the late 1920s, he worked out a theory of chemical bonds that applied quantum mechanics, a new set of theories describing the behavior of particles within atoms that he had learned about in Europe, to chemistry. Pauling's theory allowed molecules to be described in three dimensions for the first time. He described his ideas most fully in *The Nature of the Chemical Bond and the Structure of Molecules and Crystals* (1939), which fellow Nobel laureate JAMES WATSON has called "the most influential chemistry book of the century."

In the mid-1930s, Pauling became interested in the molecular structure of proteins. His first work, beginning in 1934, was on a protein called hemoglobin, which gives blood its red color. Pauling showed in 1936 that hemoglobin binds to oxygen, which it carries throughout the body. Turning to a different group of proteins, he demonstrated in 1942 that antibodies, which certain cells in the body's defense system produce, are variations of a single protein called globulin.

Meanwhile, other kinds of defense were on everyone's mind. Pauling did war-related research during World War II, for instance, inventing a new type of explosive and a substitute for human blood plasma, but he refused to work on the Manhattan Project, which developed the atomic bomb. Once atom bombs became public knowledge, he joined scientists such as Albert Einstein in protesting against their use and testing because he feared the damage that radiation could do to genes and tissues.

Pauling returned to his interest to proteins after the war. Scientists knew that proteins were made up of smaller molecules called amino acids, joined together in chains (polypeptides). Pauling believed that the chains were folded in some way, but he was unable to decide how until one day in 1948, when he was bedridden with a kidney infection. As he wrote later, he "took a sheet of paper and sketched the atoms with the bonds between them and then folded the paper to bend one bond at the . . . angle I thought it should be . . ., and kept doing this, making a helix [corkscrew shape], until I could form hydrogen bonds between one turn of the helix and the next turn." A few hours of this molecular origami revealed what came to be called the alpha helix.

Other scientists questioned Pauling's proposal at first, but experiments soon confirmed that the alpha helix represented the basic structure of protein molecules. During the next several years, Pauling built on his discovery to work out the structure of specific proteins in skin, muscle, and other tissues. He won his first Nobel Prize, a prize in chemistry awarded in 1954, for his work on the structure of proteins and other molecules.

Pauling also began investigating sickle-cell anemia in the late 1940s. In this illness, fairly common among people of African or Mediterranean descent, red blood cells with crescent shapes block tiny blood vessels, causing pain and tissue damage. Pauling showed that the blood cells of people with the disease contain abnormal hemoglobin molecules produced by a defective gene that substitutes one amino acid for another in the part of the molecule called globin. This was the first genetic disease shown to be due to a molecular defect. Pauling and coworkers H. Itano and J. Singer published their work on sickle cell anemia in 1949.

In the early 1950s, Pauling stepped up his antinuclear protests, a dangerous thing to do in the repressive political environment of the time. He was accused of being a Communist, though he always denied the charge. The U.S. government saw him as a security risk and repeatedly refused to issue him a passport, which prevented him from attending scientific meetings in other countries. He could not even obtain permission to go to the 1954 Nobel Prize ceremonies until two weeks before they took place.

Pauling biographer Anthony Serafini thinks that these travel restrictions may have contributed to Pauling's most famous mistake. In early 1953, Pauling claimed that the DNA molecule was a triple helix, but rival researchers Watson and FRANCIS CRICK at Cambridge University showed soon afterward that it was in fact a double helix. Serafini believes that if Pauling had seen the photographs that X-ray crystallographer ROSALIND ELSIE FRANKLIN presented at a key scientific meeting in spring 1952, he might not have made this error.

Despite these problems, Pauling refused to give up his political activism. In 1957, he helped to collect signatures from 11,000 scientists from 49 countries on a petition protesting the buildup of nuclear weapons and aboveground tests of nuclear bombs, which contributed substantially to bringing about a permanent ban on atomic testing in 1963. Pauling's antinuclear activities earned him his second Nobel, an unshared Peace Prize. The prize, although considered to be for 1962, was actually awarded about the time the test ban took effect.

Pauling left Caltech in 1963, partly because of criticism of his political position. He thereafter taught and researched at the Center for the Study of Democratic Institutions (1963–67), the University of California, San Diego (1967–69), and Stanford University (1969–73). In 1973, he founded his own research institution, the Linus Pauling Institute of Science and Medicine, in

Palo Alto, California, and worked there for the rest of his life.

Pauling's institute researched a new approach to preventing and treating disease that he called orthomolecular medicine. In several popular books, he claimed that large daily doses of vitamin C (ascorbic acid) and certain other vitamins and nutrients would prevent or cure everything from colds to cancer, heart disease, and mental illness. Although his proposals drew praise in some quarters—*Vitamin C and the Common Cold* won the Phi Beta Kappa Award for the best new book on science in 1970, for instance—critics said that Pauling lacked sufficient evidence to back up his claims. The merits of consuming vitamin C, at least in the doses Pauling recommended, are still debated.

Controversial though some of his ideas were, Pauling received innumerable awards, both in the United States and overseas. He became a member of the U.S. National Academy of Sciences in 1933, the youngest person to gain that honor up to that time. Other major awards included the Presidential Medal of Merit (1947), the Davy Medal of Britain's Royal Society (1947), the Dr. Martin Luther King Medical Achievement Award (1972), and the National Medal of Science (1975).

Pauling continued to do research in the 1980s, investigating subjects ranging from superconductivity (a property of some rare metals that could lead to cheaper electric power) to AIDS. Active almost to the end of his life—he was working on three books at the time of his death—Pauling died of prostate cancer at his home in Big Sur, on the California coast, on August 19, 1994. In an obituary notice, the British magazine *New Scientist* called him one of the 20 greatest scientists in history.

Further Reading

Goertzel, Ted, and Ben Goertzel. *Linus Pauling: A Life in Science and Politics*. New York: Basic Books, 1995.

Hager, Tom. *Force of Nature: The Life of Linus Pauling*. New York: Simon & Schuster, 1995.

———. *Linus Pauling and the Chemistry of Life*. New York: Oxford University Press, 1998.

Marinacci, Barbara, ed. *Linus Pauling in His Own Words: Selections from His Writings, Speeches, and Interviews*. New York: Simon & Schuster/Touchstone, 1995.

"Pauling, Linus Carl." *Current Biography Yearbook 1994*. New York: H. W. Wilson, 1994.

Serafini, Anthony. *Linus Pauling: A Man and His Science*. New York: Paragon House, 1989.

Pavlov, Ivan Petrovich

(1849–1936)
Russian
Physiologist

Ivan Pavlov received the 1904 Nobel Prize in physiology or medicine for his careful experiments on digestion, but he is best known for later experiments that defined learning and behavior in terms of "conditioned reflexes." He was born in the Russian village of Ryazan on September 27, 1849.

Pyotr Dmitrievich Pavlov, Pavlov's father, was a priest, and his mother, Varvara Ivanovna, was the daughter of a priest. Pavlov, too, was destined for the ministry, but he was more interested in nature and science, and in 1870 he left the ecclesiastical school in which he was enrolled and began to study natural science at the University of St. Petersburg. He received the equivalent of a bachelor's degree in 1875, then went on to obtain a medical degree from the Academy of Medical Surgery in 1879. He married Seraphima Karchevskaya, a teacher, in 1881, and they later had four children who survived to adulthood.

One of Pavlov's teachers at St. Petersburg interested him in physiology, especially the physiology of digestion. While still a medical student, Pavlov did research on the pancreas, a digestive organ, that not only won a gold medal but

Russian physiologist Ivan Petrovich Pavlov developed the concepts of conditioning and conditioned reflexes, central to behavioral psychology, in the early years of the 20th century. *(National Library of Medicine)*

resulted in his being asked to head a new experimental laboratory in the city in 1878. There, he studied the nerves that control the heart and the way different drugs affect them, earning a Ph.D. in 1883. He did postdoctoral research at several German universities during the next two years, then returned to St. Petersburg.

Pavlov gained greater scope for his experiments in 1890, when he became head of a new physiology department at the Institute of Experimental Medicine and a professor of pharmacology at the Military-Medical Academy, both in St. Petersburg. In 1895, he became chairman of the latter organization's physiology department as well. He kept this post until 1925 and stayed with the Institute of Experimental Medicine for the rest of his life. He proved to be a stern but able administrator, at times managing hundreds of scientists.

During the 1890s, Pavlov studied digestion by surgically creating permanent openings, or fistulas, that connected different parts of dogs' digestive systems to the surface of their bodies so that he could, for example, insert food into the openings or remove samples of digestive juices. (These experiments recreated the unusual situation, resulting from a gunshot wound to one of his patients, that had allowed American surgeon WILLIAM BEAUMONT to perform groundbreaking experiments on digestion earlier in the century.) Pavlov's experiments were unusually successful, both because of his skill and care in surgery (his operating rooms were cleaner than those in most hospitals of the time) and because, unlike other physiologists who operated on animals for experimental purposes, he waited until his dogs had recovered from their surgery before he studied them. He was thus able to observe relatively normal physiology in healthy animals.

Pavlov and his coworkers examined digestion minutely in hundreds of experiments. Perhaps most importantly, they proved that the nervous system is the chief regulator of digestion, a fact that the biographical sketch of Pavlov published by the Nobel Foundation claims "is . . . the basis of modern physiology of digestion." Pavlov published his results in *Lectures on the Function of the Principal Digestive Glands* in 1897 and received the Nobel Prize (the monetary part of which the Russian State Bank confiscated) for the work seven years later.

By the time he won the prize, Pavlov had left his research on digestion to follow up an observation made during his earlier experiments. Placing food or a dilute solution of acid in a dog's mouth made saliva flow from glands in the mouth; this was an automatic, or reflex, action in response to a direct physical stimulus. Pavlov had noticed, however, that his dogs actually began salivating as soon as they saw food or even saw the keepers who normally brought them food. This response had to be more than a simple reflex. He decided to adapt his existing techniques to learn more about these "psychic secretions."

Measuring saliva output through fistulas in dogs' salivary glands, Pavlov found that if a signal, such as a ringing bell, repeatedly occurred at the same time as or slightly before a stimulus that produced a reflex response, such as contact with food, a dog would eventually produce the response (salivation) when it heard the signal, even if food did not actually appear. He called this response a conditioned, or conditional, reflex, a term he first used around the time he won the Nobel Prize. He saw conditioned reflexes as adaptations of automatic, or unconditioned, reflexes. Both kinds of reflexes, he said, consist of a stimulus and a response. In this example, contact with food is an unconditioned stimulus and salivation an unconditioned response. The sound of the bell is a conditioned stimulus. It produces no unconditioned response by itself, but in time, because it is associated with the unconditioned stimulus, it produces a conditioned response, which in this case is also salivation.

Viewing what had formerly been termed "psychic" activity as a kind of reflex allowed Pavlov to investigate it with the same sorts of objective, physiology-based experimental techniques that he and others had used to study unconditioned reflexes. This was important to him because he considered himself to be a physiologist, not a psychologist. Indeed, he strongly distrusted psychology because, following in the wake of Sigmund Freud and others, this field at the time focused on speculations about hidden thoughts and feelings. Pavlov believed that science should concern itself only with phenomena that could be studied experimentally. "It is still open to question," he once wrote, "whether psychology is a natural science, or whether it can be regarded as a science at all."

Pavlov and his army of researchers investigated the elements of conditioned reflexes in thousands of experiments during the rest of the scientist's long life. They measured the amounts of saliva or gastric juice (liquid made by the stomach) secreted in response to various conditioned and unconditioned stimuli, investigated how many pairings of conditioned and unconditioned stimuli were necessary to produce conditioned responses under various circumstances, and so on. They showed that conditioned reflexes could be reinforced, or made more likely to occur, by additional pairings of the signal and the physiological stimulus. On the other hand, if the signal was separated from the stimulus on a number of occasions after the conditioned reflex had been established, the conditioned response eventually ceased to occur, or was extinguished.

Pavlov showed that the cerebral cortex, the "thinking" part of the brain, was necessary for establishing conditioned reflexes. He eventually concluded that all behavior, even the complex thought processes of humans, grew out of conditioned reflexes and therefore could be studied by objective, physiological techniques. These views were tremendously influential, both in his native Russia and abroad, especially in the United States. In the latter country, they led to the establishment of a psychological school called behaviorism, which dominated psychology during the first half of the century and attracted followers such as B. F. SKINNER. Beginning in the late 1920s, Pavlov himself attempted to apply his discoveries about behavior and conditioned reflexes to human mental illness, but he was not notably successful.

Pavlov received honors for his work throughout his career, including election to the Russian Academy of Sciences in 1907, the Order of the French Legion of Honor in 1915, and a special Soviet government decree signed by Lenin in 1921 praising "the outstanding scientific services of Academician I. P. Pavlov." The government was also very generous in providing financial support for his large laboratory. Pavlov, however, sometimes criticized Communism and its officials. He died of pneumonia in St. Petersburg on February 27, 1936, dictating precise observations of his physical and mental condition a mere hour before his death.

Further Reading

"Ivan Petrovich Pavlov—Biography." *Nobel Lectures, Physiology or Medicine, 1901–1921*. Available online. URL: www.nobel.se/medicine/laureates/1904/pavlov-bio.html. Last updated 2001.

Todes, Daniel Philip. *Ivan Pavlov: Exploring the Animal Machine*. New York: Oxford University Press, 2000.

———. *Pavlov's Physiology Factory: Experiment, Interpretation, Laboratory Enterprise*. Baltimore, Md.: Johns Hopkins University Press, 2001.

Pincus, Gregory Goodwin

(1903–1967)
American
Physiologist

With Min-Chueh Chang, Gregory Pincus developed a safe, effective hormone mixture that could be taken by mouth to prevent pregnancy. Pincus was born in Woodbine, New Jersey, on April 9, 1903. His father, Joseph William Pincus, was a lecturer in agriculture at Cornell University in Ithaca, New York, and Pincus also went to Cornell to study agriculture as a young man. He graduated with a B.S. in 1924 and obtained a Ph.D. from Harvard University in 1927. He did postdoctoral work on genetics and physiology in England and Germany in 1929 and 1930. He married Elizabeth Notkin in 1924, and they had three children.

Pincus taught and did research at Harvard from 1930 to 1938, at Clark University in Worcester, Massachusetts, from 1938 to 1945, and at Tufts Medical School in Boston from the late 1930s to 1950. In 1950, he became a research professor of biology at Boston University, a position he held until the end of his life. His chief affiliation, however, was with the Worcester Foundation for Experimental Biology in Shrewsbury, Massachusetts, which he cofounded with Hudson Hoagland in 1944. Pincus was first the institution's director of laboratories and then, beginning in 1956, its research director.

Pincus's main interest was in animal and human reproduction. He achieved test-tube fertilization (fertilization of eggs outside a living body) in rabbits in 1934, while at Harvard. At Clark in 1938, he worked out chemical techniques that mimicked the effects of sperm on the egg of a rabbit so effectively that, when implanted into the uterus of a doe rabbit, the egg developed into a normal offspring. This was the first known parthenogenetic, or fatherless, birth of a mammal in history. Pincus and Hudson Hoagland also developed a method for quick-freezing living sperm that allowed the sperm to be used for fertilization when thawed out later. In the late 1940s, Pincus began investigating hormones involved in reproduction.

Margaret Sanger, a pioneer crusader for a woman's right to control the size of her family through contraception, visited Pincus in 1951. At the time, not only was contraception (prevention of fertilization) hard to accomplish dependably, but the distribution, advertising, or sale of contraceptive devices was also illegal in many states because preventing pregnancy violated the beliefs of the Catholic Church and some other religious denominations. The 68-year-old Sanger nonetheless asked Pincus and his coworkers to try to develop a safe, effective human contraceptive, and wealthy Sanger supporter Katherine McCormick and the Planned Parenthood Federation (which Sanger had founded) offered financial support for the effort. Pincus agreed to take on the task.

Pincus and his coworkers at the Worcester laboratory knew that a hormone called progesterone is produced when a female mammal ovulates, or releases an egg from the ovary for possible fertilization. Among other effects, progesterone prevents further ovulation, which could cause overlapping pregnancies. (If fertilization does not take place, progesterone secretion stops after a few days.) This effect, the researchers realized, might make progesterone a good contraceptive.

Soon after Pincus and Chang, his Chinese-born assistant, began their attempts to turn progesterone into a usable contraceptive, other scientists discovered substances in Mexican yams (sweet potatoes) that mimicked progesterone and another female hormone, estrogen. These imitation hormones were inexpensive and were not harmed by digestion, which meant that they could be taken by mouth. By 1954, Pincus and Chang had developed a combination of these yam-based hormones that prevented pregnancy in animals.

Since Pincus and Chang were not medical doctors, they could not test their formula on humans. They therefore formed a research partnership with John Rock, a gynecologist (specialist in medical conditions affecting women) at a fertility clinic in the Free Hospital for Women in Brookline, Massachusetts, whom Pincus had met in 1952. Some of Rock's patients volunteered to test the new hormone pills—in secret at first, because contraception was illegal in Massachusetts—and the results were promising. After Pincus and Rock published an account of the tests in 1956, several drug companies arranged for more extensive trials in Puerto Rico, Haiti, and elsewhere. Pincus's pill produced only one unwanted pregnancy per 1,000 women, and most of the failures proved to be due to women's failure to follow the somewhat complicated instructions for taking the medication. The hormone pills caused fewer side effects than normal pregnancy.

The U.S. Food and Drug Administration (FDA) approved the pills for use in treating miscarriages and menstrual disorders in 1957 and for use as a contraceptive in 1960. They were the first practical contraceptive medication that could be taken by mouth and one of the most reliable contraceptives of any kind. Later studies showed that the quantities of hormones in the pills could be reduced, making them even safer, without changing their effectiveness.

"The Pill," as the new contraceptive was informally called, became widely used in the 1960s, causing both controversy and social change. Some religious groups opposed it, claiming that it was unnatural and promoted promiscuity. Many women, however, welcomed it as a way of avoiding the ill health that often followed pregnancy after pregnancy and the poverty that dogged families with more children than they could afford to care for. U.S. Supreme Court decisions in 1965 and 1972 nullified state anti-contraception laws and established the right of married couples, and even single people, to use contraception if they wished. A Canadian Broadcasting Corporation article said, "Few scientific inventions have impacted society as fundamentally and powerfully as the contraceptive pill. . . . It has been credited with launching the women's movement, fuelling the wild and free times of the 1960s, [and] reforming the Catholic Church."

Pincus wrote several medical books on reproduction, including *The Control of Fertility* (1965). He received awards for his work, such as the Albert Lasker Award in Planned Parenthood and the Oliver Bird Prize (both in 1960) and the Modern Medicine Award (1964). He died on August 22, 1967.

Further Reading

Asbell, Bernard. *The Pill: A Biography of the Drug that Changed the World*. New York: Random House, 1995.

"The Pill That Transformed America." *Newsweek*, Winter 1997.

"Pincus, Gregory (Goodwin)." *Current Biography Yearbook 1966*. New York: H. W. Wilson, 1966.

Prusiner, Stanley B.

(1942–)
American
Neurobiologist, Virologist, Biochemist

Stanley Ben Prusiner insisted for decades that certain proteins can reproduce without genetic material and cause deadly brain diseases. The nature of these proteins, which Prusiner calls prions, is still hotly debated, but Prusiner gathered enough evidence for their existence to earn the 1997 Nobel Prize in physiology or medicine.

Prusiner was born on May 28, 1942, in Des Moines, Iowa, and grew up in Des Moines and Cincinnati, Ohio. His parents were Lawrence Prusiner, an architect, and Miriam Prusiner. He studied chemistry at the University of Pennsylvania, graduating with honors in 1964, and obtained a medical degree from the same university in 1968. After an internship at the University of California, San Francisco, Medical Center (UCSF) and three years of postdoctoral work at the National Institutes of Health, he returned to UCSF in 1972, and he has spent the rest of his career there, first joining the faculty in 1974. He is currently professor of neurology, virology, and biochemistry at UCSF and a professor of virology at the nearby University of California, Berkeley. He is married to the former Sandy Turk, and they have two daughters.

In 1972, while serving a residency in neurology at UCSF, Prusiner treated a woman with a rare brain disease called Creutzfeldt-Jakob disease (CJD), which produces symptoms similar to those of Alzheimer's disease. The cause of CJD was unknown but was suspected of being a virus. This mysterious illness "captivated [his] imagination," Prusiner wrote later, and he "began to think that defining the molecular structure of this elusive [disease-causing] agent might be a wonderful research project."

Prusiner's reading told him that CJD closely resembled two other fatal diseases, one in animals and one in humans. The animal disease was called scrapie because sheep with the illness obsessively scrape their sides against rocks as well as staggering and showing other signs of brain degeneration. Kuru, the human disease, was known to affect only the Fore people of Papua New Guinea. In the 1950s, virologist Carleton Gajdusek had determined that kuru was probably spread by the Fore's custom of eating parts of the bodies, including the brains, of relatives during funeral rites. Both scrapie and kuru could be given to animals (chimpanzees in the case of kuru) by injecting them with brain cells from victims of the diseases. CJD, too, had occasionally been accidentally transmitted through contamination of instruments used in brain surgery. Virologists thought that all these diseases, which produce similar patterns of brain degeneration, were caused by unknown "slow viruses."

Scrapie was the easiest slow virus disease to study, since it could be given to laboratory mice. Prusiner began to investigate it in 1974. Strangely, he kept finding that the fractions of brain material that gave scrapie to mice contained protein but no nucleic acid. A few British scientists had obtained similar results in the 1960s, but most researchers refused to believe that a protein by itself could cause disease. To do so, the protein would have to be able to reproduce, and no known biochemical could reproduce itself except nucleic acid (DNA or RNA).

Prusiner nonetheless came to the conclusion that protein particles were the cause of scrapie, CJD, kuru, and several other rare brain diseases. He described these particles, which he called prions (short for *pro*teinaceous *in*fectious particles), in a 1982 paper that caused a sensation in the scientific community. He claimed that the disease-causing protein, which he calls PrPSc (scrapie prion protein), is a distorted form of a normal protein, PrPC (cellular prion protein). Later in the decade, he provided evidence that PrPSc molecules' unusual shape somehow endows them with the ability to make the PrPC molecules that they contact take on the same abnormal shape. By this means, PrPSc, in effect, reproduces. Prusiner now believes that a second protein, which he calls Protein X, may be involved in the duplication process.

In the mid-1980s, Prusiner and others announced the identification of a gene that orders the making of PrPC not only in animals with scrapie but in a wide range of healthy animal species, including humans. The fact that PrPC is so widespread suggests that the protein has a vital function in the body. Some forms of CJD and certain other rare brain diseases that Prusiner believes are caused by prions can be inherited, and he reported in 1988 that he and coworker Karen Hsiao had isolated a mutant gene from a member of a family with one such disease. The gene makes a form of PrP that differs by only one amino acid from normal PrPC. Prusiner believes that this mutant protein is one form of disease-causing prion.

Unnervingly, prion infections occasionally seem able to move from one species to another. For instance, a cattle disease called bovine spongiform encephalitis (BSE), or "mad cow disease," is thought to have originated when cattle were fed ground-up carcasses of sheep that had died of scrapie. This illness became widespread in British cattle in the late 1980s.

An international panic arose in March 1996, when the British health secretary reported 10 cases of a new type of CJD in young adults (CJD had almost never been known to affect young people) and suggested that they might have gotten the disease by eating beef from British cattle with BSE. Many countries stopped importing British beef in the wake of this announcement. In an attempt to restore confidence, Britain's government ordered the slaughter and burning of 37,000 apparently healthy cattle that had been alive before 1989, when the practice of feeding ground-up carcasses to cattle was stopped, because they might be carrying the disease. Few additional cases of "variant" CJD have been reported since then, but concerns that the disease could be spread through eating infected beef continue, especially after the finding in 2002 that, at least in mice, disease-causing prions accumulate in muscle as well as nerve tissue (edible meat is chiefly muscle).

Prusiner has described his theories and research on prions in several books, including *Prion Diseases of Humans and Animals* (1993), as well as in many papers. Debate over them continues. Although Prusiner has yet to completely isolate his mutant protein, a growing number of scientists feel that his indirect evidence for its existence and behavior is impressive. Supporters see his Nobel Prize and many other awards, which include the Gairdner Foundation Award, the Albert Lasker Medical Research Award (1994), and Israel's Wolf Prize (1996), as vindications of his "heretical" ideas. Critics, however, believe that a still-undetected virus accounts for prions' apparent disease-causing effects.

In addition to continuing to study PrPSc's structure and molecular mode of action, Prusiner's laboratory at UCSF is looking for drugs to keep this rogue protein from spreading (quinacrine, a drug currently used to treat malaria, is one possibility). Prusiner believes that his research may eventually shed light not only on the rare diseases that he believes are directly caused by prions but on more common neurodegenerative illnesses that resemble the putative prion diseases in some ways, such as Alzheimer's disease, Parkinson's disease, and amyotrophic lateral sclerosis (Lou Gehrig's disease). Even more amazing, if his ideas about prions prove to be correct, Stanley Prusiner will have discovered a new form of life.

Further Reading

Prusiner, Stanley B. "The Prion Diseases." *Scientific American*, January 1995.

———. "Stanley B. Prusiner—Autobiography." *Les Prix Nobel 1997*. Available online. URL: www.nobel.se/medicine/laureates/1997/prusiner-autobio.html. Last updated 2001.

"Prusiner, Stanley." *Current Biography Yearbook 1997*. New York: H. W. Wilson, 1997.

Purkinje, Jan Evangelista (Jan Evangelista Purkyně)
(1787–1869)
Austro-Hungarian
Histologist, Physiologist

Using recently improved compound microscopes, Jan Evangelista Purkinje made numerous discoveries about the microscopic structure of the brain, heart, skin, and other tissues and organs in the mid-19th century. Purkinje was born on December 17, 1787, in Libochovice, then in Bohemia, part of the Austro-Hungarian Empire, but now belonging to the Czech Republic. His father was an estate supervisor.

Purkinje was educated by the Piarists, an order of monks devoted to teaching children of poor families, and at first he, too, planned to join the order. He changed his mind, however, and entered Prague University, where he studied philosophy, biology, physics, and eventually medicine, meanwhile supporting himself with teaching jobs. He obtained his medical degree in 1819.

Purkinje began to make a reputation even as an undergraduate. He wrote his doctoral thesis on a phenomenon of vision—one of several biological phenomena and structures to be named after him—in which, as light dims, colors at the red end of the spectrum seem to fade more than those at the blue end that are of equal actual brightness. His experiments are considered to be some of the first in the field of experimental psychology, and his thesis contains the statement, central to neuroscience, that all subjective experiences have some physical basis that can be measured objectively.

Purkinje sent a copy of his thesis to famed German poet Johann Wolfgang von Goethe, who was also interested in science, and through it he obtained Goethe's friendship. The help of Goethe and other influential patrons in Purkinje's early years helped to make up for his poverty and the discrimination he faced as a minority Czech in a German-oriented society. (Resenting this discrimination, Purkinje was a Czech nationalist all his life.)

Purkinje worked as an assistant in Prague University's physiology department from 1819 to 1823, after which (perhaps through Goethe's influence) he became professor of physiology and pathology at the University of Breslau (now Wroclaw, Poland). Around this time, he married Mary Asmund, daughter of a professor, and they later had four children, although only their two sons lived to grow up. Purkinje's teaching methods, which focused on experimentation rather than on textbooks, were unusual for the time and drew criticism from other members of the faculty, so he often held his classes in a laboratory that he built in his home. He founded the world's first physiology department in Breslau in 1839 and the first experimental institute devoted to the subject in 1842. In 1850, he returned to Prague University as professor of physiology, and he remained there until his death on July 28, 1869.

Purkinje's chief discoveries related to the microscopic structures of tissues and organs, which he began studying in 1832. Among them are large, pear-shaped cells with numerous short branches called dendrites, which he found in 1837 in the cerebellum, the part of the brain that controls movement. In 1839, he identified unusual muscle fibers in the wall of the heart that conduct electrical impulses throughout the heart from a nerve area called the pacemaker, thus regulating the heartbeat. Both of these kinds of cells still bear Purkinje's name. He first described sweat glands in 1833 and the overall structure of the skin in 1835. In that same year, he observed the motion of tiny hairs called cilia on the outside of certain cells, and in 1836 he showed that extracts from a digestive gland called the pancreas, the function of which had previously been unknown, can dissolve or digest protein.

Purkinje's achievements reached well beyond the detailing of particular structures. He proposed in 1823 that fingerprints were unique and therefore could be used for identification. He introduced such key biological terms as *protoplasm* (eventually used to mean the internal material of cells) and *plasma* (the liquid part of the blood), and he presented the basic elements of the theory that the bodies of all organisms are made up of cells in 1837, two years before MATTHIAS JAKOB SCHLEIDEN and THEODOR SCHWANN developed this idea more fully.

In addition to discovering numerous structures under the microscope, Purkinje made improvements in techniques that helped other scientists do the same. For instance, he invented a tool that could slice even bone thinly enough to be observed on a microscope slide without damaging its cellular structure. The modern version of this device is called a microtome. He also made the first photographs taken through a microscope. Purkinje is considered to be one of the founders of histology, the discipline of studying tissues and organs under the microscope.

Further Reading

Wade, Nicholas, Josef Brozek, and Jiri Hoskovek. *Purkinje's Vision: The Dawn of Neuroscience*. Mahwah, N.J.: Lawrence Erlbaum Associates, 2001.

Ramón y Cajal, Santiago

(1852–1934)
Spanish
Neurobiologist, Histologist

Santiago Ramón y Cajal established that the brain is made up of separate nerve cells that communicate but are not in contact with each other. He was born on May 1, 1852, in the village of Petilla de Aragón in northern Spain. His father, Justo Ramón y Casasús, began life as a farmer but eventually became professor of anatomy at the University of Zaragoza. An authoritarian figure, he had similar ambitions for his son, but Santiago had more interest in art than science. Nonetheless, he finally gave in to his father's demands and studied medicine at the university, obtaining a licentiate (preliminary degree) in 1873 and an M.D. in 1879.

Ramón y Cajal served as an army surgeon in Cuba in 1874 and 1875 but was sent home after developing malaria. He was an assistant in the anatomy department of the University of Zaragoza from 1876 to 1879, when he took over the university's anatomical museum. He taught anatomy at the University of Valencia from 1884 to 1887, histology at the University of Barcelona from 1887 to 1892, and histology and pathological anatomy at the University of Madrid from 1892 to 1922. He also headed two research institutes in Madrid beginning in 1902. He married Silveria Fañanás García in 1880, and they had six children.

Ramón y Cajal's groundbreaking work on the nervous system began around 1887, when he first learned about a new way of staining nerve tissue that Italian neuroanatomist CAMILLO GOLGI had developed in 1873. Ramón y Cajal (who was often known outside of Spain simply as Cajal) improved Golgi's method and invented several additional stains that revealed different parts of nerve cells or nerve tissue. In 1913, for instance, he developed a gold stain that allowed nerve tissue in embryos and brain tumors to be studied for the first time.

Ramón y Cajal described his findings about the nervous system in *Structure of the Nervous System of Man and Other Vertebrates* (1904), which Harvard neurobiologist David Hubel calls "the most important single work in neurobiology." By showing that neurons in the cerebral cortex had somewhat different structures in different parts of the brain, he suggested the possibility that parts of the brain might have separate functions. Ramón y Cajal used his artistic talent in the meticulous and beautiful drawings that filled this book.

At the time of Ramón y Cajal's research, scientists held two competing theories about the nervous system. The reticular theory, which Golgi and most others in the infant field of

Santiago Ramón y Cajal, a Spanish neurohistologist, showed through his microscopic studies that nerve cells do not touch each other. *(National Library of Medicine)*

neuroscience espoused, stated that the system was a seamless network of interconnected fibers. Ramón y Cajal, however, was an advocate of the neuron theory, which held that the nervous system was composed of separate cells that, although they certainly communicated with one another, did not physically touch. In his opinion, he proved this theory conclusively by showing that the long fibers of nerve cells, called axons, ended in the gray matter of the brain without touching other fibers or nerve cell bodies. He never convinced Golgi, but the neuron theory is now universally accepted.

Around 1903, Ramón y Cajal began studying the damage that diseases and injuries can do to the nervous system. He also studied the system's ability to repair that damage and regenerate. He described this research in *Degeneration and Regeneration of the Nervous System* (1913). He also investigated the structure of the retina, the tissue in the back of the eye that responds to light, and the tissues of the inner ear.

In addition to his scientific works, Ramón y Cajal wrote popular articles on science (some under the pseudonym of "Dr. Bacteria"), an autobiography called *Recollections of My Life* (1917), and a book about the training and attitudes necessary for a scientific career, *Rules and Advice on Biological Investigation*. His work on the nervous system won the 1906 Nobel Prize in physiology or medicine (which he shared with Golgi) as well as the Rubio Prize (1897), the Moscow Prize (1900), the Helmholtz Gold Medal from the Royal Academy of Berlin (1905), and many other awards. In 1920, when Ramón y Cajal had become internationally renowned and a hero in Spain, King Alfonso XIII established the Cajal Institute in Madrid for his research, and Ramón y Cajal headed this institute until his death in Madrid on October 18, 1934.

Further Reading

Ramón y Cajal, Santiago. *Recollections of My Life*. 1917. Reprint, Cambridge, Mass.: MIT Press, 1989.

Knudtson, Peter M. "Painter of Neurons." *Science 85*, September 1985.

Ray, John (John Wray)
(1627–1705)
British
Naturalist, Taxonomist

John Ray set up the first systematic, modern classification of plants and animals and used the term *species* for the first time, with roughly the meaning biologists give it today. He has been called the father of British natural history.

Ray was born in the village of Black Notley, Essex, England, on November 28, 1627. His father was a prosperous blacksmith and his mother an herbalist-healer. He entered Catherine's Hall (later St. Catherine's College) in Cambridge University at the unusually young age of 16, then went on to Trinity College, where he studied classics and theology and graduated in 1648. He taught Greek, mathematics, and humanities at Trinity for 13 years, becoming a fellow of the college in 1649, a lecturer in 1651, and a junior dean in 1658. He was ordained as an Anglican minister in 1660.

After a serious illness in 1650, Ray took long walks through the Cambridge countryside as part of his convalescence. He became interested in the plants he saw and, finding that no book described them in detail, he began compiling one. This book, called *Catalogue Plantarum Circa Cantabrigium Nascentium* (*Catalogue of Plants Growing Around Cambridge*), it was published in 1660. It was the first British book to describe all the types of plants in a small area.

Ray's development of a new interest proved to be convenient. In 1662, soon after Charles II restored the British monarchy following 11 years of rule by the Puritan-led Commonwealth, the government required all ministers to sign an Act of Uniformity in which they promised to adhere to certain religious principles. Ray felt that the act violated his personal beliefs and therefore refused to sign it. As a result, he lost his position at Cambridge.

Ray then joined forces with a wealthy young man named Francis Willughby, a former student who shared his fondness for nature. The two decided to write a series of books that would contain descriptions and classifications of all known plants and animals, just as ancient authors such as ARISTOTLE had done. Ray would prepare the material on plants, while Willughby covered animals. They traveled through Europe from 1663 to 1666, observing and collecting specimens. On their return, Ray moved into Willughby's home as a tutor to his children.

Unfortunately for Ray, Willughby died unexpectedly in 1672. Provisions in Willughby's will allowed Ray to remain in the house and continue his work for a time. He married Margaret Oakeley, a young governess (teacher and nursemaid for children) in the Willughby household, in 1673, and they eventually had four daughters. In 1678, however, Willughby's widow insisted that the Rays leave, so they returned to Ray's family cottage in Black Notley, where, despite ill health, Ray continued to write.

Most of Ray's major works were on plants. They included *Catalogus Plantarum Angliae* (*Catalogue of English Plants* 1670), *Methodus Plantarum Nova* (*A New Method on Plants*, 1682), and, most importantly, *Historia Generalis Plantarum* (*A General History of Plants*), a three-volume work published between 1686 and 1704. In *Methodus Plantarum*, Ray divided plants into flowerless plants (cryptogams), flowering plants with one leaf growing from the seed (monocotyledons), and flowering plants with two such leaves (dicotyledons). He was the first to make this distinction, which botanists still use. *Historia Generalis* covered 18,600 individual types (species) of plants—essentially all that were known at the time—and provided descriptions for each that included anatomy, distribution, habitats, and medical uses. The book also offered general information about plants, such as their methods of reproduction and diseases that affected them.

Ray also completed several works on animals. He published books on birds (1676) and fish (1685) under Willughby's name, but most of the work in them was his own. Ray's most important animal book was *Synopsis Methodica Animalium Quadripedum et Serpentini Generis* (*Summary of Four-footed Animals and Reptiles*), published in 1693, which classified these animals (essentially all vertebrates) according to such diverse features as the number of chambers in their hearts and the number of toes on their feet. Two other animal books, one on insects (1710) and one on fish and birds (1713) were published after his death.

Ray made the first systematic attempt to classify plants and animals according to multiple structural features that they had in common, rather than relying on a single feature or grouping according to use. Like modern biologists, he sorted plants and animals into large groups, then into smaller groups that shared several features, and finally into unique types, or species. He saw classification not only as valuable in itself but as an approach to understanding living things' physiology and behavior. Ray's system of classification is considered to be the best one developed before that of CAROLUS LINNAEUS, and Linnaeus admired and learned from Ray's work.

In addition to his books on plants and animals, Ray wrote two books about the relationship between nature study and religion, *The Wisdom of God Manifested in the Works of the Creation* (1691) and *Three Physico-Theological Discourses* (1692). Unlike many thinkers of earlier times, he saw no conflict between religious devotion and a desire to understand nature. On the contrary, he espoused a philosophy called natural theology, which held that learning about nature—God's creation—was an excellent way to appreciate the Creator's wisdom and power.

Ray's religious books proved more popular than his strictly scientific works, which was somewhat surprising considering that certain ideas in them contradicted prevailing beliefs. To begin with, he accepted the idea that fossils were the remains of ancient plants and animals, a view that most scientists did not share until a century later. He doubted that all fossils could have been created during Noah's flood, as many theologians held, because this theory could not explain why fossils of shelled sea creatures were found on mountaintops or how the shells had come to be squeezed out of shape, which Ray realized would have required both great pressure and a long period of time. Toward the end of his life, Ray even began to doubt the common belief that no species had been added, destroyed, or changed since the beginning of the world. In questioning religious views about fossils, Ray helped to lay the groundwork for CHARLES ROBERT DARWIN's theory of evolution.

Other scientists greatly respected Ray's work. He was elected a member of the Royal Society of London, a new organization for bringing British scientists together, in 1667. In his later years, when poor health kept him from attending meetings, he continued to correspond with the group extensively. Ray died at Black Notley on January 17, 1705.

Further Reading

Raven, Charles E. *John Ray: Naturalist: His Life and Works*. Reprint, New York: Cambridge University Press, 2000.

Waggoner, Ben. "John Ray (1627–1705." 1996. UC Berkeley Paleontology Museum Evolution Pages. Available online. URL: www.ucmp.berkeley.edu/history/ray.html. Accessed 2003.

Röntgen, Wilhelm Conrad (Wilhelm Conrad Roentgen)

(1845–1923)
Prussian/German
Physicist

By discovering X rays, Wilhelm Conrad Röntgen gave a substantial gift not only to physics, his own field, but to medicine as well. Röntgen was born in Lennep, Prussia (later a part of Germany), on March 27, 1845. His father, Friedrich Röntgen, was a well-to-do merchant and manufacturer of cloth. When Wilhelm was three years old, the family moved to Apeldoorn, Holland, where the family of his mother, the former Constanze Frowein, lived. Wilhelm grew up in Apeldoorn, where he enjoyed taking long walks in the countryside and building mechanical devices.

Röntgen entered a technical school at Utrecht in 1862, but he was expelled two years later after being blamed for a caricature of a teacher that another student had actually drawn.

Fortunately, he learned that the Polytechnic University in Zurich, Switzerland, would accept students who passed an entrance exam, whether or not they had any other credentials. Röntgen entered the university as a student of mechanical engineering, graduating in 1868. He earned a Ph.D. in physics from the University of Zurich in 1869.

After obtaining his doctorate, Röntgen became an assistant to one of his former professors, August Kundt, and followed him to the Universities of Wurzburg and Strasbourg. During his rather peripatetic career, Röntgen taught physics at the Universities of Wurzburg (1870–72 and 1888–99), Strasbourg (1872–74 and 1876–79), Wurttemberg (1875), Giessen (1880–87), and Munich (1900–23). In 1872, he married Anna Bertha Ludwig, the daughter of the owner of a café he frequented in Strasbourg. They had no children, but they adopted Bertha's young niece in 1887.

On November 8, 1895, Röntgen was in his laboratory at Wurzburg, investigating certain newly discovered phenomena related to electricity. British physicist William Crookes had invented a device called a Crookes tube, which was a sealed tube, drained of air to create a near-vacuum, in which two electrodes were placed. When the electrodes were connected to an electric current, a stream of electrons flowed from the cathode (negatively charged electrode) to the anode (positively charged electrode), and the tube glowed brightly with what Crookes called cathode rays. Röntgen, one of many physicists investigating this phenomenon, wanted to know whether any other kinds of radiation were produced at the same time as the cathode rays. He therefore darkened his laboratory and covered his Crookes tube with heavy cardboard to block the bright rays so that he would have a better chance of seeing anything else that might appear.

When Röntgen turned on the current that afternoon, he was surprised to see a faint greenish glow coming from a table about a yard away from the Crookes tube. The glowing object proved to be a small screen coated with a compound called barium platinocyanide. This substance was known to fluoresce, or glow, when cathode rays struck it, but no one had ever reported that the rays could affect it from such a distance. Röntgen suspected that something other than cathode rays was causing the glow.

Excitedly, Röntgen began experimenting with this new phenomenon. (When a reporter later asked him what he had thought at the moment of his great discovery, he replied, "I did not think. I investigated.") He turned the current off and the glow disappeared, only to reappear when he turned it on again. The glow

German physicist Wilhelm Conrad Röntgen discovered X rays in 1895, and physicians and surgeons were soon using his "new kind of ray" to examine bones and other structures inside the body. *(National Library of Medicine)*

dimmed to invisibility when he moved the screen two yards away from the tube. He placed various objects, ranging from a sheet of paper to a thick book and lead laboratory weights, between the tube and the screen and saw that only the lead weights stopped the glow. He found that the rays could expose, or darken, photographic film, just as ordinary light would, and that any objects that could block the rays left their outlines on the film. Most amazing of all, when he held a lead weight between the tube and the glowing screen, he saw on the screen not only a shadow shaped like the weight but the faint outline of his own thumb and finger—with the bones clearly visible inside, as though his flesh had become transparent. Cathode rays produced none of these effects, confirming Röntgen's belief that he had found a new kind of radiation.

At first, Röntgen told his laboratory colleagues and even his wife only that he had "discovered something interesting." On December 22, however, he brought Bertha into the laboratory to show her what he had found. He had her place her hand on a photographic plate, then stand still for 15 minutes while he aimed the rays at it. She waited again as he developed the film. She shuddered when she saw the finished picture, which clearly showed the bones of her hand and her wedding ring. That ghostly vision was the first X ray photograph of a part of the human body.

Concerned that someone else would make the same discovery he had and publish it before him, Röntgen quickly described it in a paper called "On a New Kind of Rays." He gave the rays the name of X rays, with the X standing for "unknown." He persuaded the Physical and Medical Society of Wurzburg, to which he belonged, to publish the paper (complete with photographs, including the one of Bertha's hand) in its journal on December 28.

Röntgen sent copies of his article to a number of fellow physicists, and some passed it on to their own friends. One of those friends was the son of a newspaper publisher in Vienna, Austria, who showed it to his father. The father, who knew news when he saw it, published a front-page story about the new rays on Sunday, January 5, 1896. Telegraphs transmitted the news all over the world, and many more front-page stories followed. A few months later, an article in an American magazine claimed that "in all the history of scientific discovery there has never been . . . so general, rapid, and dramatic effect . . . as has followed" the announcement of Röntgen's discovery.

Physicians and surgeons realized almost immediately that Röntgen's rays could let them see inside a body without cutting it open. Robert Jones, a British surgeon, used the rays to locate a bullet in a boy's wrist in February 1896, the first known use of X rays in medicine. Shortly afterward, *McClure's Magazine* writer H. J. W. Dam called X rays (or Röntgen rays, as they were often termed) "a greater blessing to humanity than even the Listerian antiseptic system of surgery." Later scientists would expand and improve the use of X rays in diagnosis and use them to treat certain illnesses, but they would also learn that high doses or frequent exposure to the rays could cause dangerous health problems such as burns and cancer.

Röntgen himself was more irritated than pleased by his sudden fame. Barrages of questions from reporters and fellow scientists made work impossible for a few weeks. Even having the chance to demonstrate the rays before the German emperor's court and being awarded the Order of the Crown, Second Class, did not entirely soothe his feelings. Once the worst of the publicity died down, he resumed his regular research, but he never made another major discovery. Nonetheless, he won the British Royal Society's Rumford Medal in 1896 and the first Nobel Prize in physics in 1901. Röntgen died of intestinal cancer in Munich on February 10, 1923.

Further Reading

Glasser, Otto. *Wilhelm Conrad Roentgen and the Early History of the Roentgen Rays*. Reprint, Novato, Calif.: Norman Publishing, 1993 reissue.

"Wilhelm Conrad Röntgen—Biography." *Nobel Lectures, Physics 1901–1921*. Available online. URL: www.nobel.se/physics/laureates/1901/rontgen-bio.html. Last updated 2002.

Rosenberg, Steven A.

(1940–)
American
Immunologist, Surgeon

Steven Aaron Rosenberg is a pioneer in attempts to make the body's defense system attack cancers. He was born on August 2, 1940, to Abraham and Harriet (Wendroff) Rosenberg, Polish immigrants who lived in the Bronx, part of New York City, and ran a luncheonette. The family lost many European relatives in the Holocaust, and Rosenberg has said that growing up with stories about these losses made him want to help people.

Rosenberg decided to be a doctor when he was only seven years old. In 1963, he completed a special six-year program at Johns Hopkins University in Baltimore, Maryland, that allowed him to earn both a bachelor's degree and an M.D. He published his first research papers on cancer while still in medical school. He took advanced training at Harvard University in both surgery and research, two medical specialties normally considered to have little in common, and obtained a Ph.D. in biophysics from Harvard around 1964. Meanwhile, when doing his internship (postmedical training) at Peter Bent Brigham Hospital in Boston, he fell in love with the emergency room's head nurse, Alice O'Connell, and they married in 1968. They have three daughters. Rosenberg became chief of surgery at the National Cancer Institute, part of the National Institutes of Health (NIH) in Bethesda, Maryland, in 1974. He still holds this position and is also chief of the institute's tumor immunology laboratory.

Rosenberg first became interested in the immune system's effects on cancer in 1968, when he examined a 63-year-old man at a veterans' hospital in West Roxbury, Massachusetts. The man, then suffering from a medical problem unrelated to cancer, mentioned that 12 years before, he had had a huge tumor in his stomach and secondary growths elsewhere in his abdomen. Doctors at the time had expected him to die within months—but the cancers had simply gone away by themselves. Such spontaneous remissions had been reported occasionally, but Rosenberg had never seen one. Suspecting that the immune system held the key to this seeming miracle, he added immunology to his mix of specialties.

Immunology, however, had nothing to do with Rosenberg's first appearance in the media, which occurred in 1985. Called in as a consultant when then-President Ronald Reagan developed colon cancer, he became the presidential medical team's chief spokesperson. He won praise, both for his part in the cancer treatment and for his handling of the ensuing publicity. A *U.S. News & World Report* writer called Rosenberg's first press conference "the best performance by a doctor since Marcus Welby [a beloved physician in a popular fictional TV series] hung up his stethoscope."

Rosenberg made headlines again later in 1985, and this time his research was the reason. He had learned that, although the immune system does attack cancer cells, it is much less effective at doing so than at destroying bacteria or viruses. In 1976, he and coworker Mike Lotze had synthesized a natural growth factor called interleukin-2 (IL-2), which made certain cells in the immune system, termed T cells, capable of multiplying in test tubes. It also seemed to encourage the cells to attack tumors. In the early 1980s, Rosenberg gave people with advanced melanoma (a fast-spreading, deadly cancer) IL-2, or else he removed T cells from their blood, treated the cells with

IL-2 in the laboratory, and then reinjected them. He reported in the December 1985 *New England Journal of Medicine* that the combination of IL-2 and newly activated cells, which he named lymphokine-activated killer cells (LAK), had caused tumors to shrink by 50 percent or more in 11 of 25 patients.

The treatment was time consuming, expensive, and often caused severe side effects. In most cases, furthermore, the tumors eventually grew back, killing the patients. Nonetheless, the results were impressive for such an early test. One woman's tumors even vanished permanently (she was still cancer-free in 2001). Reporters picked up the story, and, although Rosenberg repeatedly warned that his results were preliminary, their accounts made some readers think that cancer was all but cured. Other researchers confirmed some of his findings, although their results were not quite as spectacular as his, and reduced the treatment's side effects by lowering the dosage of IL-2.

By 1988, Rosenberg had changed to a different type of T cells called tumor-infiltrating lymphocytes (TILs), which specifically homed in on cancer cells. Again, he extracted these cells from blood, treated them with IL-2, and reinjected them. This treatment also produced promising results in early tests on patients with advanced cancer, shrinking 40 percent of tumors temporarily, and it produced fewer side effects than the earlier procedure. Nonetheless, it was not as effective as Rosenberg had hoped, and he wanted to learn more about what happened to the treated TILs after they were put back into patients' bodies. Working with fellow NIH researcher and gene therapy pioneer W. FRENCH ANDERSON in 1989, therefore, he incorporated a marker gene into the cells that made them resistant to a certain antibiotic. He expected this characteristic to help him identify and track the TILs after they were reinjected.

Rosenberg's treatment was not gene therapy because the added gene was not expected to help his patients, but it was the first artificial insertion of a gene into cells given to human beings. He reported in 1990 that the marker genes had done the patients no harm and had allowed him to track the TILs to the tumors. This preliminary success opened the way for true gene therapy, which both Anderson and Rosenberg explored. In 1990, Rosenberg added a gene that made a protein called tumor necrosis factor (TNF) to his TILs in the hope of making them better tumor fighters. The therapy was not usually effective, however, perhaps because the cells often kept the inserted genes from functioning.

Rosenberg has received the Public Health Service Medal (1981) and a share of the Armand Hammer Cancer Prize (1985) for his work. He continues to explore and refine his immunological approach to cancer treatment, which he calls adoptive immunotherapy. For instance, he is attempting to make "cancer vaccines" containing antigens from patients' own cancer cells or from cancer-related genes that he has identified. He reported in 1999 that 42 percent of 31 patients receiving the vaccines, which are given along with IL-2, showed some tumor shrinkage. "I have seen with my own eyes . . . widely metastatic [spreading to many parts of the body], invasive, bulky cancers . . . disappear completely with immune manipulations," he said in 2001. He is also experimenting with ways to make added immune cells more effective. He reported in late 2002, for example, that killer T cells multiplied in "staggering" numbers after being reinjected and had long-term effects on tumors in some patients if the patients' immune systems were temporarily disabled by drugs before the injections.

Further Reading

Newman, Judith. "I Have Seen Cancers Disappear." *Discover*, May 2001.

Rosenberg, Steven A., with John M. Barry. *The Transformed Cell: Unlocking the Mysteries of Cancer*. New York: Putnam, 1992.

"Rosenberg, Steven A." *Current Biography Yearbook 1991*. New York: H. W. Wilson, 1991.

Ross, Ronald
(1857–1932)
Indian
Epidemiologist, Physician

Ronald Ross proved that malaria is transmitted by the bite of mosquitoes, and he worked out part of the life cycle of the microscopic parasite that causes the disease. He was born on May 13, 1857, in Almora, a military post in the foothills of the Himalayas that was then part of India (it is now in Nepal), to Campbell Ross, a British army officer, and his wife.

Ross spent his early years in India, then was sent to relatives in England when he was about eight years old so that he could go to British schools. Although chiefly interested in literature (he wrote poetry, novels, and dramas all his life), Ross, at his father's urging, took medical training at St. Bartholomew's, a famous London hospital, and joined the Indian Medical Service in 1881. He served in various parts of India during the 1880s.

During a leave in England in 1889, Ross took a course in bacteriology, which taught him how to use a microscope. He also met and married Rosa Bloxam; they later had two sons and two daughters. In 1894, during a second British trip, he met Patrick Manson, another physician who had worked in the tropics. Manson had shown in 1876 that a tropical disease called elephantiasis, caused by a parasitic worm, was transmitted by mosquitoes—the first disease proved to be spread by an insect. He told Ross about his belief that malaria, a much more common and often deadly illness, was transmitted in much the same way. (French army surgeon Alphonse Laveran had identified a microscopic parasite that invades red blood cells as the cause of malaria in 1880, but he had not found out how the disease spread.) Manson thought people caught malaria by drinking water containing dead mosquitoes that had been infected with the parasite. By then retired

Ronald Ross, a British physician born in India, helped to show in the late 1890s that mosquitoes transmit the parasite that causes the serious blood disease malaria. *(National Library of Medicine)*

in England, he asked Ross to test this theory when Ross returned to India.

Ross, who had already begun to study malaria in 1892, was eager to do the experiments, but fate frustrated him for several years. Only one out of many kinds of mosquito carries the malaria parasite, and that type did not live in the places to which the medical service sent him. Nonetheless, he persisted, trapping hundreds of mosquitoes, feeding them on malaria victims, and dissecting them under a microscope in the hope of finding the parasite in their bodies. He wrote often to Manson, describing his frustrations and occasional progress.

Finally, in Secunderabad on August 20, 1897—which Ross referred to afterward as

"Mosquito Day"—he tiredly turned his microscope on the stomach of the last mosquito in a batch of 38 and spotted Laveran's parasites. Thrilled, he celebrated his discovery in verse:

> I know this little thing
> A myriad [million] men will save.
> O Death, where is thy sting,
> Thy victory, O Grave!

Ross still did not know, however, how the parasites moved from the mosquito's stomach into the blood of their next victim. After failing repeatedly to answer this question for human malaria, he followed Manson's advice to study the disease in birds. On June 25, 1898, he showed that mosquitoes that had fed on birds with malaria could give the disease to healthy birds by biting them. Over the next few weeks, he tracked the parasites from the mosquitoes' stomachs to the insects' blood and, finally, on July 4, to their salivary glands. From there, they would be injected along with the mosquito's saliva when the mosquito took its next blood meal. On July 9, Ross wrote to Manson, "I think I may now say QED"—that he had proved exactly how malaria was transmitted.

Manson made sure that a paper describing Ross's work appeared in the *British Medical Journal* later in July. French, Italian, and German researchers were also trying to determine how malaria was spread, and Manson and Ross wanted England to have the honor of the discovery. In fact, the Italian team, led by Giovanni Battista Grassi, published similar research at almost the same time, making the same demonstration with human malaria that Ross had made with bird malaria.

Dissatisfied with his advancement in the Indian Medical Service, Ross resigned in 1899 and, with Manson's help, joined the new Liverpool School of Tropical Medicine. He remained there until 1912, serving as the Sir Alfred Jones Professor of Tropical Medicine from 1902. He then left to become Physician for Tropical Diseases at King's College Hospital in London. He was also malariology consultant to the British War Office during World War I. In 1926, with funding from admirers, he founded the Ross Institute and Hospital for Tropical Diseases in Putney, a London suburb, and he directed this institute for the rest of his life. During his career in England, he worked on plans for controlling mosquitoes and preventing malaria, identified the parasites that cause another mosquito-borne disease called African fever, and devised mathematical models that showed how malaria spreads in populations. He led expeditions to West Africa to study and set up prevention programs for malaria and African fever.

In 1902, Ross was awarded the Nobel Prize in physiology or medicine, and he was knighted in 1911. Nonetheless, he argued endlessly in his later years with Grassi and others, even including his old mentor, Manson, about who deserved credit for which part of the malaria discovery. Historians have tended to give Ross the greater credit because he published his work first, although it was Grassi who first identified *Anopheles* as the specific type of mosquito that carried human malaria and traced the malaria parasite's life cycle completely. Ross died in Putney on September 16, 1932, after a long illness.

Further Reading

Bynum, W. F., and Caroline Overy, eds. *The Beast in the Mosquito: The Correspondence of Ronald Ross and Patrick Manson*. Amsterdam and Atlanta, Ga.: Rodopi Bv. Editions, 1998.

de Kruif, Paul. *Microbe Hunters*. New York: Harcourt Brace, 1926.

Gibson, Mary E., and Edwin R. Nye. *Ronald Ross, Malariologist and Polymath*. New York: Palgrave Macmillan, 1997.

Ross, Ronald. *Memoirs, with a Full Account of the Great Malaria Problem and Its Solution*. London: John Murray, 1923.

Rous, Peyton
(1879–1970)
American
Virologist, Pathologist

Although it took more than 25 years for his work to be accepted, Francis Peyton Rous was the first person to show that a virus could cause cancer. He also contributed to the technology of blood transfusions and to understanding the liver and gall bladder.

Rous was born in Baltimore, Maryland, on October 5, 1879. His father, Charles Rous, a grain exporter, died when Peyton was still a child, and his mother, the former Frances Wood, raised him. Rous attended Johns Hopkins University in Baltimore, earning a B.A. in 1900 and an M.D. in 1905. After two years as an instructor in pathology at the University of Michigan, he was invited to join the then-new Rockefeller Institute of Medical Research in New York (later Rockefeller University) in 1909. He stayed there for the rest of his life, becoming a full faculty member in 1920. He married Marion de Kay in 1915, and they had three daughters.

The research for which Rous is best remembered began around 1910, when a farmer brought him a hen with a fleshy tumor called a sarcoma in its breast. Rous, who was already studying cancer in animals, was intrigued because the farmer said that other chickens in his flock had similar tumors, which suggested that the cancer might be caused by something that could be passed from chicken to chicken. When Rous ground up the chicken tumor, filtered the mixture to remove cells and bacteria, and injected the remaining liquid into the breasts of other hens, they, too, developed tumors. This was the first time a scientist had transmitted a cancer from one animal to another.

Because Rous's filter was supposed to have removed all known microorganisms, he suspected that the cancer was caused by a virus, an agent whose existence was still just a theory. (Scientists in the late 19th and early 20th centuries had shown that certain diseases, including serious human illnesses such as rabies and yellow fever, could be transmitted by liquids that had passed through the type of filter Rous had used, yet microscopes of the day revealed no microbes in these fluids. The scientists therefore assumed that the disease-causing agents, which they called viruses, were too small to see.) He proposed this idea in a 1912 paper called "Transmission of a Malignant New Growth by Means of a Cell-Free Filtrate." No one had ever found a microorganism that could cause cancer, however, and most researchers thought that such a thing was impossible. They claimed that either some cancer cells had slipped through Rous's filter or the chicken growths were not true cancers.

Rous was still convinced that his idea was right, but at the time he could find no way to prove it. He therefore turned to research on other subjects. Beginning in 1915, he and coworkers J. T. Turner and Oswald H. Robertson studied the preservation of blood. Rous and Turner developed a solution that could keep blood from spoiling, and Robertson applied it in establishing the first blood banks, which were used in 1917 on the Belgian front during the last part of World War I.

During the 1920s, Rous and other coworkers investigated the functions of the liver and a small associated organ, the gall bladder, in which a liver secretion called bile is stored. They learned about the process by which bile helps to digest fats and explored diseases that affect the liver and bile system.

Meanwhile, evidence to support Rous's half-forgotten cancer virus theory finally began to appear. British scientists confirmed his work with chicken sarcomas in 1925, and in the mid-1930s, other researchers showed that a type of chicken leukemia was also caused by a virus. Later in the decade, scientists using newly developed electron microscopes saw Rous's virus, now called the Rous sarcoma virus, for the first time.

Once the existence of these tumor-causing viruses was verified, researchers began to hunt for other viruses that might cause cancer in animals. They found several in the 1950s, and in 1978, ROBERT GALLO discovered the first virus shown to cause a human cancer.

Rous himself returned to the study of tumor viruses in 1934, working with a virus that caused small benign (noncancerous) tumors in rabbits. He demonstrated that certain chemicals, such as coal tar, could make these tumors develop into cancers and suggested that cancer formation (carcinogenesis) involves two stages, initiation and promotion, which may be triggered by different things. In this case, the virus was the initiator and coal tar was the promoter.

Rous nominally retired from the Rockefeller Institute in 1945, but in fact he continued to do research there until his 90th birthday. He developed several techniques useful in virology, including improved ways to isolate cells and grow viruses. He suggested that genetics might play a role in cancer and other diseases and recommended that all laboratory animals used in a particular experiment be as genetically similar as possible.

Rous was very old when his work on cancer finally received the honors it deserved. He won the Lasker Award from the American Public Health Association in 1958, a prize from the United Nations in 1962, the National Medal of Science in 1965, and, finally, when he was 87 years old, the Nobel Prize in physiology or medicine (which he shared with Charles B. Huggins, another cancer researcher) in 1966. Rous died on February 16, 1970.

Further Reading

"Rous, (Francis) Peyton." *Current Biography Yearbook 1967*. New York: H. W. Wilson, 1967.

S

⊠ Sabin, Albert Bruce
(1906–1993)
Russian/American
Virologist

Albert Sabin developed a vaccine against the virus that causes polio (poliomyelitis or infantile paralysis) that for many years replaced an earlier vaccine developed by JONAS SALK. Sabin was born to Jacob and Tillie (Krugman) Sabin on August 26, 1906, in Bialystok, a town then belonging to Russia but now part of Poland. The Sabins came to the United States in 1921 to escape religious persecution and settled in Paterson, New Jersey, where Jacob Sabin took a job in the textile industry.

An uncle offered to pay Albert Sabin's tuition at New York University's College of Dentistry if Sabin would join his dental business afterward, but Sabin found medical research more interesting than filling teeth. The uncle stopped his support when Sabin quit the dental program, but Sabin continued his education with scholarships and obtained a bachelor of science degree from the university in 1928 and an M.D. in 1931. In 1935, he married Sylvia Tregillus, and they had two children. (Sabin married Jane Weller soon after his first wife's death in 1966, but the couple quickly divorced. He married Brazilian-born Heloisa Dunshee de Abranches in 1972.)

Sabin began working with viruses while still in medical school and isolated a new kind of pneumonia virus. After completing his internship and residency, he worked at the Rockefeller Institute from 1935 to 1938. He became an associate professor of research in pediatrics at the University of Cincinnati (Ohio) College of Medicine and a fellow of the affiliated Children's Hospital Research Foundation in 1939. During World War II, he served in the U.S. Army Medical Corps, eventually attaining the rank of lieutenant colonel. He developed vaccines against two serious tropical illnesses, dengue fever and Japanese encephalitis, and these discoveries protected soldiers fighting in the Pacific. He returned to the University of Cincinnati in 1946 and became a professor of research pediatrics.

From almost the beginning of his research career, Sabin's chief interest was polio, which brought death or permanent paralysis to thousands of children and adults each summer. (In 1952 alone, the disease killed 3,000 American children and infected nearly 58,000 more.) He had begun studying this disease while still at the Rockefeller Institute. Sabin made several basic discoveries about the virus, including the fact that it could live in the digestive system as well as in the nervous system and probably was spread by mouth.

Sabin and rival researcher Jonas Salk of the University of Pittsburgh raced to develop a vaccine for polio in the early 1950s. Salk's vaccine used killed viruses, but Sabin believed that a polio vaccine could convey long-lasting immunity only if it was made, as vaccines traditionally were, from living microorganisms weakened naturally or artificially so that they could not cause serious illness. He therefore looked for naturally weak strains of polio virus. By 1955, he had created a vaccine incorporating three such strains and tested it on a small number of volunteers, including himself and his family. He then began testing it on volunteer prisoners.

The powerful Foundation for Infantile Paralysis supported both Salk's and Sabin's research, but Sabin did not receive as much money or receive it as soon as Salk did. Perhaps because of this, the Salk vaccine was ready for mass testing first. Once Salk's vaccine was pronounced safe and effective in April 1955, interest in Sabin's all but disappeared in the United States. Sabin, however, convinced the World Health Organization (WHO) that his vaccine had several advantages over Salk's. It could be given by mouth, so it was easier and safer to administer than the Salk vaccine, which had to be injected. Sabin's vaccine was also cheaper and easier to store, an advantage in developing countries. Because it produced an actual infection, it created longer, perhaps lifetime, immunity, and it induced immunity very quickly, which could be important if the vaccine was given during an epidemic. Also, because the weakened viruses were excreted in waste and spread just like natural poliovirus, they infected and therefore gave immunity to people who had never taken the vaccine.

Beginning in 1957, the WHO sponsored mass tests in which some 80 million people in the Soviet Union and several other countries were given Sabin's vaccine. Like Salk's, it proved safe and effective. After seeing the results of these tests, U.S. authorities approved the Sabin vaccine in 1960. It was given to about 100 million people in the United States between 1962 and 1964 alone, and by the end of the 1960s it had almost completely replaced the Salk vaccine. The rate of polio infection in the United States, which the Salk vaccine had already lowered from 135 cases per million in the early 1950s to 26 cases per million people by 1960, fell to 2.4 per million by 1965. The last natural case of polio in the country was reported in 1979, and, thanks to a massive WHO-sponsored vaccination program, polio had almost been eradicated worldwide by the start of the 21st century.

The Sabin vaccine's very success, however, eventually worked against it. In a tiny number of cases—about six children (usually babies) a year in the United States—the virus in the vaccine somehow became able to cause paralytic polio. Americans found that small risk acceptable when natural polio was a major threat, but it became less so after the natural disease had been eradicated. In 2000, therefore, the U.S. Advisory Committee for Immunization Practices recommended that only the Salk vaccine be used. Sabin's vaccine is still preferred in some parts of the world, however.

While his polio vaccine was saving lives and limbs, Albert Sabin continued his research career. He was a Distinguished Service Professor at the University of Cincinnati from 1960 to 1970, spent three years (1970–72) in Israel as president of the Weizmann Institute of Science, and then returned to the United States and became a Distinguished Research Professor of Biomedicine at the Medical University of South Carolina. He kept this position until 1982, then became a senior expert consultant at the Fogarty International Center for Advanced Studies in the Health Sciences, part of the National Institutes of Health, in 1984. He worked at the Fogarty Center full-time until 1986 and part-time until 1988, when worsening health forced him to retire.

During the later part of his career, Sabin identified several disease-causing microorganisms, including a new kind called toxoplasma. He also studied genetic factors in resistance to viruses, investigated cancer-causing viruses, and developed a sprayable vaccine against measles, which kills thousands of children in developing countries each year. He supervised mass tests of the vaccine in Mexico and Brazil in the early 1980s. Later in the 1980s, he investigated possible vaccines for AIDS, although he eventually concluded that such a vaccine could not be made because the AIDS virus changes so quickly.

Sabin never received the public acclaim that Salk did. Perhaps for that very reason, however, his reputation was higher than Salk's in the scientific community, and he won more awards. For example, he was elected to the U.S. National Academy of Sciences in 1951, an honor Salk never received. Sabin also won the U.S. Legion of Merit (1945), the National Medal of Science (1970), and the Medal of Freedom and Medal of Liberty (both in 1986). Several other countries also gave him prizes or medals. Sabin died of congestive heart failure in Washington, D.C., on March 3, 1993.

Further Reading

Blume, Stuart, and Ingrid Geesink. "A Brief History of Polio Vaccines." *Science*, June 2, 2000.

Bowen, Ezra. "Albert Sabin: The Doctor Whose Vaccine Saved Millions from Polio Battles Back from a Near-Fatal Paralysis." *People Weekly*, July 2, 1994.

"Sabin, Albert B(ruce)." *Current Biography Yearbook 1958*. New York: H. W. Wilson, 1958.

Salk, Jonas
(1914–1995)
American
Virologist

Before the mid-1950s, epidemics of poliomyelitis, or infantile paralysis, brought terror to the United States each summer. Polio, as this highly contagious viral disease was usually called, killed 3,000 American children and infected nearly 58,000 more in 1952 alone, leaving many permanently paralyzed. Jonas Edward Salk became a national hero for creating a vaccine that helped to end this threat.

Salk was born on October 28, 1914, in New York City, the oldest son of Daniel and Dora (Press) Salk, immigrants of Polish-Jewish ancestry. Daniel Salk worked in the city's garment industry. The family moved to the Bronx shortly after Jonas's birth. Although the Salks were poor, they were determined that their children

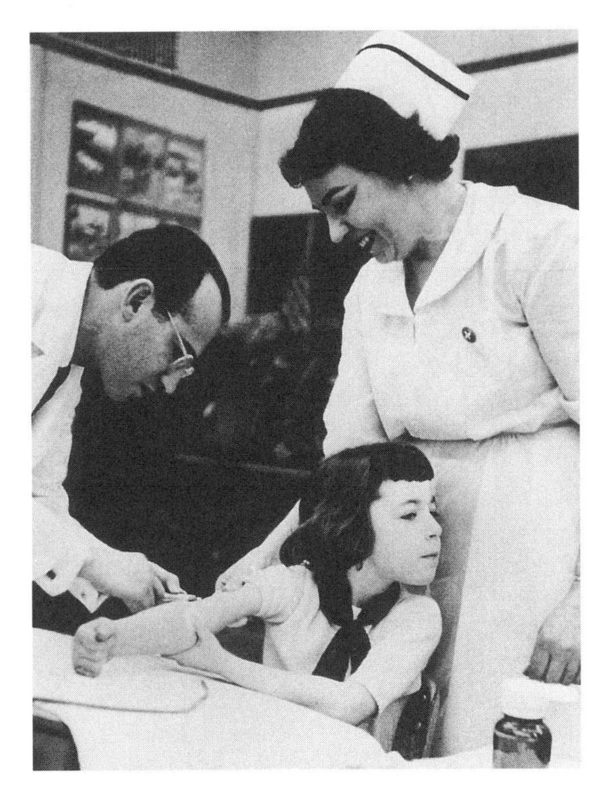

This shot may hurt, but it is protecting the little girl from paralysis or death from polio (poliomyelitis), a virus-caused disease that was once epidemic in the United States and elsewhere; Jonas Salk, the inventor of the polio vaccine that went into widespread use in 1955, is shown at left, administering the injection. *(National Library of Medicine)*

would obtain an education. In 1933, when Jonas graduated from the City College of New York with a B.Sc. in biology at the age of 19, he became the first member of his family to obtain a college degree.

Salk had originally intended to study law, but while an undergraduate he became more interested in biology and medicine. He earned an M.D. from the New York University College of Medicine in 1939. In that same year, he married Donna Lindsay, a social worker, and they had three sons. The couple divorced in 1968, and two years later Salk married painter Françoise Gilot, former companion of Pablo Picasso and mother of two of Picasso's children.

While Salk was a medical student, one of his professors, Thomas Francis Jr., interested him in virus diseases. Salk followed Francis to the University of Michigan in 1942 and held various junior positions in the epidemiology department of the university's School of Public Health. There, in 1943, Francis and Salk developed a vaccine for influenza that protected thousands of American soldiers during the last part of World War II. This vaccine was the first to be made from killed rather than living, weakened viruses.

Salk moved to the University of Pittsburgh Medical School in 1947, becoming an associate research professor of bacteriology and director of the school's virus research laboratory. He was made a full research professor two years later. Turning his attention from influenza to polio, he decided that a polio vaccine, like his flu vaccine, should use killed viruses, even though most other researchers thought such vaccines could not stimulate the immune system strongly enough to be effective.

Basil O'Connor, head of the National Foundation for Infantile Paralysis (popularly known as the March of Dimes), saw reports of some of Salk's early research and decided to fund his vaccine work. Several other groups, most notably one led by ALBERT BRUCE SABIN at the University of Cincinnati, were also trying to develop polio vaccines in the early 1950s, but O'Connor's support and Salk's own ambitious drive helped him finish his vaccine first. Small-scale tests of the vaccine's safety (first on children who already had polio and then on about 100 people who did not, including Salk and his family) were begun in 1952 and reported successful in 1953. Then, in 1954, O'Connor's foundation launched a massive test involving almost two million children, who proudly called themselves the "Polio Pioneers"—the largest field test in medical history up to that time.

In a nationwide radio broadcast on April 12, 1955, officials announced that the gigantic test had proved the Salk vaccine "safe, effective, and potent." Church bells rang and factory sirens blared to celebrate the news. Sharon Begley wrote later in *Newsweek* that by taking away families' reluctance to enter crowds because of the fear of catching polio, "Salk's announcement sparked a tectonic shift in the way people thought and lived." Four million doses of the Salk vaccine were administered by May 7.

Salk's vaccine was not without problems. One early batch that accidentally contained live virus caused the disease in 200 young people in June 1955. More general problems were that the Salk vaccine had to be injected, whereas Sabin's weakened-virus vaccine, first mass distributed in the United States in the early 1960s, could be taken by mouth. Sabin's vaccine was also easier to store and conferred lifetime immunity, which Salk's did not. Because of these advantages, Sabin's vaccine replaced Salk's in many parts of the world for decades. By 2000, however, Salk's vaccine had again become the preferred one in the United States because it is slightly safer than the Sabin vaccine, in which the weakened virus occasionally reverted to its dangerous form and caused a few cases of polio each year.

Like opinions of his vaccine, feelings about Salk himself varied considerably. Some people saw him as a national hero, a modest man who had refused to profit from his vaccine by patenting it ("Could you patent the sun?" he is reported to have said). Others portrayed him as ruthlessly ambitious and criticized him for seeking publicity and not giving sufficient credit to his predecessors or coworkers. The rivalry between Salk and Sabin, in particular, produced bitter feelings for decades. "A monument to the conquest of polio faithful to the facts would consist of not one man in a white lab coat but two of them glaring at each other," Wilfrid Sheed wrote in *Time* in 1999. "But since the public usually prefers one hero to two, and since Salk did get there first, he got the monument."

Salk's is the name that most people remember in connection with the vaccines that reduced the number of polio cases in the United States by 96 percent by 1961, brought natural cases of polio to an end in the United States in 1979, and had almost eradicated the disease worldwide by 2002. On the other hand, although Salk received the Medal of Freedom in 1977, he was not elected to the prestigious U.S. National Academy of Sciences as Sabin was, and he won relatively few other awards. Several commentators, including Sheed, believe that this situation arose because of the scientific community's ongoing disapproval of Salk.

During the rest of his career, Salk did research on cancer, autoimmune diseases, and connections between the immune system and the nervous system. He achieved a lifetime ambition in 1963 by opening the Salk Institute for Biological Studies, a research institute in La Jolla, California, which has since become world famous. He wrote several philosophical books on human evolution during the 1970s, including *Survival of the Wisest* (1973). Beginning in the early 1980s, he focused his research on AIDS, trying unsuccessfully to develop a vaccine that could be given to people already infected with HIV to prevent development of the full-blown disease. He was working on this vaccine the day before he died of congestive heart failure, at age 80, on June 23, 1995.

Further Reading

Blume, Stuart, and Ingrid Geesink. "A Brief History of Polio Vaccines." *Science*, June 2, 2000.

MacPherson, Stephanie Sammartino. *Jonas Salk: Conquering Polio*. Minneapolis, Minn.: Lerner, 2001.

Sheed, Wilfrid. "Virologist Jonas Salk." *Time*, March 29, 1999.

Smith, Jane S. "An Icon Named Salk." *Life*, April 1990.

———. *Patenting the Sun*. New York: William Morrow, 1990.

Sanger, Frederick

(1918–)
British
Biochemist, Molecular Biologist

Frederick Sanger used traditional methods of biochemistry as well as new techniques of molecular biology to work out the sequence of smaller molecules within the large, complex molecules of proteins and nucleic acids. These achievements, which New York University Medical Center biochemist G. Nigel Godson says "single-handedly engineered two revolutions in biology," made Sanger the first person to win two Nobel Prizes in chemistry.

Sanger was born on August 13, 1918, in Rendcombe, Gloucestershire, England, to Frederick Sanger, a prosperous physician, and the former Cicely Crewdson, heiress of a wealthy cotton manufacturer. Sanger attended St. John's College at Cambridge University, graduating in 1939. Perhaps inspired by his father, he began by studying medicine, but he soon changed his major to natural science. The only course he did exceptionally well in was biochemistry, and he decided to make this his specialty. During World War II, he was a conscientious objector and stayed in Cambridge to work toward his Ph.D. in biochemistry, which he obtained in 1943. He married Joan Howe in 1940, and they later had two sons and a daughter.

Sanger has remained at Cambridge all his life, beginning as a Beit Memorial Fellow for Medical Research (1944–51). In 1951, he joined the Medical Research Council, and in 1962 he became director of the division of protein and nucleic acid chemistry at the council's Laboratory of Molecular Biology in Cambridge.

Sanger's first major research project involved trying to determine the sequence of amino acids within a protein. By the mid-1940s, scientists knew that proteins are made up of 20 kinds of amino acids, linked by chemical bonds into long chains called polypeptides. They knew that amino acid molecules are arranged in a different order within each type of protein, but they did not know how to find out what that sequence was.

On the advice of his mentor, Albert Chibnall, Sanger decided to study the pancreatic hormone insulin. Insulin, Chibnall pointed out, had several advantages as a subject: It was a small molecule as proteins went, its atomic composition was known, it was important in medicine, and it could be obtained fairly pure in large quantities from slaughtered cattle. Beginning in 1944, Sanger used chemicals to break insulin molecules into fragments and then developed a new technique using another chemical (still called Sanger's reagent) to label the amino acid at the end of each fragment. Then, using another new technique (developed by others) called paper chromatography, which sorts molecules by size, he and coworker Hans Tuppy showed that insulin consists of two polypeptide chains, one 21 amino acids long and one 30 amino acids long.

Sanger worked out the sequence of amino acids in the 30-acid chain by 1950 and the sequence in the shorter chain, which proved to be the more difficult of the two, several years later. He went on to show how the chains were linked together and to demonstrate slight differences in composition among the insulins of cattle and four other animal species. He was awarded the 1958 Nobel Prize in chemistry for this research, which marked the first time that the sequence of amino acids within a protein molecule had been determined. Other scientists applied Sanger's methods to discover the amino acid sequences in many other proteins, an important step toward learning how cells make and use the proteins.

Around 1960, thanks in part to his friendship with fellow Cambridge researcher FRANCIS CRICK, the codiscoverer of DNA's molecular structure, Sanger turned his attention to nucleic acids. Crick and others had established that the order of four small molecules called bases, or nucleotides, within nucleic acid molecules carries the genetic "code" of instructions for making proteins, specifying which amino acids should be assembled in which order to create each protein molecule. Determining the order of bases in a particular nucleic acid molecule, however, was a lengthy and laborious process.

In 1964, Sanger and Kjeld Marcker worked out the base sequence in a simple type of RNA molecule called transfer RNA. Sanger then turned to messenger RNA, a more complex form of this nucleic acid. Adapting the techniques he had used on insulin, he determined the sequence in a messenger RNA molecule from a virus. He refined his methods during the early 1970s, eventually allowing the sequence in a particular molecule to be discovered in days instead of years, and applied them to DNA as well as RNA. In 1977, he announced that he and his coworkers had determined the order of the 5,375 bases in the DNA of a simple virus called Phi X174—the first time that the sequence of the complete genome of a living thing had been worked out.

Sanger won a second Nobel Prize in chemistry in 1980 for his work with nucleic acids. (He shared the prize with American molecular biologist WALTER GILBERT, who had developed another method of sequencing nucleic acids at about the same time, and with biochemist PAUL BERG, who had developed a way to combine genetic material from different types of living things.) Sanger has also received both the Royal Medal and the Copley Medal from Britain's Royal Society, as well as Canada's Gairdner Foundation Award and the Albert Lasker Medical Research Award in the United States. He was made a Commander of the Order of the British Empire in 1963. By 1993, Sanger had retired to devote time to his family and his favorite hobbies, gardening and sailing.

Further Reading

Douglas, Kate. "The Quiet Genius Who Decoded Life." *New Scientist*, October 8, 1994.

Ruthen, Russell. "Revealing the Hidden Sequence." *Scientific American*, October 1993.

"Sanger, Frederick." *Current Biography Yearbook 1981*. New York: H. W. Wilson, 1981.

Schleiden, Matthias Jakob

(1804–1881)
German
Botanist

With THEODOR SCHWANN, Matthias Jakob Schleiden developed the theory that all living things are made up of microscopic units called cells, a basic concept in biology. Schleiden was born in Hamburg on April 5, 1804, the son of a respected physician. He trained as a lawyer, earning his degree from Heidelberg University in 1827, and established a practice in Hamburg, but he soon found that he did not really enjoy this work. An uncle interested him in botany, so he studied botany and medicine at the Universities of Göttingen, Berlin, and Jena. He earned a degree from Jena in 1839 and taught botany at that university for 23 years.

While he was in Berlin, Schleiden met Scottish botanist Robert Brown, who in 1831 had identified a structure in plant cells that he called the nucleus. This meeting may have inspired Schleiden to turn his own attention to cells. British microscopist ROBERT HOOKE had first seen and named cells around 1665, but Hooke had really observed only the stiff cell walls left after plant cells died. By Schleiden's time, biologists knew a little more about cells, including the fact that living cells contain a semiliquid substance then called protoplasm (now termed cytoplasm), but they disagreed about how important cells were.

Using an improved type of compound microscope developed in the 1820s and 1830s, Schleiden studied cells in plants. He eventually concluded that all parts of plants were made of cells or cell products and that cells were the basic units of structure in plants and probably animals as well. He stressed cells' almost independent existence, writing in a book published in 1838 that "cells are organisms, and entire animals and plants are aggregates of these organisms, arranged according to definite laws." He also noted several new features of cells, including the movement, or streaming, of protoplasm. He correctly claimed that the nucleus was involved in cell reproduction, although his description of the way reproduction took place was mistaken.

Schleiden met physiologist Schwann in Berlin in 1838 and persuaded him to make his own studies of cells, with the result that Schwann extended the cell theory to include animals the following year. Most biologists hailed this theory as the fundamental advance that it was, although they disagreed about many of its details.

Schleiden continued his career in botany but made no other major discoveries. In 1862, he moved to the University of Dorpat in Estonia, but he returned to Germany in 1864 and from that time on taught privately in Frankfurt and other German cities. He died on June 23, 1881, in Frankfurt.

Further Reading

Baker, John Randal. *The Cell Theory: A Restatement, History, and Critique*. New York: Garland Publishing, 1988.

Schwann, Theodor
(1810–1882)
German
Physiologist

Theodor Schwann extended his friend MATTHIAS JAKOB SCHLEIDEN's cell theory to cover animals as well as plants. He also made discoveries about digestion and other body processes. He was born in Neuss, now in Germany, on December 7, 1810. First educated at the Jesuit College of Cologne, he went on to study medicine at the Universities of Bonn, Würzburg, and Berlin, graduating with a degree from the latter in 1834. He then became an assistant to physiologist Johannes Müller at the Museum of Anatomy in Berlin. He did all his significant work between 1834 and 1838, when he held this position.

With botanist Matthias Jakob Schleiden, German physiologist Theodor Schwann, shown here, showed in the late 1830s that all living things are made up of microscopic units called cells. *(National Library of Medicine)*

In 1836, Schwann extracted pepsin, an enzyme essential for protein digestion, from the walls of animal stomachs. This was the first enzyme to be extracted from animal tissue. He also identified cells that make up the waxy sheaths around the long fibers called axons that extend from nerve cells outside the brain; these supporting cells are still called Schwann cells. He coined the term *metabolism* to refer to the chemical changes that take place in living tissue and pointed out that a fertilized egg is a single cell from which an entire, complex organism grows, a basic principle of embryology.

Scientists of the time believed that putrefaction, or decay of animal tissue, was either a strictly chemical process or was caused by microscopic living things spontaneously generated from nonliving matter. In 1837, however, Schwann repeated and improved on earlier experiments by Italian researcher LAZZARO SPALLANZANI to demonstrate that putrefaction was caused by something in air that could be destroyed by heat but not by air itself. French chemist LOUIS PASTEUR would extend this finding in the early 1860s by showing that microorganisms living on dust in the air cause putrefaction.

Schwann met Schleiden, a botanist, in 1838, and Schleiden told him about his theory that microscopic structures called cells are the basic units of which all plants are made. Plant cells had been known since British microscopist ROBERT HOOKE first saw them in 1665, but they had been impossible to study in detail until compound microscopes were improved in the 1820s and 1830s. Hooke, in fact, had seen only the stiff cell walls left after plant cells died.

Excited by Schleiden's idea, Schwann began studying animal tissues, especially of embryos, with the cell theory in mind. He soon concluded (as Schleiden had also suspected) that the theory applied to animals as well as plants. In *Mikroskopische Untersuchungen über die Ueberreinstimmung in der Struktur und dem Wachstum der Tiere und Pflanzen* (*Microscopical Researches on the Similarity in the Structure and Growth of Animals and Plants*), published in 1839, Schwann stated that "everything alive has cellular origin" and demonstrated that even tissues like bone, in which no cells can be seen in adults, originate from cells.

Each cell, Schwann said, is a partly independent living thing, capable of carrying out its own metabolism, but in multicellular organisms the life of individual cells is subordinate to the life of the whole organism. Schwann pointed out that almost all cells have basic parts in common, including nuclei, a semiliquid substance that fills the cell body (then called protoplasm), and outer membranes. Schwann's work, added to Schleiden's 1838 book on cells in plants, demonstrated the unity of all living things. The cell theory quickly gained wide acceptance as a central concept in biology.

Schwann also studied the process of fermentation, in which cells break down sugar or other carbohydrates to produce carbon dioxide and other substances such as alcohol. In 1836 and 1837, he provided evidence that living microorganisms called yeasts carried out the most common kind of fermentation. Powerful German chemists Justus von Liebig and Friedrich Wöhler criticized Schwann's work severely, claiming that fermentation was strictly a chemical process and did not require living things. Pasteur would prove Schwann right some 25 years later, but at the time, the chemists' criticism made him feel that he had no future as a scientist in Germany. Depressed, Schwann moved to the University of Louvain in Belgium in 1839. He taught anatomy there until 1848 and after that at the University of Liège, but he became increasingly devoted to religious mysticism and made little further contribution to science. He died on January 11, 1882, in Cologne.

Further Reading

Baker, John Randal. *The Cell Theory: A Restatement, History, and Critique*. New York: Garland Publishing Co., 1988.

Semmelweis, Ignaz Phillipp

(1818–1865)
Austro-Hungarian
Physician

Although he was mistaken about what actually caused the illness, Ignaz Semmelweis showed how the deadly disease puerperal (childbed) fever could be prevented. Unfortunately, his fellow physicians did not listen to his advice.

Semmelweis was born in Buda (in those days, what is now Budapest was two separate cities, Buda and Pest), in the part of the Austro-Hungarian Empire that is now Hungary, in 1818. His father was a grocer. He studied medicine at the University of Pest and then the University of Vienna, Austria, obtaining his M.D. in 1844. A Hungarian outsider in the German-dominated world of Vienna, Semmelweis failed to obtain the medical jobs he wanted. Instead, in 1847 he became an assistant to Johann Klein, the head of one of the two obstetrics wards in the Vienna General Hospital.

Obstetrics, the medical specialty of helping women give birth, was a depressing field at the time because so many women died soon after their labor—especially if they had their babies in a hospital. The cause of their deaths was puerperal fever, a massive infection. Some physicians thought the disease was due to foul-smelling air, while others blamed it on seasonal weather.

Semmelweis noticed that a far higher proportion of women died in Klein's ward, which was staffed by students from the University of Vienna's medical school, than in the hospital's other obstetrics ward, where patients were cared for by midwives, women trained only to assist with births. A tragic accident early in 1847 suggested why this might be so. Jacob Kolletschka, a colleague and friend of Semmelweis's, cut his finger during an autopsy and died within a few days of an infection that strongly resembled puerperal fever. When Semmelweis in turn performed an autopsy on Kolletschka, he found that the internal signs of the disease were also similar.

This clue brought Semmelweis's attention to a key difference between the staff members of the two obstetrics wards: Midwives never performed autopsies, whereas medical students carried them out all the time. Indeed, the students often went straight from the autopsy room to the ward without washing their hands or their instruments. Semmelweis concluded that "invisible cadaver particles" on the hands and clothing of students and physicians carried puerperal fever. The doctors unknowingly gave the disease to their patients when they touched the patients' damaged tissue. He believed that not only dead bodies but any source of pus, such as infected living patients or soiled sheets or bandages, contained the particles.

Beginning in May 1847, Semmelweis ordered everyone in Klein's ward to wash their hands in a disinfectant chlorine solution to rid themselves of the invisible particles before touching patients. Within a year, the death rate in the ward dropped from 18.3 percent of patients to 1.2 percent, comparable to the 1.3 percent in the midwives' ward. Some of the hospital's younger physicians were impressed by these results, but many others, including Klein, found Semmelweis's insistent manner annoying and his ideas ridiculous. They credited the drop in deaths to seasonal or other variations in the disease. Semmelweis, in turn, did little to promote his work in scientific circles or obtain additional support for his theory through animal experiments. Instead, after failing to win a renewal of his assistantship, he left Vienna abruptly in mid-1850 and returned to Pest.

In 1851, Semmelweis became the director of obstetrics at Pest's St. Rochus Hospital. There, he set up the same procedures he had tried to establish at Vienna and achieved the same amazing results—a death rate from puerperal fever of only 0.85 percent of patients. He was made a professor of obstetrics at the city's university in 1855, and his ideas came to be accepted locally, but elsewhere they were rejected, misunderstood, or forgotten.

Suddenly, after years of silence, Semmelweis began to advertise his ideas with a vengeance around 1858. He sent letters to influential obstetricians all over Europe, asking their opinions of his theory, and wrote vituperative responses to any who failed to agree with him, calling them not only fools but murderers. This same strident tone, as well as repetitious writing and confused reasoning, permeated his single book on the subject, *The Etiology [Cause], Concept, and Prevention of Puerperal Fever,* which was published in 1860. Partly because of these stylistic flaws, most physicians either ignored the book or attacked it.

During these few years, Semmelweis's behavior became more and more erratic and aggressive, until finally, on July 31, 1865, his wife, Maria, and some of his former friends in Vienna committed him to an insane asylum in that city. He died there just two weeks after his commitment, on August 13, 1865. Ironically, his death was caused by the same infection against which he had crusaded. Until recently, historians had believed that his infection, like Kolletschka's, had begun with a cut he received during hospital work done just before his commitment. A new examination of autopsy records, however, has suggested that it arose in injuries resulting from a beating in the asylum.

Semmelweis's ideas were ahead of their time—but only by a little. In 1867, less than a decade after he published his book, British surgeon JOSEPH LISTER concluded that microorganisms cause wound infections. Like Semmelweis, Lister greatly reduced deaths from such infections by a program that combined increased cleanliness with chemical treatment to kill the "germs." Furthermore, Lister, a more determined and less abrasive man than Semmelweis, eventually convinced others to follow his lead. LOUIS PASTEUR, the French chemist whose research on microorganisms had inspired Lister, identified the specific bacterium that causes puerperal fever, a type called streptococcus, in 1878. Pasteur shared Semmelweis's conviction that physicians unconsciously spread the disease from woman to woman, and he said that it could be prevented by such methods as boiling instruments and bandages and washing with disinfectants. By the time these advances occurred, however, Ignaz Semmelweis, who had in part foreseen them, was long forgotten.

Further Reading

Carter, K. Codell, and Barbara R. Carter. *Childbed Fever: A Scientific Biography of Ignaz Semmelweis*. Westport, Conn.: Greenwood Publishing Group, 1994.

Nuland, Sherwin B. *Doctors: The Biography of Medicine*. New York: Random House, 1988.

Sherrington, Charles Scott
(1857–1952)
British
Neurobiologist

Charles Scott Sherrington revealed much about how nerve cells communicate with muscles and with each other and helped to found the field of neurophysiology. He was sometimes called the "WILLIAM HARVEY of the nervous system" because his discoveries revolutionized scientists' picture of the brain as much as the earlier British physician's had changed their view of the blood circulation.

Sherrington was born in the Islington district of London on November 27, 1857. His father, James Sherrington, died when Sherrington was a child. His mother, Anne Brooks, soon married Caleb Rose Jr., a physician, archaeologist, and classics scholar who interested young Sherrington in medicine. Sherrington took part of his medical training at St. Thomas's Hospital in London and part at Cambridge University, where he studied physiology. He obtained his medical degree in 1885. In 1892, he married Ethel Wright, and they had one son.

After three years of postdoctoral research in continental Europe, Sherrington returned to Britain to teach physiology at St. Thomas's. He became a professor at the University of London and the head of the Brown Institute for Advanced Physiology and Pathological Research, the university's veterinary research institution, in 1891. In 1895, he joined the University of Liverpool as a professor of physiology and then, in 1913, became Waynflete Professor of Physiology at Oxford University. He kept this position until his retirement in 1936.

When Sherrington began his research on the nervous system around 1881, some scientists, chiefly Italian neurohistologist CAMILLO GOLGI, believed that the system was a completely interconnected network, an idea called the reticular theory. Others, led by SANTIAGO RAMÓN Y CAJAL, an equally eminent Spanish neurohistologist, thought that it was made up of separate nerve cells that did not touch—the so-called neuron theory. During the 1880s, Sherrington found evidence for both points of view and combined them in a theory of his own. He said that Ramón y Cajal was correct that the brain and nerves were made of separate cells that physically did not touch. Nonetheless, the cells communicated in a way that made them function like a network. This communication took place across microscopic gaps that Sherrington termed synapses.

Sherrington studied communication between nerves and muscles during the 1890s. He showed that muscles contain not only motor nerve endings, through which the nervous system orders them to contract or relax, but also sensory endings, which communicate sensations back to the nervous system. The sensory endings include a type that Sherrington was the first to describe: proprioceptive endings, which sense conditions within the muscles, such as the effects of body position and movement. Sherrington also studied communication between the brain and spinal cord and mapped areas of the brain that control sensation and movement in different parts of the body.

Sherrington's research revealed that most automatic activities, or reflexes, were the result of many muscles acting together. Using what he called an "electric flea," for instance, he investigated the scratching reflex in dogs in 1906. He showed that 19 muscles took part in the reflex itself and another 17 kept the dog upright during the activity. He also studied the knee jerk reflex in humans and animals. He pointed out that in a reflex, some muscles always contract while others relax, resulting in coordinated activity. He termed this idea reciprocal innervation; it is also known as Sherrington's law.

Sherrington described his ideas about the nervous system in *The Integrative Action of the Nervous System* (1906), which the *Cambridge Dictionary of Scientists* calls "a classic of neurology." His work led to improvements in brain surgery and in understanding of diseases of the nervous system as well as the system's basic method of functioning. He also published a respected general textbook on physiology, *Mammalian Physiology: A Course of Practical Exercises* (1919), which went through many editions.

The nervous system was not Sherrington's only area of interest. During World War I, for instance, he headed the British government's Industrial Fatigue Board. As part of an attempt to determine how long a workday was safe for factory employees, he personally worked in a munitions plant for three months.

Sherrington also did bacteriological research. He investigated cholera epidemics in Spain and Italy in the early 1880s and, while at the Brown Institute, studied antisera, a then-new treatment for certain bacterial diseases that used the liquid part of the blood of animals made immune to the diseases. In 1893, while testing the process of making antiserum for diphtheria from a horse, he learned that a young relative had fallen severely ill with the disease, which at the time was often fatal. He quickly bled the immunized horse and gave the serum to the child, saving its life. This was the first use of diphtheria antiserum in England. Sherrington also worked on antisera for cholera and tetanus.

Sherrington's neurological studies won many honors, most prominently the 1932 Nobel Prize in physiology or medicine, which he shared with fellow British neurophysiologist EDGAR DOUGLAS ADRIAN. Sherrington also received the Royal Society's Royal Medal (1905) and Copley Medal (1927) and was president of the society from 1920 to 1925. He was made a Knight Grand Cross of the British Empire in 1922 and given the Order of Merit in 1924.

A man of many interests, Sherrington had hobbies ranging from skydiving to poetry. His weekend parachute jumps from the tops of London hospitals often attracted large crowds. His poetic side came through in his scientific writing as well as in actual verse (of which he published a book in 1925), as when, for instance, he described the nervous system as an "enchanted loom, where millions of flashing shuttles weave a dissolving pattern." He expressed his philosophical beliefs in *Man on His Nature* (1940). Sherrington died of heart failure in Eastbourne on March 4, 1952.

Further Reading

Eccles, John C. *Sherrington, His Life and Thought*. New York: Springer Verlag, 1979.

Sherrington, Charles Scott. *The Integrative Action of the Nervous System*. 1906. Reprint, Manchester, N.H.: Ayer Co., 1973.

"Sir Charles Scott Sherrington—Biography." *Nobel Lectures, Physiology or Medicine 1922–1941*. Available online. URL: www.nobel.se/medicine/laureates/1932/sherrington-bio.html. Last updated 2001.

Simpson, George Gaylord
(1902–1984)
American
Paleontologist, Evolutionary Biologist, Taxonomist

Along with ERNST MAYR and several other mid-20th-century biologists, George Gaylord Simpson helped to unite paleontology with genetics, taxonomy, and evolutionary biology to create a "new synthesis" of understanding about evolution. Simpson was born in Chicago on June 16, 1902, but grew up in Denver, Colorado.

On summer trips into the Rockies with his father, Joseph Simpson, a land developer and miner, Simpson became interested in geology. He first studied literature and humanities when he entered the University of Colorado at Boulder, thinking he might become a poet, but he changed his major to geology and transferred to Yale University in 1922. After earning a Ph.B. (bachelor of philosophy) degree from Yale in 1923, he stayed at Yale for graduate work in geology and paleontology and earned a Ph.D. in 1926. He secretly married Lydia Pedroja in 1923, but, although the couple eventually had four daughters, the marriage was unhappy and ended in divorce in 1938. Immediately afterward, Simpson married Anne Roe, a psychologist. This marriage was much more successful.

Simpson's Ph.D. thesis concerned the taxonomy and evolutionary relationships of mammals in the Mesozoic period, based primarily on surveys of fossils in U.S. museums. In a year of postdoctoral study at the British Museum of Natural History, he extended his catalog to include holdings in European museums. Joe Cain of University College, London, author of a biographical sketch of Simpson, calls the total work, issued in 1945, "a long-awaited synthesis" and "a major success, with information from around the world . . . masterfully brought together." Simpson's classification of fossil mammals is now considered standard.

On his return to the United States in 1927, Simpson joined the American Museum of Natural History in New York as assistant curator of vertebrate paleontology. He became curator of fossil mammals and birds and chairman of a new department of paleontology and geology in 1944. He also taught paleontology at Columbia University from 1945 to 1959. During his time at the museum, Simpson led numerous field expeditions, mostly in North America. Two of his expeditions went to Patagonia, at the southern tip of South America, and he described these in a travel book called *Attending Marvels* (1934). During World War II, beginning in 1942, he used his fluency in several languages as an intelligence officer on the Mediterranean front.

Simpson returned to the Museum of Natural History after the war. Much of his work during the next 15 years focused on fossils of the American Southwest and on the migration of ancient mammals between North and South America. In 1959, he became Alexander Agassiz Professor of Vertebrate Paleontology at Harvard University's Museum of Comparative Zoology. Increasing ill health led him to move to Tucson, Arizona, in 1967, although he kept his Harvard professorship until 1970. He was associated with the University of Arizona from 1967 until he retired in 1982.

Simpson's main area of study was the evolution of mammals. He was especially interested in biogeography, or the distribution and range of species. He traced migrations of key species over long periods of evolutionary time and analyzed them according to principles of population genetics. He also developed a theory to explain the different rates and patterns of changes in form that species showed during evolution. He made a special study of horses and showed that, rather than following a single line of descent, the horse family had developed on a "zigzag course" involving several branching lines. He stressed the importance of having classification of species reflect evolutionary history, an idea that by no means all taxonomists shared. He also defended the importance of paleontology and evolutionary theory against "new" biologists who preferred to stress genetics and molecular biology.

Innumerable books, papers, and essays, both popular and scientific, poured from Simpson's pen. *Tempo and Mode in Evolution* (1944), *The Major Features of Evolution* (1953), and *Principles of Animal Taxonomy* (1961) were among his most highly regarded scientific books. His best-known popular books include *The Meaning of Evolution* (1949), which not only explained recent developments in the theory of evolution but described Simpson's ideas about ethics and future human evolution, which he believed that humans themselves could and should guide. He even wrote one novel, *The Dechronization of Sam Magruder*, which was published in 1996, after his death. He won many awards, including the Dumont Medal of the Geological Society of Belgium and the British Royal Society's Darwin Medal.

Simpson did not form friendships easily, and he had vigorous arguments with colleagues such as fellow Harvard professor STEPHEN JAY GOULD. Most of those colleagues, however, respected him highly. Gould, for instance, wrote that "George Gaylord Simpson, in the impact of his ideas and by the power of his writing, . . . was the most important paleontologist since GEORGES CUVIER." Simpson died from complications of pneumonia on October 6, 1984, in Arizona.

Further Reading

Laporte, Leo F. *George Gaylord Simpson: Paleontologist and Evolutionist*. New York: Columbia University Press, 2000.

Simpson, George Gaylord. *Concession to the Improbable: An Unconventional Autobiography*. New Haven, Conn.: Yale University Press, 1978.

"Simpson, George Gaylord." *Current Biography Yearbook 1964*. New York: H. W. Wilson, 1964.

Skinner, B. F.

(1904–1990)
American
Psychologist

Burrhus Frederic Skinner aroused controversy by claiming that the study of human as well as animal psychology should be limited to observable behavior initiated in response to the environment. He was born on March 20, 1904, in Susquehanna, Pennsylvania. His father, William Skinner, was a lawyer. In a biographical sketch published in 1967, Skinner described his mother, the former Grace Burrhus, as a "bright and beautiful" woman with "rigid standards of what was 'right.'" He called his childhood "warm and stable" and wrote that, although Susquehanna was "a rather dirty railroad town," it was located in "a beautiful river valley." He enjoyed walking through the countryside as well as inventing and constructing mechanical devices.

At Hamilton College, a small men's college in Clinton, New York, Skinner majored in English, thinking he might become a writer. After two years of failed writing attempts following his graduation in 1926, however, he decided to investigate behavior through science instead of literature and therefore went to Harvard University to study psychology. He earned his M.A. in 1930 and his Ph.D. in 1931. He married Yvonne (Eve) Blue in 1936, and they later had two daughters.

Skinner stayed at Harvard until 1936, when he moved to the University of Minnesota, Minneapolis. He taught there until 1945, rising to the rank of associate professor in 1939. He headed the psychology department at Indiana University, Bloomington, from 1945 to 1947. In 1948, he returned to Harvard as a professor of psychology and remained there until his retirement, becoming Edgar Pierce Professor of Psychology in 1958.

Building on the teachings of IVAN PETROVICH PAVLOV and John B. Watson, Skinner studied animal behavior that could be observed and manipulated in the laboratory, an approach that Watson called behaviorism. Like Pavlov, Skinner saw behavior as something that occurred in response to stimuli and could be changed by changing the stimuli. However, he moved beyond Pavlov in seeing behavior not merely as modified ("conditioned") reflexes but as general responses to the environment ("operant conditioning").

Pavlov had concentrated on the stimuli that first produced behavior, but Skinner focused on the events that followed behavior and their effects on the behavior's continuation. "Behavior is followed by a consequence, and the nature of the consequence modifies the organism's tendency to repeat the behavior in the future," he once wrote. If the behavior was rewarded, or positively reinforced—for instance, by the appearance of food—the behavior would be repeated often. On the other hand, an unpleasant consequence, or lack of any identifiable consequence, would cause the behavior to become less frequent and, eventually, stop occurring.

In the late 1930s, the inventive Skinner adapted an ice chest to create the so-called Skinner box, which proved to be an invaluable tool for investigating and measuring laboratory animals' interaction with their environment. Rats or pigeons in the soundproofed, climate-controlled box learned to press a lever or carry out other behaviors to obtain food rewards. Skinner described the box (which other behaviorists soon adopted) and experiments he had done with it in his first major book, *The Behavior of Organisms*, published in 1938.

Skinner attempted to put his techniques to practical use during World War II by training pigeons to peck at the center of a target, intending them to become living guidance systems for antiaircraft missiles or torpedoes. The U.S. Office of Scientific Research and Development sponsored this research, which began in 1940 and was dubbed "Project Pigeon," but Skinner's pigeons never actually went into service.

Skinner aroused public controversy for the first time in 1945 with an article in the *Ladies' Home Journal* that described another of his inventions, the Air Crib. The crib, which he had designed for his younger daughter, Deborah, was air conditioned, soundproofed, temperature controlled, and completely enclosed, with a window of safety glass. Skinner intended it to be a safe environment in which parents could let babies sleep or play without constricting clothing for relatively long periods. Some manufacturers attempted to market it, but it never became popular, perhaps because critics saw it as disturbingly similar to the Skinner box.

At Harvard during the 1950s, Skinner studied relationships between behaviors and different schedules of reinforcement, including fixed schedules, in which a reward dependably appears after a certain number of behaviors, and variable schedules, in which the reward arrives at random intervals. In *Schedules of Reinforcement* (1957), he and coworker Charles B. Ferster showed that different kinds of schedules produce different patterns of behavior. They also described how animals could be trained to perform behaviors nothing like their natural ones by a process they called shaping, or "the method of successive approximations." In this training technique, any behavior even slightly similar to the desired one is rewarded at first. The rewards are then successively limited to behaviors more and more closely resembling the goal activity. Skinner's most famous application of this method was teaching pigeons to play Ping-Pong.

In the late 1950s, Skinner concluded that children, like his pigeons, learned best when subject matter was broken down into small steps and the mastering of each step was followed by reward. In the case of students, the reward was being informed that they had made a correct answer. This, he pointed out, was very different from the standard approach used in schools. He designed "teaching machines" that used his method, which he called programmed learning, and described the machines and their use in *The Technology of Teaching* (1968).

Like many of Skinner's other inventions, teaching machines and programmed learning stirred hot debate. Supporters praised the machines' ability to let students work through material independently, at their own pace, and the encouragement that frequent feedback provided. Critics, on the other hand, found Skinner's highly structured lessons rigid, mechanistic, and boring. Programmed learning enjoyed a wave of popularity when computers began to be used in schools in the 1970s and 1980s, but it never became a dominant force in education. Aspects of Skinner's approach nonetheless have been incorporated into many educational and training programs.

Skinner stirred up even more trouble when he attempted to apply behaviorism to human psychology. He maintained that, in people as well as animals, only observable behavior was a fit subject for scientific study. Furthermore, he said, human behavior could and should be shaped to meet society's needs. (He was accused of being totalitarian, but he always stressed that conditioning by means of rewards was far more effective and therefore preferable to conditioning through punishments.) He first expressed this idea in a novel, *Walden Two* (1948), which described an ideal society based on behaviorist principles. He carried it even further in *Beyond Freedom and Dignity* (1971), which claimed that the concepts named in the title were both illusory and unnecessary. Not surprisingly, many people disagreed.

Skinner's retirement from Harvard in 1974 did not mean the end of his career. He wrote numerous books after that time, including a three-volume autobiography—*Particulars of My Life* (1976), *Shaping of a Behaviorist* (1979), and *A Matter of Consequences* (1983)—and a book of advice on dealing with the difficulties of old age (coauthored with Margaret Vaughan). His work ended only with his death from leukemia on August 18, 1990.

Skinner won numerous awards for his work, including the American Psychological Association's Distinguished Scientific Contribution Award (1958) and Gold Medal (1971), a Career Award from the National Institute of Mental Health (1964), the National Medal of Science (1968), and the Joseph P. Kennedy Jr. Foundation Award (1971). Although many psychologists, both in Skinner's own day and since, regarded his ideas as extreme, he helped to bring the methods of psychology into closer conformity with the experimental approach used in other fields of biology. His work provided techniques that have proved useful in areas ranging from animal training to treatment of mental illness.

Further Reading

Bjork, Daniel W. *B. F. Skinner: A Life*. New York: Basic Books, 1993.

Richel, Marc N. *B. F. Skinner: A Reappraisal*. Mahwah, N.J.: Laurence Erlbaum Associates, 1993.

Skinner, B. F. "B. F. Skinner . . . A Brief Autobiography." Available online. URL: http://ww2.lafayette.edu/~allanr/autobio.html. Accessed 2003.

"Skinner, B(urrhus) F(rederick)." *Current Biography Yearbook 1979*. New York: H. W. Wilson, 1979.

Snow, John

(1813–1858)
British
Epidemiologist, Physician

Although he was known in his own time chiefly as one of the first physicians to use anesthesia regularly, John Snow is remembered today for his pioneering work in another medical specialty, epidemiology. He used statistics and maps to show how the dangerous epidemic disease cholera was spread and suggested ways of preventing it, even though he did not know its true cause.

Snow was born to William Snow, a poor farmer, and his wife, Frances, near York, England, on March 15, 1813. Possibly with financial help from his uncle, a well-to-do seller of books and art, he obtained a basic education and became an apprentice, or assistant, to a surgeon in Newcastle at age 14. In 1836, he began studying at the Hunterian College of Medicine and the Westminster Hospital in London. He obtained a surgeon's and apothecary's (druggist's) license in 1838 and a medical degree from the University of London in 1844.

In 1831, while Snow was still an apprentice, he treated an outbreak of cholera among coal miners in the town of Killingworth, part of a nationwide epidemic of the disease. (Cholera, which originated in India, reached England for the first time in that year.) He saw victims of the illness "brought up from . . . the coal-pits . . . fast approaching to a state of collapse" and die within days or even hours from diarrhea and vomiting that leached most of the water from their bodies. This disturbing experience may have interested Snow in combating the disease.

British physician John Snow, one of the first epidemiologists, provided convincing evidence during an 1854 epidemic of cholera that this disease is transmitted by contaminated water. *(National Library of Medicine)*

Snow began to study cholera seriously in 1849, when another widespread epidemic swept through England. At the time, many scientists thought that cholera and other contagious diseases were caused or spread by "bad air," especially the foul air rising from the piles of garbage, including human waste and decaying animal corpses, that infested the poverty-stricken tenements of cities like London. Snow, however, began to suspect that water rather than air was the source of cholera. The disease's symptoms showed that it affected the digestive tract, he said, and he believed that it spread when waste from cholera victims contaminated drinking water. Snow published his ideas in a pamphlet called *On the Mode of Communication of Cholera* in 1849. The Institute of France awarded it a prize, but it failed to convince most physicians and city officials in Britain.

A third cholera epidemic in 1854 gave Snow a chance to test his theory "on the grandest scale." Most homes in the south of London received piped-in drinking water from one or the other of two water companies. Both companies took their water from the Thames, parts of which were heavily polluted by sewage. The Lambeth Water Company drew its water from a spot upriver from the main sewage dump and therefore had relatively clean water. The Southwark and Vauxhall Company, on the other hand, took its water from below the sewage dump, so Snow expected the water to be polluted. Britain had recently begun to collect information about births, deaths, and social and health conditions, so records of cholera cases already existed. Snow realized that if he found out which homes used each water company and compared this information with the list of homes that contained cholera victims, he should be able to determine whether cholera was associated with polluted water.

With the cooperation of William Farr, the official in charge of the government registry of births and deaths, Snow and an army of assistants collected their water data in July and August 1854. He then made his comparison with the cholera records and found that users of Southwark and Vauxhall water had 71 cases of cholera per 10,000 homes—whereas Lambeth customers had only 5 cases per 10,000 homes—strong support for his theory.

Snow gained even clearer evidence from a cholera outbreak that occurred in early September in a poor part of London called Golden Square. This outbreak, which he called "the most terrible . . . which ever occurred in this kingdom," killed more than 500 people in 10 days in an area just 1,500 feet across. People in places like Golden Square pumped their drinking water from community wells rather than having it piped in, and when Snow visited Golden Square, he immediately became suspicious of a well in Broad Street because it was less than three feet downhill from a pool into which waste was dumped. If the brickwork of the well was cracked, sewage from the pool could easily leak into it.

When Snow conducted a quick survey, he found that 59 out of 77 cholera-stricken households in the area had obtained their drinking water from the Broad Street well. Most households or institutions that had been spared, on the other hand, took their water from other sources. Snow even learned of two women living five miles away who had had water specially sent to them from Broad Street because they liked its taste: They had contracted cholera, even though no one else near them was sick.

Snow described his concerns about the Broad Street well to a neighborhood government group, the Board of Guardians of St. James Parish, on September 7 and asked that the handle of the well's pump be removed so that people could no longer use it. Although, according to one physician witness, "not a member of his own profession, not an individual in the parish believed Snow was right" in his basic ideas about the way cholera was spread, the guardians agreed that contamination might be a factor and therefore did as he asked. The outbreak ceased shortly after the handle of the Broad Street pump was taken off.

Henry Whitehead, a local minister, did an independent study of the Golden Square outbreak in early 1855. He found evidence that the outbreak was already dying down by the time Snow spoke to the guardians and guessed that it probably would have ended soon on its own. Nonetheless, he concluded, the guardians' action probably saved lives by preventing a second outbreak. Whitehead discovered the case that had probably started the first outbreak, a sick baby whose mother had thrown the dirty water from washing its soiled diapers into the pool near the Broad Street pump. The woman's husband developed cholera on the same day the pump handle was removed, and his waste, too, went into the pool. That could have started the cycle of contamination and disease all over again if the well had still been in use. In addition, a city engineer who checked the area at Whitehead's request found broken and stained underground brickwork between the pool and the well, providing further evidence that the pool could have leaked into the well.

Snow described the Golden Square outbreak in a book-length revision of his cholera pamphlet, published in 1855. The book included information about many other outbreaks during the 1849 and 1855 pandemics (epidemics affecting many countries at once), not only in Britain but also in Russia, France, and India. It ended with 12 recommendations for preventing cholera and other epidemic diseases that Snow thought were spread by water. This book, which stressed using statistics about large numbers of people in a community to track epidemic dis-cases, was one of the first major contributions to the new specialty of epidemiology. Furthermore, although many physicians and others concerned with public health still clung to the "bad air" theory of disease, Snow's book inspired them to add installation of sewers and purification of drinking water to their list of desired reforms. Carrying out these changes helped to prevent the spread of several epidemic digestive diseases transmitted by contaminated water, including typhoid fever as well as cholera.

Snow did not speculate much about what actually caused cholera, but he suggested that it was a "poison, which has the property, under suitable circumstances, of reproducing its kind." In this he was very close to the truth, even though no one yet suspected that microorganisms could cause disease, an idea that would be proposed by French chemist LOUIS PASTEUR and German bacteriologist ROBERT KOCH in the 1860s and 1870s. In 1883, Koch proved that a comma-shaped bacterium caused cholera.

Unfortunately, John Snow did not live to see that discovery. He had been one of the first British physicians to use ether anesthesia, first popularized by WILLIAM THOMAS GREEN MORTON in the United States in late 1846, and administration of anesthesia became his specialty. He gave another anesthetic gas, chloroform, to Queen Victoria during the birth of two of her children in 1853 and 1857. Snow also did research on anesthesia, sometimes testing new anesthetics on himself. Damage from these experiments may have contributed to his death from a stroke in London on June 16, 1858, when he was only 45 years old.

Further Reading

Brody, Howard, et al. "Map-making and Myth-making in Broad Street: The London Cholera Epidemic, 1854." *Lancet*, July 1, 2000.

Frerichs, Ralph R. "John Snow." Available online. URL: www.ph.ucla.edu/epi/snow.html. Accessed 2002.

Spallanzani, Lazzaro

(1729–1799)
Italian
Naturalist

Lazzaro Spallanzani made several significant discoveries in biology, including a clear demonstration that microorganisms cannot be created from nonliving matter. He was born in Scandiano, a village near Modena in northern Italy, on January 12, 1729. His father was a well-to-do lawyer who, after Spallanzani attended a Jesuit college in Reggio, persuaded him to study law at the University of Bologna. Spallanzani obtained a law degree, but the profession apparently never really appealed to him. He may have been ordained as a priest around 1757, but if so, he did not often practice this calling either. Instead, his cousin, Laura Bassi, a professor of physics and mathematics at the University of Bologna (Italy, unlike the rest of Europe at the time, allowed women to hold academic positions), encouraged his interest in natural science, and that became his chief activity.

Spallanzani began teaching at the Reggio college in 1757. He was a professor of physics and mathematics at San Carlo College in Modena from 1760 to 1769 and then professor of natural history at the University of Pavia for the rest of his life. He also headed the Pavia university's museum and traveled widely to gather specimens for it.

Some of Spallanzani's first work concerned digestion. Lowering perforated boxes containing food into animals' stomachs—and his own—on strings and then removing them, he showed that liquid produced by the stomach, which he was the first to call gastric juice, dissolved the food. This demonstration disproved the widely held belief that the stomach digested food by heating or cooking it.

Spallanzani also studied reproduction. Among other things, he was the first person to perform artificial insemination in the laboratory.

(Arab breeders had been doing it with horses for centuries.) By making tiny trousers and putting them on sexually excited male frogs, he succeeded in capturing a few drops of clear liquid which, when added to unfertilized frog eggs, caused tadpoles to develop. Development did not occur if he filtered the liquid and thereby (although he did not realize it) removed the male sex cells. He artificially inseminated a dog as well.

Spallanzani's interest in reproduction led to his best-known experiments, which provided a convincing disproof of the idea that microorganisms could arise spontaneously from nonliving matter. People in early times had believed that, for instance, wormlike maggots could be created from rotting meat, but in 1668 another Italian, Francesco Redi, had demonstrated that maggots would not appear if flies were prevented from laying eggs on the meat. This experiment convinced most scientists that spontaneous generation (creation of living things from nonliving material) did not apply to living things large enough to be seen with the naked eye. Many, however, still thought that microorganisms, those mysterious creatures that only a few microscopists such as ANTONI VAN LEEUWENHOEK had seen, might still arise in this way.

In 1748, an English priest named John Needham had claimed to have produced living microorganisms from flasks of mutton broth or seeds in water boiled for up to an hour and sealed with corks and gum. The British Royal Society and the influential French scientist GEORGES-LOUIS BUFFON accepted Needham's work, but Spallanzani did not, and he set out to show that microorganisms, like larger living things, could arise only from parents like themselves. He put mixtures like Needham's in glass flasks and boiled them for an hour. No living microorganisms remained in the liquids after this treatment, but if he either left the flasks open to the air or sealed them with Needham's method, within a few days the broths looked under the microscope "like lakes in which swim fishes of all sizes, from whales to minnows." If he sealed the flasks by melting their narrow necks in a flame before boiling them, however, no microorganisms appeared in them even after weeks or months.

Spallanzani published accounts of his experiments on digestion and reproduction, including the spontaneous generation experiments, in a work called *Preface to Studies on Animal Reproduction* in 1768. His experiments convinced most scientists that even microbes could not be spontaneously generated, although some belief in spontaneous generation of microorganisms persisted until about a hundred years later, when French chemist LOUIS PASTEUR performed experiments that were similar to Spallanzani's but even more persuasive.

Anatomy and physiology were among Spallanzani's interests as well. In 1771, for instance, he became the first person to show the tiny blood vessels called capillaries, which connect arteries with veins, in a warm-blooded animal, the chick embryo. (Italian microscopist MARCELLO MALPIGHI had first seen capillaries in frogs in 1661.)

Spallanzani sometimes carried his curiosity about animal life to bizarre extremes. He proved that snails could regenerate their heads and that male frogs could continue mating even after their heads were cut off. He demonstrated that blinded bats not only could fly but could even catch insects and avoid colliding with fine silk threads. He dried out tiny water animals called tardigrades, or water bears, so completely that they appeared dead, kept them in that condition for up to four years, and then revived them by adding water to them. This experiment caused a sensation when he reported it in 1775 because of the religious implications of the apparent resurrection.

In addition to his biological investigations, Spallanzani did experiments in chemistry, physics, meteorology (the study of weather and climate), and geology (he was one of the first people to study volcanoes systematically). His tireless explorations, which he described in numerous letters and papers, earned him the honor of election to membership in Britain's Royal Society in 1768 and won the patronage of learned men such as Voltaire as well as several European rulers. They ended only with his death from a stroke in Pavia on February 11, 1799.

Further Reading

Epstein, Samuel. *Secret in a Sealed Bottle: Lazzaro Spallanzani's Work with Microbes*. New York: Putnam, 1979.

Sperry, Roger Wolcott

(1913–1994)
American
Neurobiologist

Roger Sperry showed that the left and right hemispheres of the brain have different functions and that the basic "wiring" of the nervous system is determined by chemical and genetic factors. He was born on August 20, 1913, in Hartford, Connecticut, the older of Francis and Florence (Kraemer) Sperry's two sons. Sperry's father, a banker, died when Roger was 11 years old, and his mother went to business school and became an assistant to the local high school principal to support her family.

Sperry majored in English at Oberlin College in Ohio, graduating with a B.A. in 1935, but an undergraduate psychology course inspired him to change to that field in graduate school. He earned an M.A. in psychology from Oberlin two years later, then entered the University of Chicago to do Ph.D. studies under Paul Weiss. Weiss had proposed that neurons' function was determined primarily through experience and learning and that changing a nerve's connection to other nerves therefore would change its function. To test Weiss's theory, Sperry surgically crossed the nerves that controlled the muscles of a rat's hind legs. If Weiss's idea was correct, the nerves should eventually have "reeducated" themselves so that they would stimulate the muscles to which they had been newly attached. Even after a long period of recovery, however, stimulating the rat's right foot always produced a response in the left foot, showing that the nerves had retained their original function.

Sperry also tested Weiss's theory in amphibians, which, unlike mammals, can regenerate cut nerves. He cut a salamander's optic nerves and reimplanted the animals' eyes upside down. If Weiss's idea was correct, the nerves should "learn" to take the eye displacement into account when they grew back. Sperry found, however, that long after the nerves had regenerated, the salamanders moved away from a lure instead of toward it, indicating that functionally the nerves had remained reversed. This suggested that nerve connections and functions are innate and cannot be changed. Sperry earned a Ph.D. in zoology in 1941 for proving his famous professor wrong.

Meanwhile, around 1940, Sperry theorized that brain and nerve connections are determined before birth by substances within the embryo, which guide each nerve to its appropriate attachment point in the body—what he called "a kind of probing chemical-touch system." The placement of these chemical markers, in turn, is controlled by the genes. The specific functions of the nerves are also determined at this time, he concluded. He continued to develop and research this theory, called the chemoaffinity or chemospecificity theory, for the next 20 years. The theory came to be accepted as a basic principle of developmental neurobiology, and researchers have found that other tissues in the embryo develop in the same way.

Sperry did postdoctoral research at Harvard under another well-known psychology professor, Karl Lashley, and followed him to the Yerkes Laboratories of Primate Biology, then in Orange Park, Florida, in 1942. At first, as a World War II-related project for the Office of Scientific Research and Development, Sperry investigated surgical repair of nerve injuries. Then, after the war was over, he tested Lashley's ideas about electric fields in the brain, which, somewhat like Weiss's theory, pictured neuronal circuits as interchangeable. Once again, Sperry showed that his mentor was in error.

Sperry became an assistant professor of anatomy at the University of Chicago in 1946 and remained there until 1954, when he became Hixon Professor of Psychobiology at the California Institute of Technology (Caltech) in Pasadena. He held this position until his death. His first projects at Caltech were concerned with vision and memory. Normally, one set of nerves from each eye goes to the brain hemisphere on the same side of the body and another set crosses over to connect to the opposite hemisphere. Sperry cut the crossover point in cats so that each cat's right eye was connected only to its right hemisphere and its left eye only to its left hemisphere. He then covered one of the cat's eyes with a patch and trained the animal to recognize the difference between a circle and a square. When he switched the patch to the other eye, the cat still remembered the distinction, which meant that its memory had been transferred from one hemisphere to the other. This result interested Sperry in communication between the two halves of the brain and launched his best-known experiments.

At the time, biologists knew that most nerves crossed over, so that the right hemisphere of the brain received input from, and controlled, the left half of the body and vice versa. They knew little about whether or how the hemispheres communicated, however, or whether the corpus callosum, a thick band of nerve tissue that connects the two hemispheres in mammals, played any role in such communication. Lashley, for one, thought that the corpus callosum's only function was to keep the two hemispheres from sagging into each other. When Sperry cut the corpus callosum in cats and monkeys, however, he and coworker Ronald Meyers found that knowledge gained by training that involved one eye or one side of the body was no longer transferred to the other side of the body. As he put it in a January 1964 *Scientific American* article, "The split-brain animal behaved . . . as if it had two entirely separate brains."

In the 1960s, Sperry became able to study split-brained humans as well because surgeons began cutting the corpus callosum in certain people with uncontrollable epilepsy in order to keep the abnormal electrical activity that caused their seizures from spreading across their whole brain. The two brain hemispheres have identical functions in animals, but (by studying people whose brains had been damaged by strokes or injuries) neurobiologists had learned that this is not true in humans. The brain centers responsible for speech, logical reasoning, and numerical calculation, for instance, were known to exist only in the left hemisphere. Researchers knew much less about the right hemisphere, and most thought it inferior to the "dominant" left hemisphere.

Overall, Sperry found, split-brained people showed normal intelligence and functioning. Striking differences appeared, however, when he and coworker Michael S. Gazzaniga began testing each side of the brain separately, as Sperry had done with his cats. If split-brained people saw an object with just the left eye, they could not name what they had seen because the language-oriented left hemisphere had received no information about it, but they could easily use their left hand to pick out a matching object from a pile of similar-sized items hidden behind a screen. Gazzaniga wrote in 1981, "No one was prepared for the riveting experience of observing a split-brain patient generating integrated activities with the mute right hemisphere that the language-dominant left hemisphere was unable to describe or comprehend. That was the sweetest afternoon."

After numerous experiments of this type, Sperry and Gazzaniga concluded in the late 1960s that both hemispheres are equally important but have different functions. The right brain specializes in spatial relationships, music, and social and emotion-related activities such as identifying expressions on faces. This research completely changed scientists' views of brain function and had implications for understanding and treatment of both physical brain damage and mental illness. Educators also drew from it the inference that nonverbal aspects of learning are just as important as verbal ones.

Sperry received numerous awards for his split-brain research, most notably a share of the Nobel Prize in physiology or medicine in 1981. He received half the prize and two other brain scientists, Torsten Wiesel and David Hubel of Harvard University, divided the other half. Sperry also won the Albert Lasker Medical Research Award and Israel's Wolf Prize in 1979.

Brain research was by no means Sperry's only interest. He was a star athlete in college and later enjoyed hobbies ranging from folk dancing to art and paleontology. He married Norma Deupree in 1949, and they had a son and a daughter. Late in his life, he turned to the philosophy of science, attacking what he felt was the excessively mechanistic view of many mainstream scientists. In *Science and Moral Priority*, published in 1983, he argued that science and religion should cooperate in establishing a new system of ethics and values. Sperry died on April 17, 1994.

Further Reading

Erdmann, Erika, et al. *Beyond a World Divided: Human Values in the Brain-Mind Science of Roger Sperry*. New York: Random House, 1991.

"Sperry, Roger W(olcott)." *Current Biography Yearbook 1986*. New York: H. W. Wilson, 1986.

⊠ Starling, Ernest Henry

(1866–1927)
British
Physiologist

Ernest Starling greatly increased biologists' understanding of the way the heart and other organs work and are controlled. He was born on April 17, 1866, in London. He attended King's College, London, from 1880 to 1882 and then studied medicine at Guy's Hospital, graduating in 1889.

Starling was a lecturer in physiology at Guy's and a part-time researcher at University College, part of London University, from 1889 to 1899. In the latter year, he became Jodrell Professor of Physiology at University College, a position he held until his retirement in 1922. During World War I, he was director of research at the Royal Army Medical Corps College, where he investigated antidotes for poison gases used in the war. He was also head of the Royal Society's Food Committee and scientific adviser to the Ministry of Food from 1917 to 1919. He was Foulerton Research Professor of Physiology at the Royal Society from 1923 until his sudden death in Kingston, Jamaica, during a Caribbean vacation on May 2, 1927.

Starling's first research was on blood pressure and the exchange of fluid between tissues and the body's tiniest blood vessels, the capillaries. He showed in 1896 that a balance, or equilibrium, exists between hydrostatic pressure, which pushes fluids from the capillaries into the tissues, and osmotic pressure, which pushes fluids from the tissues into the capillaries. He found that a protein called albumin is important in controlling this exchange, which helps to determine blood pressure.

At University College, Starling and William Bayliss studied the way the nervous system controls the digestive system. They found that nerves stimulate constant waves of muscular action, called peristalsis, which flow down the digestive tract and move food through it. In 1902, they isolated a substance that the duodenum, the part of the small intestine just below the stomach, secretes when food and juices from the stomach pass into it. They showed that this chemical, which they named secretin, travels through the blood to the pancreas, another digestive organ, and stimulates it to release substances necessary for digestion. The compounds were released even when all nerves to the pancreas were cut, proving that secretin, not the nervous system, produced this effect.

Starling and Bayliss coined the term *hormone* in 1905 to describe chemical messengers like secretin, which are made in one part of the body and affect organs or tissues in another part. Secretin was the first hormone to be identified as such, although adrenalin (epinephrine), another biochemical discovered in 1901 by JOKICHI TAKAMINE, was later also classified as a hormone. The identification of hormones as a class marked the beginning of endocrinology, a medical specialty that deals with hormones and the organs that produce them, which are called endocrine glands.

Starling's best-known work was on the heart. In 1918, he showed that the heart muscle can adjust the strength of its beat to deal with variations in blood flow caused by exercise and other factors without relying on the nervous system or changing the frequency (rate) of the beats. The more blood that flows into the heart during the relaxation phase of the heartbeat, called diastole, the more strongly the heart muscle will contract during systole, the contraction or pumping phase, thus increasing the heart's output to match the increased input. This rule, called Starling's Law of the Heart, was eventually generalized into a basic law governing all muscle, which states that the more a muscle fiber is stretched during relaxation, the harder it will contract. Starling also showed that because of this law, a heart weakened by disease must increase in size in order to do the same amount of work it did when it was healthy. A slowly enlarging heart is therefore a sign of heart disease.

Starling was also a pioneer in maintaining organs in the laboratory for research purposes. He did his heart research on animal hearts and lungs separated from the rest of the body, and in 1924 he also succeeded in keeping a mammalian kidney alive for a short period. He wrote a standard textbook on physiology, *Principles of Human Physiology*, which was first published in 1912. He received surprisingly few awards for his contributions to physiology, possibly because of his negative comments about British leaders during World War I.

Further Reading

Abbott, David, ed. *The Biographical Dictionary of Scientists: Biologists*. New York: Peter Bedrick Books, 1983.

Stevens, Nettie Maria

(1861–1912)
American
Geneticist

Nettie Stevens was one of two scientists who discovered what determines whether a living thing will be male or female. She was born in Cavendish, Vermont, to Ephraim and Julia (Adams) Stevens on July 7, 1861. Her mother died when she was a child, and her father, a carpenter, and stepmother, Ellen Thompson, raised her and her sister, Emma. After earning a credential from Westfield Normal School in Massachusetts in 1883, she spent the first part of her adult life as a teacher and librarian. Eventually, however, she decided on a career in research. She enrolled at Stanford University, in California, in 1896, earning a bachelor's degree in 1899 and a master's degree in physiology a year later.

Stevens first studied single-celled sea creatures, but while doing doctoral research at Bryn Mawr College, a women's college in Pennsylvania, she met genetics pioneer THOMAS HUNT MORGAN, who was teaching there, and became interested in genetics herself. She earned her Ph.D. in 1903 and then joined the faculty of Bryn Mawr, rising to the rank of associate professor and becoming beloved as a teacher. She once told a student, "How could you think your questions would bother me? They never will, so long as I keep my enthusiasm for biology; and that, I hope, will be as long as I live."

Geneticists were just beginning to associate the "factors" that controlled inheritance of traits, first described by Austrian monk GREGOR MENDEL in 1866, with threadlike bodies called chromosomes in the nucleus of cells. In her most important research, Stevens observed that although all unfertilized eggs of the common mealworm contained the same 10 chromosomes, that was not true of the insect's sperm. One chromosome in some sperm cells, which she called X, resembled one seen in the egg. In other sperm cells, this chromosome was replaced by another, smaller one, which she termed Y. She speculated that if an egg was fertilized by a sperm carrying an X chromosome, the resulting offspring would be female. If the egg was fertilized by a sperm carrying a Y, the offspring would be male.

Most other geneticists doubted Stevens's theory when she published it in 1905, but when a better-known scientist, Edmund B. Wilson of New York's Columbia University, made the same finding independently shortly afterward, it came to be accepted. For many years, only Wilson was given credit for the discovery, but science historians now agree that Stevens should be equally honored. Stevens's and Wilson's discovery not only showed how gender is determined but also confirmed the link between chromosomes and inheritance.

In 1905, the year she published her groundbreaking chromosome research, Stevens won the Ellen Richards Prize for outstanding scientific research by a woman. She went on to find chromosome differences in other insects, such as aphids, that were similar to what she had found in mealworms. She also proved that inheritance of gender followed the rules that Mendel had worked out and made the important discovery that chromosomes exist in pairs. Edmund Wilson wrote that she was "not only the best of the women investigators, but one whose work will hold its own with that of any of the men of the same degree of advancement." Unfortunately, Stevens died of breast cancer in Baltimore, Maryland, on May 4, 1912, at the age of 51, limiting her late-blooming career to only 12 years.

Further Reading

Ogilvie, Marilyn Bailey, and Clifford J. Choquette. "Nettie Maria Stevens (1861–1912): Her Life and Contributions to Cytogenetics." *Proceedings of the American Philosophical Society*, August 1981.

Veglahn, Nancy J. *Women Scientists*. New York: Facts On File, 1991.

Sutherland, Earl Wilbur, Jr.

(1915–1974)
American
Biochemist

Earl Wilbur Sutherland discovered the "second messenger" chemical through which hormones affect cells. He was born in Burlingame, Kansas, on November 19, 1915. He earned a B.S. from Washburn College in Topeka in 1937 and an M.D. from Washington University Medical School in St. Louis in 1942. He was an army surgeon during World War II. He married in 1963 and had two sons and two daughters.

Returning to Washington University as a lecturer in pharmacology after the war, Sutherland began working with CARL FERDINAND CORI and GERTY THERESA RADNITZ CORI on liver metabolism. He and coworker Ted Rall showed that the hormones adrenalin (epinephrine) and glucagon control the cycle by which glycogen (a complex carbohydrate) and glucose (a simple sugar) are alternately built up and broken down by activating and deactivating an enzyme called liver phosphorylase.

After rising through faculty ranks at Washington University Medical School to become an associate professor of biochemistry, Sutherland became professor of pharmacology and director of the pharmacology department in the medical school at Western Reserve (now Case Western

Reserve) University in Cleveland, Ohio, in 1953. He held this post for 10 years and then was professor of physiology at Vanderbilt University in Nashville, Tennessee, for another 10. He taught at the University of Miami's medical school from 1973 until his death.

Continuing to study the action of adrenalin and glucagon in the liver, Sutherland identified a chemical in the mid-1950s that was essential to that action. It came to be called cyclic adenosine 3',5'-monophosphate, or cyclic AMP. Sutherland and others went on to find cyclic AMP not only in the liver but in heart, brain, muscle, and, indeed, all animal cells. Biochemists had believed that hormones affected cell processes directly, but Sutherland found that they instead increase the amount of cyclic AMP in the cells they influence. Cyclic AMP, in turn, adds or subtracts phosphorus atoms from proteins like liver phosphorylase, thereby turning the action of these proteins on or off.

Sutherland called cyclic AMP the body's "second messenger" (hormones are the first messengers) and said that it "affects everything from memory to toes." Work by many scientists has confirmed that it is one of the most essential compounds in cell chemistry. Sutherland won the 1971 Nobel Prize in physiology or medicine for discovering it and was also elected to the National Academy of Sciences. He died suddenly of massive bleeding in Miami on March 9, 1974.

Further Reading

Raju, Tonse N. K. "1971: Earl Wilbur Sutherland, Jr. (1915–74)." *Lancet*, September 11, 1999.

Swammerdam, Jan
(1637–1680)
Dutch
Entomologist, Histologist

Jan Swammerdam made some of the most careful microscope observations of his time, as well as detailed studies of the life cycles of many kinds of insects. He is considered one of the founders of entomology. He was born in Amsterdam, Holland, on February 12, 1637. His father was a well-to-do apothecary (pharmacist) whose hobby was collecting unusual natural objects. The senior Swammerdam's home museum of fossil, animal, and insect "curiosities" interested his son in natural history, and the boy soon began building an insect collection of his own.

Swammerdam hoped that Jan would become a priest, but instead the young man studied medicine in Leiden and Paris, obtaining a degree from Leiden in 1667. He never actually worked as a physician, however, but instead returned to Amsterdam and lived on an allowance in his father's house. There, he concentrated on doing what interested him most: studying insects and examining specimens of all kinds under the microscope.

Like fellow Dutch microscopist ANTONI VAN LEEUWENHOEK, Jan Swammerdam made his own microscopes. Although less powerful than Leeuwenhoek's, Swammerdam's were easier to use. In dissecting insects and tissues, he employed tools so tiny that he had to observe them under the microscope while sharpening them.

Swammerdam carefully studied the anatomy and life cycle of such insects as the mayfly and the honeybee, correcting mistaken ideas that even scientists had believed since ancient times. For instance, many people thought that insects' bodies were simply fluid-filled bags with no internal structure, but Swammerdam showed that insects possessed organs just as "higher" animals did. He identified the largest bee in a honeybee colony—the supposed "king"—as a fertile female, the "drones" as males, and the worker bees as sterile females. He classified insects into four major groups depending on the amount of metamorphosis, or major bodily change, that they underwent during their lives, a division basically the same as the one biologists still use.

Swammerdam also made discoveries under his microscope that did not relate to insects. For instance, he was the first to see red blood cells, or at least to describe them in detail. He spotted them in the blood of a frog in 1658, when he was only 21 years old. He showed that a dead frog's leg muscle could be made to contract when the nerve in it is stimulated, a prelude to the later electrical experiments of LUIGI GALVANI and others, and that muscle does not change in volume when it contracts. He was one of the first scientists to show that female mammals produce eggs, essentially similar to the eggs of birds, even though the eggs do not have a hard shell and develop inside the female's body rather than being deposited outside.

A tone of sadness pervades Swammerdam's description of the mayfly: "It is born into the world, it is a worm, it sheds its skin twice, it becomes an adult, it lays eggs, it grows old and dies at last—all in the brief period of five hours." Swammerdam's own life was also short and rather tragic. Around the time he finished his medical studies, he caught malaria, and probably as a result of this chronic illness, both his physical and his mental health were unstable for the rest of his life. In his later years, he argued increasingly with his elderly father, who finally cut off his allowance around 1672 and ordered him to use his physician's training to earn a living. Instead, Swammerdam joined a religious cult led by a woman named Antoinette Bourignon and abandoned science entirely. He died in Amsterdam on February 15, 1680, when he was just 43 years old.

Only one book by Swammerdam, *A General History of Insects*, was published during his lifetime (in 1669). Fortunately, however, the meticulous notes and drawings he had made were not lost. They passed from hand to hand until famed Dutch physician and teacher Hermann Boerhaave finally published them as a two-volume work called *Biblia Naturae* (*The Book of Nature*) in 1737. This large, beautiful set of books gained for Swammerdam a fame he had never enjoyed in life.

Further Reading

Cobb, Matthew. "Reading and Writing the Book of Nature." *Endeavour*, September 2000.

Swammerdam, Jan. *The Book of Nature*. Edited by Thomas Floyd. 1737. Reprint, Ayer, Mass.: Ayer Co., 1978.

Szent-Györgyi, Albert

(1893–1986)
Austro-Hungarian/Hungarian
Biochemist

Albert von Nagyrapolt Szent-Györgyi identified vitamin C (ascorbic acid) and revealed the chemical reaction that makes muscles contract. He was born in Budapest, Hungary (then part of the Austro-Hungarian Empire), on September 16, 1893. His father, Nicolaus, was a wealthy landowner and, according to Szent-Györgyi, cared mostly about farming, but his mother, the former Josefine Lenhossek, came from a family of scientists and interested him in both science and the arts. Szent-Györgyi did some of his earliest research in the laboratory of her brother, a professor of anatomy at the University of Budapest.

Szent-Györgyi began studying medicine at the University of Budapest in 1911, but World War I temporarily halted his studies. He served on the Russian and Italian fronts, was wounded, and won a silver medal for bravery. After being sent home, he returned to medical school and obtained his degree in 1917. In that same year, he married Cornelia Demény, daughter of Hungary's postmaster-general; they later had one daughter. Following his graduation and marriage, he spent 10 years doing biochemical research at universities in various parts of Europe and the United States.

The work for which Szent-Györgyi was to win the 1937 Nobel Prize in physiology or medicine began while he was at the University of Groningen in the Netherlands, continued at Britain's Cambridge University, and concluded after he returned to Hungary in 1930 and became professor of medical chemistry at the University of Szeged. His studies of oxidation in cells at Groningen reconciled competing theories that two eminent German biochemists, Otto Warburg and Heinrich Wieland, held about the process and provided information that HANS ADOLF KREBS later used in working out the Krebs or citric acid cycle, a vital part of the reactions through which cells obtain energy.

These oxidation studies, in turn, led to Szent-Györgyi's discovery, or in a sense rediscovery, of ascorbic acid (vitamin C). The path to that achievement began when he noticed that certain fruits, such as apples and bananas, turn brown when damaged, but some others, such as lemons, do not. He suspected that the brown color was caused by an oxidation reaction in the first group of fruits and that the second group contained something that blocked this reaction. His suspicion became stronger when he found that an oxidation process that took place immediately when certain chemicals were mixed in a test tube took half a second longer when carried out in tissues from the second group of fruits. He used this delay as a quick, simple test to help him identify the mystery substance.

Hungarian-born Albert Szent-Györgyi helped to identify ascorbic acid (vitamin C) and the proteins that make up muscle fibers. *(National Library of Medicine)*

After only two weeks, Szent-Györgyi crystallized small amounts of this antioxidant, or reducing agent, from orange juice, cabbage, and the outer layer, or cortex, of the adrenal gland, which also could carry out this oxidation reaction. He turned his discovery into a thesis that earned a Ph.D. from Cambridge in 1927. He called his new compound hexuronic acid, but he soon began to suspect that it was identical to a chemical that two German scientists had isolated in 1907 and named vitamin C. Vitamin C had been shown to cure scurvy, which could be prevented or cured by eating oranges or lemons.

At about the same time Szent-Györgyi found hexuronic acid, an American scientist, Charles C. King, and his coworkers at the University of Pittsburgh isolated the same compound and, also suspecting that it was the substance that cured scurvy, began testing it on guinea pigs given the disease experimentally. J. L. Svirbely, a young American of Hungarian descent who had worked with King, provided a lucky link between the two laboratories when he visited Szent-Györgyi in Szeged in 1931. Szent-Györgyi had not yet tested hexuronic acid to find out whether it really was vitamin C (he found vitamins "theoretically uninteresting"), but when Svirbely offered to settle the question by means of the guinea pigs, Szent-Györgyi was happy to let him do so. The substance cured the animals, proving that it was indeed vitamin C or, as Szent-Györgyi suggested calling it, ascorbic ("no-scurvy") acid. Szent-Györgyi went on to discover that Hungarian red pepper, or paprika, could provide abundant amounts of the vitamin for further experiments.

Both the Szent-Györgi–Svibely team and King's group, which had also confirmed that hexuronic acid was vitamin C, wrote accounts of their work in 1932. Szent-Györgyi, who published slightly ahead of the other group, was awarded the Nobel Prize for this discovery as well as for his work on cellular oxidation. He described both in a well-received book called *Oxidation, Fermentation, Vitamins, Health and Disease*, published in 1939.

By the time this book appeared, Szent-Györgyi had stopped his work on oxidation and vitamin C and was beginning a new line of research on muscle. Biologists had believed that muscle consisted almost exclusively of a single protein called myosin. Szent-Györgyi showed in 1940, however, that what appeared to be one protein was actually two, the second of which he named actin. He and coworkers Ilona Banga and F. Bruno Straub made artificial muscle fibers of these two proteins and demonstrated that the fibers contracted when a third chemical called adenosine triphosphate (ATP), a donor of energy in biochemical processes, was added. "Seeing this artificial bundle contract was the most exciting moment of my scientific career," he wrote later. *Current Biography Yearbook 1955* calls Szent-Györgyi's description of contraction as a result of the interaction of actin, myosin, and ATP "the first workable theory of muscle physiology." Szent-Györgyi described his research on muscles in *Chemistry of Muscular Contraction* (1947).

In between prizewinning discoveries, Szent-Györgyi led an adventurous and sometimes perilous life. Having survived battle in World War I, he went on to take part in the anti-Nazi underground movement during World War II, using his scientific travels as a cover for smuggling documents to British offices in neutral countries. He narrowly escaped the Gestapo, or Nazi secret police, several times.

Peril continued when Szent-Györgyi returned to Hungary in 1945 and became professor of biochemistry at the University of Budapest's medical school. He entered Hungarian politics and was elected to the country's legislature. A pacifist, he hoped to cooperate with Hungary's new Soviet government, but he soon found that impossible and therefore emigrated to the United States in 1947. He opened a small laboratory, the Institute for Muscle Research, within the Marine Biological Laboratory at Woods Hole, Massachusetts, and there continued his experiments on muscle biochemistry. He received the Albert Lasker Award from the American Heart Association in 1954 for the contribution that his muscle work made to the understanding of heart disease.

Szent-Györgyi began studying cancer in the 1960s and 1970s, a personal crusade for him because the disease had killed both his first wife and his daughter. (He had married again, to Marta Borbiro, a Woods Hole coworker, in 1949.) His approaches included studying the movements of electrons in cells, which he called quantum biology, and further examination of biochemicals that affect oxidation, including compounds related to vitamin C. He persuaded two wealthy donors, Franklin and Tamara Salisbury, to open the National Foundation for Cancer Research at Woods Hole in 1975, and he served as scientific director of this organization from its founding until his death.

"The Prof," as his many friends called him, never lost his taste for adventure. He enjoyed sports, sailing, and fishing at his home in Penzance Point, Massachusetts. In addition to scientific books, he wrote a book of philosophy, *The Crazy Ape* (1970), in which he decried the tendency to make war and urged scientists to lead humanity toward a more cooperative attitude. Szent-Györgyi died on October 22, 1986.

Further Reading

Moss, Ralph W. *Free Radical: Albert Szent-Györgyi and the Battle over Vitamin C*. New York: Paragon House, 1987.

"Szent-Györgyi, Albert (von Nagyrapolt)." *Current Biography Yearbook 1955*. New York: H. W. Wilson Co., 1955.

Weber, George. "A Genius for Discovery." *Saturday Evening Post*, May–June 1984.

T

Takabe, Tetsuko

(1947–)
Japanese
Biochemist, Geneticist

Tetsuko Takabe tries to reduce food shortages and environmental damage by using genetic engineering to make plants more tolerant of high salt levels in soil. Born in Aichi Prefecture, Japan, on February 20, 1947, she was a poor student until her father and her fourth grade teacher convinced her that she could achieve something for herself and help her country as well if she studied harder. Before long, she became famous in her village for spending all her time with her books. Her favorite subjects were mathematics and physics.

Takabe was accepted as a science student at Nagoya University, one of Japan's seven large national universities, in 1965. When she arrived, she was startled to find journalists waiting to interview her. She learned that she was the highest scorer on the tests that students had to take before they entered the university—the first time a woman had achieved that honor. Such distinctions did not help her, however, after she earned her bachelor's and master's degrees (in 1969 and 1970, respectively) and began seeking a job. Very few industries or universities would employ women scientists in those days.

Nagoya University's School of Agriculture finally hired Takabe as a technician in 1971. Her supervisor, Takashi Akazawa, encouraged her to pursue her studies on the biochemistry of photosynthesis. After she obtained a Ph.D. in plant biochemistry from the university in 1975 and joined its faculty as an assistant professor in 1979, Akazawa helped her arrange to do postdoctoral work at Cornell University in the United States for two years (1980–82). She then returned to Nagoya, where she became an associate professor in 1989.

Takabe decided to specialize in agriculture because, she says, "I wanted to contribute . . . directly to human life." Most of her research has centered on attempts to make crop plants more tolerant of salt. When humans clear and cultivate land in dry areas, evaporation draws salty groundwater to the surface. Salt therefore accumulates in the soil unless proper irrigation is practiced. Most plants do not grow well in salty soils, so such soils remain barren, reducing the amount of land that can be used to grow food crops and increasing erosion and desertification. Making plants, especially crop plants, more salt tolerant thus potentially benefits the environment as well as increasing food supplies.

Takabe found that some naturally salt-tolerant plants make a compound called glycinebetaine, which protects their cells against the effects of excess salt. At the time she began her research, the gene that controls production of this substance in plants was not known, but an equivalent gene in bacteria had been identified. Takabe and coworker Sachie Kishitani therefore used genetic engineering techniques to insert the glycinebetaine gene from the common bacterium *E. coli* into rice. They found, however, that the bacterial gene usually did not function in the plants. They and their coworkers spent several years modifying the gene, then tried putting it into rice once more.

The new transgenic rice did make glycinebetaine, and it grew better than ordinary rice in both salty soils and dry conditions. Nonetheless, it was not strong enough to survive in heavily salted soils. Takabe therefore has been working to isolate the genes that plants themselves use to make glycinebetaine and other salt-protective compounds. She has cloned and studied about 500 such genes. She hopes eventually to make not only crop plants but trees salt tolerant, because trees planted in salty soil could help to prevent erosion and desertification.

In 1997, Takabe won the Saruhashi Prize, which Japanese geochemist Katsuko Saruhashi and the organization she founded, the Association for the Bright Future of Women Scientists, award yearly to an outstanding Japanese woman scientist. Partly as a result of this award, Takabe thinks, she was promoted to the rank of full professor at Nagoya in 1999. She is married to coworker Teruhiro Takabe, and they have one daughter.

Takabe says that, although women scientists in Japan still advance much more slowly than their male counterparts, the situation is "changing slowly." She tries to speed that change by mentoring younger women. "I have brought up female scientists and watched them earn their Ph.D.," she writes. "They are enjoying both life and science actively."

Further Reading

Kozai, Yoshihide, et al., eds. *My Life: Twenty Japanese Women Scientists*. Tokyo: Uchida Rokakuho, 2001.

Takamine, Jokichi

(1854–1922)
Japanese/American
Biochemist

In addition to developing several patented processes in agricultural and industrial biochemistry, Jokichi Takamine was one of two scientists to isolate adrenaline (epinephrine), the first hormone isolated in pure form from a natural source. Takamine was born on November 3, 1854, in Takaoka, Japan, now a part of Toyama Prefecture. He earned a chemical engineering degree from the College of Science and Engineering of the Imperial University in Tokyo in 1878 and then did postgraduate study at Anderson's College in Glasgow, Scotland.

When Takamine returned to Japan in 1882, he began working at the Imperial Department of Agriculture and Commerce, focusing on brewing processes and paper manufacture. He was soon promoted to head of the chemistry division. He was also appointed vice commissioner of the country's patent office in 1885. He left these jobs in 1887 to start his own chemical company, the Tokyo Artificial Fertilizer Company, which became the first business to make superphosphate fertilizers in Japan. In 1889, he also isolated a starch-digesting enzyme from a fungus that grew on rice, which proved useful in brewing.

Takamine was invited to come to the United States in 1890 and introduce his patented enzyme techniques to the American distilling industry. He married an American woman and decided to remain in the United States permanently. After working on enzyme production at several factories, he set up his own laboratory in Clifton, New Jersey, to develop his methods further.

Impressed with the processes Takamine had developed for extracting enzymes, the drug company Parke-Davis asked him to try to extract the active substance from the medulla, or interior part, of the adrenal glands, two small glands located on top of the kidneys in mammals. The mystery substance was known to affect blood pressure and had potential uses in medicine. In 1901, Takamine crystallized a chemical from animal adrenal glands that he called adrenaline. John Jacob Abel, a biochemist with whom Takamine had worked, isolated the same substance independently at about the same time and named it epinephrine. Both names are still used.

Around 1905, British biochemist ERNEST HENRY STARLING identified certain biochemicals as "messengers" that are made in one part of the body and travel through the bloodstream to affect organs or tissues in another part. He named this group hormones. Biochemists realized that Takamine's adrenaline belonged to this class.

Using the process Takamine had developed, Parke-Davis began producing adrenaline commercially and marketing it under the name Takadiastase. Takamine remained associated with Parke-Davis for the rest of his career, but he also continued to run his own company. He aided the development of industries in Japan as well. The Japanese government awarded him the Third Class Order of Merit in 1922. Takamine died on July 22, 1922.

Further Reading

St. Francis Xavier University Multicultural Science Education Research Project. "Takamine, Jokichi." Available online. URL: www.upei.ca/~xliu/multiculture/taka.htm. Accessed 2001.

Temin, Howard Martin

(1934–1994)
American
Virologist

Howard Temin discovered that certain viruses contain an enzyme that allows them to insert their genes directly into a cell's genome, a feat formerly thought to be impossible. For this discovery, he shared the 1975 Nobel Prize in physiology or medicine with DAVID BALTIMORE, who found the same enzyme independently at about the same time, and Renato Dulbecco, another researcher who studied cancer-causing viruses.

Temin was born in Philadelphia, Pennsylvania, on December 10, 1934, the second son of Henry and Annette Temin. His father was an attorney, and his mother was active in civic affairs related to education. Temin decided on a biology career in high school after spending summers at the Jackson Laboratory in Bar Harbor, Maine, in a special program that Baltimore also attended. He graduated from Swarthmore College in Pennsylvania in 1955, earning a B.A. with honors.

Temin began graduate work in experimental embryology at the California Institute of Technology (Caltech) in Pasadena, but meeting Dulbecco there inspired him to change his focus to animal virology. He obtained a Ph.D. from Caltech in 1959 and then, after an additional year of postdoctoral work in Dulbecco's laboratory, moved to the McArdle Laboratory for Cancer Research at the medical school of the University of Wisconsin, Madison. "My first laboratory was in the basement," he recalled in the autobiographical sketch he wrote for the Nobel Foundation, "with a sump in my tissue culture lab and with steam pipes for the entire building in my biochemistry lab." He spent the rest of his career at the university, rising to become American Cancer Society Professor of Viral Oncology and Cell Biology in 1974. Temin married Rayla Greenberg, a population geneticist, in 1962, and they had two daughters.

Most of Temin's early research, beginning in his Caltech days, was done with the Rous sarcoma virus, which causes cancer in chickens. Temin shared Dulbecco's conviction that the ability of this and other cancer-causing viruses to change cells into a form that multiplied endlessly was connected with the viruses' reproduction. Several such viruses, including Rous sarcoma virus, had been found to carry their genes in the form of RNA rather than the usual DNA, and Temin proposed that these viruses could somehow carry out a process he called reverse transcription, in which they copied their genes into the DNA of a cell's genome. There, as what Temin termed a provirus, the viral genes would be copied along with the cell's own genes each time the cell reproduced.

Most virologists in the 1960s rejected Temin's theory because they believed that RNA could not copy itself into DNA. Temin, however, proved his claim in 1970 by isolating from Rous sarcoma virus a unique enzyme called reverse transcriptase, which allowed the "backwards" copying to take place. RNA viruses containing reverse transcriptase, which include the AIDS virus (HIV) as well as many cancer-causing viruses, came to be known as retroviruses. The enzyme has proved useful in genetic engineering as well as in showing how retroviruses affect cells.

Temin went on to propose in 1971 that reverse transcription can sometimes occur in cells not infected by retroviruses, for instance, during embryonic development. Indeed, he claimed, retroviruses probably had evolved from cellular genes that could reproduce by means of reverse transcriptase. Temin's theory that certain genes in normal cells could reproduce by reverse transcription was confirmed when scientists found such genes in fruit flies and yeast in the mid-1980s, although the role of such genes in embryonic development proved to be minor.

During the later part of his career, Temin continued to investigate the way retroviruses reproduce, form DNA and integrate it with cellular genomes, and cause cancer. He also studied factors in blood serum that stimulate cells (both normal and virus-infected) to multiply. When HIV was discovered in the early 1980s, he did research on it as well, and he spent his last years attempting to devise a vaccine for AIDS. Temin, whose awards included the Warren Triennial Prize, the Gairdner International Award, and the Albert Lasker Medical Research Award, as well as the Nobel Prize, died on February 9, 1994.

Further Reading

Cooper, Geoffrey M., et al. *The DNA Provirus: Howard Temin's Scientific Legacy*. Washington, D.C.: American Society of Microbiology Press, 1995.

Temin, Howard. "Howard Martin Temin—Autobiography." *Les Prix Nobel* 1975. Available online. URL: www.nobel.se/medicine/laureates/1975/temin-autobio.html. Last modified 2002.

Theophrastus

(ca. 372–ca. 287 B.C.)
Greek
Botanist, Taxonomist

The ancient Greek writer and philosopher Theophrastus was the first to describe and classify plants systematically. CAROLUS LINNAEUS, who adapted and improved Theophrastus's classification system more than a thousand years later, called him "the father of botany."

Theophrastus was born in Eresus, on the Greek island of Lesbos, around 372 B.C. His father was a fuller, or processor of wool cloth. Theophrastus studied under the renowned philosopher Plato at Plato's Academy in Athens and there met ARISTOTLE, another philosopher, whose close friend he became. The two spent three years (344–342 B.C.) together on Lesbos, and Theophrastus probably helped Aristotle study marine life there. When Aristotle established his own philosophical school, the Lyceum, in Athens in 335 B.C., Theophrastus became his chief assistant.

Theophrastus is said to have written more than 200 books on science, literature, politics, law, and other subjects, but only a few have survived. Of these, the most important for biology are two multivolume works on botany, *Historia Plantarum* (*The History of Plants*) and *Plantarum Causae* (*The Etiology [Causes] of Plants*). *History* describes and classifies plants, while *Etiology* contains information about plant physiology.

The nine-volume *History* covers more than 500 types of plants, virtually all that were known at the time. Theophrastus's plant descriptions, which he modeled after descriptions of animals in Aristotle's writings, include geographical location and uses as well as structural features. He is thought to have obtained samples of and information about foreign plants from his and Aristotle's students, some of whom came from distant parts of Greece, and from soldiers who had traveled with the army of Alexander the Great as far away as India. The six-volume *Etiology* describes plant reproduction and growth from seeds, the effects of soil and climate on plants' geographical distribution, plant diseases, and cultivation techniques.

Theophrastus divided plants into trees, shrubs, undershrubs (small plants with woody stems), and herbaceous plants. He listed the basic features of plants as roots, stems, branches, and twigs, and he classified plants within each large group according to similarities in these features. (He considered leaves, flowers, and fruit to be less important because they were not permanent parts of the plant.) He recognized important distinctions that biologists still use, including the divisions between flowering plants (angiosperms) and cone-bearing plants (gymnosperms) and between monocotyledons and dicotyledons (two types of flowering plants). He recognized the relationship between flowers, fruits, and seeds. Much of his information is still considered to be basically accurate.

Theophrastus, friend and heir of the famous ancient Greek philosopher Aristotle, described and classified plants much as Aristotle did for animals; Theophrastus is often called "the father of botany." *(National Library of Medicine)*

Aristotle made Theophrastus his heir. When Aristotle was forced to retire and leave Athens in 323 B.C. after the death of his unpopular former patron, Alexander the Great, Theophrastus took over the Lyceum. He enlarged the school and made it even more prosperous than it had been in Aristotle's time, reportedly drawing in some 2,000 students during his 35-year tenure. He headed the school until his death around 287 B.C. in Athens.

Further Reading

Fortenbaugh, William W., ed. *Theophrastus of Eresus: On His Life and Work*. New Brunswick, N.J.: Transaction Publications, 1985.

Tinbergen, Niko
(1907–1988)
Dutch/British
Zoologist, Ethologist

With KONRAD LORENZ and KARL VON FRISCH, Nikolaas Tinbergen, or Niko, as he preferred to be called, founded the biological specialty of ethology, which studies animal behavior in the wild. The three men shared the Nobel Prize in physiology or medicine in 1973.

Tinbergen was born in The Hague, the capital of the Netherlands, on April 15, 1907, the second son of Dirk and Jeannette (van Eek) Tinbergen. (His older brother, Jan, also won a Nobel Prize, in economics.) In an autobiographical essay for the Nobel Foundation, Tinbergen described his father, an elementary school teacher, as "intellectually stimulating" and his mother as "warm [and] impulsive." He collected and studied plants and animals as a child, but school bored him, and he had no definite plans for a career until his father arranged for him to spend the summer of 1925 at Vogelwarte Rossitten, a famous bird observatory in Germany. That experience inspired him to enter the State University of Leiden as a biology major.

Tinbergen found his Ph.D. project in a colony of large wasps called digger wasps, bee killers, or bee wolves that he spotted near his parents' coastal cottage one summer. Each wasp in the colony had a separate underground nest, and Tinbergen wondered how the wasps found their way to their own nests. Employing the mixture of patient observation and ingenious experimentation that would mark all his work, he painted individual wasps with quick-drying enamel to identify them and then changed various factors in their environment to see what would disturb their navigation. He determined that the wasps used visual landmarks as their chief cues. The thesis in which he described this work was only 32 pages long, the shortest ever submitted at Leiden. Nonetheless, after much debate among the faculty, the university granted his degree in 1932.

By this time, Tinbergen had married Elizabeth Rutten, a chemistry student at the university (they later had five children), and the two had obtained positions in a state-sponsored scientific expedition to Greenland. During the 14-month expedition, in 1932 and 1933, they lived with an isolated group of Eskimos and learned about their culture. The athletic Tinbergen also spent many hours in the wild, observing a variety of animals including birds called snow buntings. He wrote a book on the behavior of these birds that was published in 1939.

Tinbergen joined the Leiden faculty in 1935 and taught courses in animal behavior that centered around one of his favorite animals, a fish called the stickleback. Using dummy fish to trigger responses from live ones, he showed students how changes in appearance that mark different stages in the fishes' social and reproductive behavior—red belly patches in males aggressively defending their territory or swollen bellies in females ready to lay eggs, for instance—affect other sticklebacks' reactions.

Tinbergen met Austrian zoologist Konrad Lorenz at a conference on instinct in 1936, and the two, as Tinbergen later wrote, "clicked" immediately. In contrast to behaviorists such as B. F. SKINNER, both men felt that most animal behaviors are instinctive, or genetically determined, rather than learned. They also shared the belief that, although behavior might be manipulated experimentally, it usually should be studied in animals' natural environment, not in a laboratory. Tinbergen and Lorenz became close friends and collaborated on several studies during the late 1930s, developing a new way of studying animal behavior that came to be known as ethology.

World War II interrupted Tinbergen's research. After Germany seized control of the Netherlands in 1940, he was sent to a prison camp for two years because he protested the firing of three Jewish professors at Leiden. He returned to the university after the war, becoming a professor of experimental zoology in 1947. He and his students studied the functions of color in animals during the late 1940s. Then, in 1949, he accepted an invitation to start a center for the study of animal behavior at Britain's prestigious Oxford University. There, he wrote his first major book, *The Study of Instinct* (1951), which, like most of his later writings, was addressed to laypeople as well as scientists.

Although Tinbergen studied many kinds of animals during his career, his name was most strongly associated with herring gulls. He began observing these noisy, squabbling birds' tightly packed colonies (which he called "cit[ies] of thieves and murderers") in the Netherlands and continued to do so in England. He learned much new information about their behavior, discovering, for instance, that gull chicks make their parents regurgitate fish by pecking a red spot near the tip of the adults' beaks.

Tinbergen also offered a theory to explain why two male gulls, meeting at the edge of their respective territories, often alternated aggressive displays with the odd behavior of pulling up grass, an action normally associated with nest building. He proposed that the gulls, torn between the urges to fight (defending their own territory) and flee (the normal response after straying into another male's territory), instead exhibited "displacement" behavior that involved neither action. They used this behavior as a substitute for actual fighting, thus avoiding possible injury or death. Tinbergen also showed that gulls' elaborate courtship displays defused aggression between the male and female as well as stimulating sexual activity. He described his gull research in his best known book, *The Herring Gull's World* (1953).

In his later years, Tinbergen drew on what he had learned from animals to comment on human behavior. Like Lorenz, he wrote about human aggression, which both men considered to be instinctive and therefore unavoidable, although Tinbergen had more conviction than Lorenz that it could be redirected into safe channels.

Tinbergen and his wife also studied a mysterious condition called autism, which appears in early childhood and severely limits ability to interact with other people. The Tinbergens believed that autistic children's repetitive actions were displacement behaviors produced by the conflict between the children's abnormal fear of others and their desire to connect with them. Like some other theorists at the time, they thought the condition resulted from poor parental treatment and could be overcome by patience and affection. These ideas, which the Tinbergens described in *Early Childhood Autism: An Ethological Approach* (1972), were Tinbergen's most controversial by far. They are generally disregarded today because most psychiatrists now feel that autism is an organic brain disorder rather than a psychological problem.

In addition to the Nobel Prize, Tinbergen received the Jan Swammerdam Medal in 1973 and was elected to both the British Royal Society (1962) and the Netherlands Academy of Sciences (1964). He was respected as a teacher and writer as well as for his scientific work. Former students such as British ethologist Desmond Morris praised his blend of laboratory and field methods: "[Most biologists] wear a white coat or Wellington boots [for work in the outdoors], one or the other," Morris once said. "Tinbergen does both. In my book, that makes him the most important man in his field this century." In addition to *The Study of Instinct* and *The Herring Gull's World*, Tinbergen's books included the partly autobiographical *Curious Naturalists* (1958) and the two-volume *The Animal in Its World: Explorations of an Ethologist, 1932–1972* (1972, 1973). Reviewers called his essays about animals "delightful" and "loving." Niko Tinbergen retired from Oxford in 1974 and died on December 21, 1988.

Further Reading

Tinbergen, Nikolaas. *Curious Naturalists*. 1958. Reprint, Amherst: University of Massachusetts Press, 1984.

———. *The Herring Gull's World: A Study of the Social Behavior of Birds*. 1953. Reprint, Guilford, Conn.: Lyons Press, 1989.

———."Nikolaas Tinbergen—Autobiography." *Les Prix Nobel 1973*. Available online. URL: www.nobel.se/medicine/laureates/1973/tinbergen-autobio.html. Last updated 2001.

"Tinbergen, Niko(laas)." *Current Biography Yearbook 1975*. New York: H. W. Wilson, 1975.

Tonegawa, Susumu

(1939–)
Japanese/Swiss
Molecular Biologist, Immunologist, Neurobiologist

Susumu Tonegawa won the 1987 Nobel Prize in physiology or medicine for solving one of the greatest mysteries of the immune system. He was born in Nagoya, Japan, on September 6, 1939. His father, Tsutomu Tonegawa, worked as an engineer for a textile company with several factories in southern Japan, and the family moved frequently as the senior Tonegawa was assigned to different factories. Susumu remembers enjoying the rural surroundings of the towns where the factories were located. When he and his older brother were teenagers, however, their parents sent them to live with an uncle in Tokyo so they could receive a better education.

Tonegawa became interested in chemistry during high school and majored in that subject at Kyoto University, graduating in 1963. In his senior year, his focus changed to molecular biology. He earned a Ph.D. in that field from the University of California, San Diego, in 1968. He then did postdoctoral work at the nearby Salk Institute, but his visa expired in 1970, and he had to remain outside the United States for two years before he could obtain a renewal. His mentor at the institute, Renato Dulbecco, suggested that he apply to the new Basel Institute for Immunology in Switzerland. Although Tonegawa knew very little about immunology, the institute accepted him, and he began working there in 1971.

Filling in the gaps in his knowledge of the body's defense system, Tonegawa learned that certain cells in the blood, called B cells, make and release proteins termed antibodies when the cells detect foreign proteins (antigens) on the surfaces of invaders such as bacteria or viruses. Each kind of invader has different antigens, which the antibodies must fit as precisely as a key fits into a lock. B cells therefore potentially must be able to make millions of different kinds of antibodies, and immunologists had no idea how the cells accomplish this feat. Some thought that instructions for making all the different antibodies must be carried in the cells' genes, but that seemed improbable because human cells contain only about 100,000 genes, and most of them have nothing to do with antibodies. Other researchers believed that the genetic code for antibodies must be developed through mutations occurring after birth, but that, too, seemed unlikely because such mutations do not happen often.

Immunologists knew that an antibody molecule is made up of two sections, the light chain and the heavy chain. Each chain, in turn, has a constant region and a variable region. In 1965, two researchers at the University of Alabama School of Medicine had proposed that, even though each chain in an antibody can be considered to be a single molecule, the genetic code for making it might be carried in two separate stretches of DNA, one for the constant region and one for the variable region. Indeed, there might be several genes for the variable region, although only one of these would be involved in making any particular antibody. Most immunologists doubted this theory because it required two things not then known to exist: split genes (separated stretches of DNA that carry instructions for making a single molecule) and a mechanism to bring these pieces close enough together to produce a protein.

Several laboratories, including Tonegawa's, tested this proposal in the early 1970s. In 1976, Tonegawa and his chief coworker, Nobumichi Hozumi, reported that instructions for the constant and the variable regions of each antibody chain are carried on separate pieces of DNA, just as the theory predicted. They went on to find that these two DNA stretches are far apart in the genomes of embryonic mice but, by the time the animals' B cells are mature, the gene parts have moved much closer together. As Tonegawa wrote in a 1985 article in *Scientific American*, "Antibody genes offer dramatic evidence that DNA is not an inert archive but can be altered during the life span of an individual."

With scientists from other laboratories, Tonegawa's group worked out the sequence of bases in the DNA segments that code for an antibody light chain in adult mouse B cells. They then used the sequence for the light chain's variable region to produce a protein, but their protein proved to be shorter than the natural one. They concluded that the code for the missing piece of the protein must be contained on still another gene segment. Christine Brack, a scientist in Tonegawa's laboratory, eventually found this segment, and the group named it J, for "joining." (Another molecular biologist, Philip Leder, independently discovered the J segment at about the same time.) Meanwhile, LEROY HOOD, then at the California Institute of Technology in Pasadena, found that the gene for the heavy chain consisted of three parts like the ones in the light chain, plus a fourth one that he called D, for "diversity."

Each of these gene parts exists in many forms (there are 150 possible forms for the heavy chain variable region, for instance), and each of these forms can combine with any of the others. Antibody genes have been found to mutate more often than most other genes as well. All this variation is enough to produce up to a billion different antibodies with just a handful of genes. As Tonegawa has said, "It's like when GM [General Motors] builds a car they want to meet the specific needs of many customers. If they custom-make each car, it is not economical, so they make different parts, then they assemble it in different ways, and therefore one can make different cars." Tonegawa won the 1987 Nobel Prize for solving this basic puzzle of immunology. He was the first Japanese to win a medical Nobel.

Tonegawa returned to the United States in 1981 and became a professor at the Center for Cancer Research, part of the Massachusetts Institute of Technology (MIT). There, in addition to continuing research on antibody genes, he studied a different kind of immune system cell called the T cell.

In the 1990s, Tonegawa changed his field of research completely. Leaving the immune system behind, he began combining genetics, molecular biology, physiology, and behavioral techniques to investigate the cellular mechanisms underlying learning, memory, and sense perception in mammals. He has done much of his work with "knock-out" mice, in which particular genes are knocked out, or turned off, in the whole animal or in particular tissues, to see what changes their absence produces. For instance, he has found a gene whose absence makes a mouse unable to remember the spatial layout of groups of objects or the sequence of events in time. Some of his research sheds light on the ways in which heredity and environment interact to shape brain development in infants. Some may help physicians understand and perhaps treat Alzheimer's disease, schizophrenia, and other illnesses in which the ability to learn and remember is damaged.

In addition to the Nobel Prize, Tonegawa has received awards that include the Louisa Gross Horwitz Prize of Columbia University (1982), Canada's Gairdner Foundation Award (1983), the Albert Lasker Medical Research Award (1987), and several medals from the

Japanese government. He is married to Mayumi Yoshinari (his second marriage), and they have three children. He is presently Whitehead Professor of Biology and Neuroscience at MIT and director of the university's Center for Learning and Memory. He is also an investigator at the Howard Hughes Medical Institute in Chevy Chase, Maryland.

Further Reading

Marx, Jean L. "Antibody Research Garners Nobel Prize." *Science*, October 23, 1987.

"Susumu Tonegawa—Autobiography." *Les Prix Nobel 1987*. Available online. URL: www.nobel.se/medicine/laureates/1987/tonegawa-autobio.html. Last updated 2002.

Yount, Lisa. *Asian-American Scientists*. New York: Facts On File, 1998.

Tsui, Lap-Chee

(1950–)
Chinese/Canadian
Geneticist, Molecular Biologist

In 1989, Lap-Chee Tsui and his coworkers isolated the defective gene that causes cystic fibrosis, the most common inherited disease among Caucasians. Tsui was born in Shanghai, China, on December 21, 1950, and grew up in the village of Dai Goon Yu on the Kowloon side of Hong Kong. His parents were Jing-Lue and Hui-Ching (Hsue) Tsui. His interest in biology may have been stirred in childhood when he and his friends collected tadpoles, fish, and silkworms (a kind of caterpillar).

At first, Tsui planned to be an architect, but while at the Chinese University of Hong Kong he decided to study biology instead. He earned a B.Sc. in 1972 and a master's degree from the same university two years later. He obtained a Ph.D. from the University of Pittsburgh in 1979, then did postdoctoral work, which included his first research in genetics, at several institutions, ending with the Hospital for Sick Children in Toronto, Ontario, Canada, in 1981. He has remained at the hospital ever since. He is married to the former Lan Fong Ng, and they have two sons.

Tsui's research on cystic fibrosis started in 1982. This disease strikes about one in every 3,000 children in the United States and one in every 2,000 in Canada. It produces a number of problems, of which the most severe is abnormally thick mucus that makes breathing difficult and leads to constant, lung-damaging infections. Many people with the disease die before age 30. At the time Tsui began his studies, cystic fibrosis was known to be inherited, but the defective gene that causes it had not been located, and the exact nature of the defect was still a mystery. Developing treatments was therefore difficult.

In the early 1980s, finding a particular gene's location on one of humans' 23 pairs of chromosomes required a combination of lengthy, tedious work and good luck. It has been compared to locating a single house in a large city with no idea of the address. As a result, few such genes had been found. Tsui's group began their hunt for the cystic fibrosis gene by studying a large number of families in which the disease was common, looking for certain forms of known, variable stretches of DNA that, although they themselves had nothing to do with the illness, were commonly inherited along with it. Such DNA segments are called markers. In 1985, Tsui's laboratory and two others announced the finding of several markers on chromosome 7 that were usually inherited with the cystic fibrosis gene.

Once markers thought to be on either side of a disease-causing gene were discovered, researchers usually attempted to "walk" along the DNA strand between the markers, testing innumerable genetic fragments for correlation with the disease. Tsui, however, felt that this process was too slow. In late 1987, therefore, he enlisted the help of Francis Collins, then at the University of Michigan, Ann Arbor, who had developed a process called "jumping" that was five to 10 times faster than "walking." Using a mixture of walking and jumping, Tsui and Collins came to focus on a stretch of DNA that was identical in several animal species (which suggested that it carried instructions for making an important protein) and overlapped a known gene affecting sweat glands and other tissues that were abnormal in cystic fibrosis. They then began trying to find out whether the order of bases in this stretch, which would determine the protein's exact composition, differed between normal people and people with the disease.

On May 9, 1989, Richard Rozmahel, a researcher in Tsui's laboratory, told him that a particular group of three bases, representing a single amino acid in the unknown protein, was present in this DNA segment in cells from healthy people but missing in cells from people with cystic fibrosis. When further tests showed that this set of bases was absent in 70 percent of people with the disease, Tsui and Collins became convinced that they had indeed found the cystic fibrosis gene. They published their work in the prestigious American journal *Science* on September 8, generating immense excitement.

Now that the cystic fibrosis gene had been discovered, the next step was to find out what its protein did in the body. Tsui and others found that the protein, which came to be known as CFTR (cystic fibrosis transmembrane conductance regulator), controls the passage of chloride ions through the membranes of certain cells. These ions, in turn, affect the movement of water in and out of the cells. When the CFTR protein is defective, lung cells cannot put enough water into the mucus they make, so the mucus becomes destructively thick.

Awards that Tsui has received for his groundbreaking work on cystic fibrosis include the Gairdner Foundation International Award (1990), the Canadian Medical Association's Medal of Honor (1996), the Canadian Medical Research Council's Distinguished Scientist Award (2000), and the Canada Council's Killam Prize (2002). He has been made an Officer of the Order of Canada and a member of both the Royal Society of Canada and the Royal Society of London. Tsui was an International Scholar of the Howard Hughes Medical Institute from 1991 to 2001. Today, he is chief geneticist, head of the genetics and genomic biology program, and Sellers Professor of Cystic Fibrosis Research in the Department of Genetics at the Research Institute of the Hospital for Sick Children. He is also a University Professor in the Department of Molecular and Medical

Working at the Hospital for Sick Children in Toronto, Ontario, Canada, Chinese-born Lap-Chee Tsui codiscovered the gene that causes the common and deadly inherited disease called cystic fibrosis in 1989. *(Lap-Chee Tsui)*

Genetics at the University of Toronto Medical School and a professor at the university's Institute of Medical Science.

Tsui and his coworkers continue to learn about how both the normal and the defective forms of the cystic fibrosis protein work in the body. They are also identifying other mutations, including mutations in genes other than CFTR, that can produce or affect the severity of cystic fibrosis or conditions related to it. In addition, as part of the Human Genome Project, they are studying chromosome 7 as a whole and trying to learn the function of other genes on it. Their work on cystic fibrosis has greatly improved physicians' ability to determine whether a couple will have a child with the disease and, they hope, will eventually lead to improved treatment or even a cure.

Further Reading

Marx, Jean L. "The Cystic Fibrosis Gene Is Found." *Science*, September 1, 1989.

Pines, Maya. "Stalking a Lethal Gene." Howard Hughes Medical Institute. Available online. URL: www.hhmi.org/genetictrail/a110.html. Posted 1997.

Shell, Barry. "Lap-Chee Tsui, Molecular Geneticist." Great Canadian Scientists. Available online. URL: www.science.ca/scientists/scientistprofile.php?pID=19. Posted 2001.

Varmus, Harold E.

(1939–)
American
Molecular Biologist, Geneticist, Virologist

With J. MICHAEL BISHOP, Harold Eliot Varmus revealed that cancer arises from altered forms of genes that play essential roles in normal cells. After this achievement, which earned both men the 1989 Nobel Prize in physiology or medicine, and other basic research, Varmus began a second career in scientific administration, serving as head of the National Institutes of Health (NIH) from 1993 to 1999 and presently heading the Memorial Sloan-Kettering Cancer Center in New York.

Varmus was born in Oceanside, Long Island, New York, on December 18, 1939, and grew up in nearby Freeport. His father, Frank, was a physician, and his mother, Beatrice, was a psychiatric social worker. Planning to follow his father's example, Varmus took premedical courses at first when he attended Amherst College in Massachusetts, but he soon became more interested in literature. He edited the school paper and changed his major to English, in which he earned a bachelor's degree in 1961. He took a master's degree in 17th-century literature from Harvard University in 1962, but by then he had come to feel that friends in medical school were "more engaged with the real world," so he decided to become a physician after all. He earned his M.D. from Columbia University's College of Physicians and Surgeons in New York in 1966.

After considering careers in psychiatry and international medicine, Varmus was introduced to basic biomedical research in 1968, when the Public Health Service assigned him to work at the NIH as an alternative to military service during the Vietnam War. The work appealed to him so much that he abandoned all thought of regular medical practice. In 1969, soon after marrying Constance Casey, a reporter (they later had two sons), he moved to the University of California, San Francisco (UCSF). He joined Bishop's laboratory in the microbiology department in 1970 and remained at the university until 1993, rising through the faculty ranks to professor of microbiology and immunology in 1979 and, starting in 1982, professor of biochemistry and biophysics as well. He was American Cancer Society Professor of Molecular Virology from 1984 to 1993.

Bishop, who later said that Varmus's arrival "changed my life and career," was studying viruses that cause cancer in animals. One of these, the

Rous sarcoma virus, existed in both a common form, which caused cancer in chickens, and an unusual form, which could not cause the disease. Other researchers had found that the two forms differed in only a single gene, which the cancer-causing form possessed and the harmless form lacked. They named this key gene *src*, short for "sarcoma," the type of cancer that the virus caused.

Scientists went on to find similar genes in other cancer-causing viruses. Robert Huebner and George Todaro of the National Cancer Institute, part of NIH, called them oncogenes, from the Greek word for cancer. Huebner and Todaro proposed that these viruses inserted oncogenes into the genomes of the cells they infected. Sometimes, however, instead of causing cancer, the genes were passed down through generations of cells in a form that remained harmless unless a chance mutation returned them to their cancer-causing form.

In 1976, working with Dominique Stehelin and Deborah Spector, Bishop and Varmus turned Huebner and Todaro's idea on its head. They not only found a gene resembling *src* in normal chicken cells but proved that it was a chicken gene rather than a virus one. They theorized that, instead of having inserted the *src* gene into chicken cells in the distant past, the viruses had picked up the gene from the cells. The gene had then become modified in a way that made it cause cancer when the viruses later reintroduced it. Spector went on to find similar genes, which Bishop and Varmus called proto-oncogenes, or cellular oncogenes, in normal cells from fish, birds, and mammals, including humans. "Cancer may be part of the genetic dowry [inheritance] of every living cell," Bishop has said.

The discovery of cellular oncogenes not only won the Nobel Prize for Varmus and Bishop but completely changed scientists' understanding of cancer. For one thing, it showed that all cancer, whether inherited or triggered by something in the environment such as a virus or radiation, is ultimately due to changes in genes. Furthermore, the fact that cellular oncogenes were so widespread in nature suggested that they had very important functions in normal cells. Indeed, during the early 1980s, Bishop, Varmus, and other researchers found that oncogenes direct cells to make proteins that cause the cells to reproduce. Normally, these genes are active only at certain times in a living thing's existence—before birth, for instance, or when new cells are needed to heal a wound—but sometimes they are damaged in a way that activates them at the wrong time or leaves them "turned on" permanently, causing uncontrolled reproduction and cancer.

Varmus received many awards besides the Nobel Prize, most shared with Bishop, for his cellular oncogene research. They included the Albert Lasker Medical Research Award in 1982 and the Armand Hammer Cancer Prize, the Alfred P. Sloan Prize from the General Motors Cancer Foundation, and Canada's Gairdner Foundation International Award, all in 1984. Varmus also won the National Medal of Science in 2002.

Varmus continued his virus and oncogene research during the 1980s and early 1990s, for instance, studying mice that had been given extra copies of oncogenes through genetic engineering. He also pursued other projects, including research on the hepatitis B virus, which causes a serious liver disease, and on genes that affect the brain's development before birth. After 1984, although he and Bishop continued to work together at times, he had his own laboratory. Varmus also began serving on advisory committees for both biotechnology companies and government agencies, including the one that settled the contentious question of what to name the virus that had recently been found to cause AIDS.

Varmus's career changed direction in 1993, when Health and Human Services Secretary Donna Shalala nominated him to head the NIH, the world's largest basic research center. In spite of his eminence as a scientist, the nomination surprised some observers because Varmus's only administrative experience had been managing his 25-person laboratory at UCSF. When President Bill Clinton supported the nomination and Congress confirmed it, however, many scientists were delighted to see the NIH placed in the hands of someone with such an extensive background in basic research.

During his time as director of the NIH, Varmus became highly respected for his ability to settle conflict and build consensus by consulting many different groups. In a period when most government agencies were losing funding, he managed to raise NIH appropriations from $10.3 billion in 1993 to $17.9 billion in 2000—a rate of growth much greater than that in almost any other category of federal spending. He also attracted prestigious scientists to the NIH, improved the morale of those already working there, planned the replacement of several aging buildings, restructured peer review and grant awarding processes, and encouraged the use of new technology. Some advocates of research on particular diseases criticized his emphasis on basic science, but he pointed out that such science could have unexpected payoffs. Understanding gained through the study of cancer-causing retroviruses, for instance, greatly speeded efforts to develop treatments for AIDS, which proved to be caused by a similar virus. In 1999, the year Varmus left the NIH, Shalala said that his appointment "may turn out to be the most important legacy of the Clinton administration."

Varmus became the head of the Memorial Sloan-Kettering Cancer Center in New York, the oldest and largest private cancer research institution in the country, in 2000. There, as at NIH, he still finds time for some basic research in addition to his administrative duties. His chief work during the last decade has been in reproducing human cancers of the breast, brain, lung, and other tissues in genetically altered mice. He is also developing or improving programs to provide free access to scientific literature on the Internet and to train and employ scientists in developing countries. In 2001, he received the National Science Foundation's Vannevar Bush Award for lifetime achievement in science and public service.

Further Reading

Fallows, James. "The Political Scientist." *New Yorker*, June 7, 1999.

Varmus, Harold E. "Harold E. Varmus—Autobiography." *Les Prix Nobel 1989*. Available online. URL: www.nobel.se/medicine/laureates/1989/varmus-autobio.html. Last updated 2002.

"Varmus, Harold E." *Current Biography Yearbook 1996*. New York: H. W. Wilson, 1996.

Venter, J. Craig
(1946–)
American
Molecular Biologist

Craig Venter and his coworkers have combined technologies to work out the sequence of bases in the genomes of several kinds of organisms, including human beings, and developed new ways to mine and manage genetic information. Venter was born on October 14, 1946, in Salt Lake City, Utah, but grew up in San Mateo, California, near San Francisco. His father was an accountant, his mother a painter.

As a teenager, Venter was more interested in surfing than studying, but his work in the medical corps in Vietnam in the late 1960s made him take life more seriously. On his return to the United States, he signed up for premedical courses at San Mateo Community College, intending to become a physician in developing nations, but after a year he decided to go into research instead and transferred to the University of California, San Diego. He completed a combined B.A.-Ph.D. program in six years, earning his bachelor's degree in biochemistry in 1972 and his doctorate in physiology and pharmacology in 1975.

Skipping the usual years of postdoctoral research, Venter became an assistant professor of pharmacology and therapeutics at the State University of New York (SUNY), Buffalo, complete with his own laboratory, as soon as he finished his degree. He spent eight years there, eventually becoming a full professor. He was also an associate chief cancer research scientist at the Roswell Park Memorial Institute in Maryland. At these institutions, he studied proteins called receptors, which allow heart and brain cells to respond to hormones and other biochemicals. He married Claire Fraser, a molecular biologist and geneticist, in 1981.

Venter became the section chief in neurology at the National Institute of Neurological Disorders and Stroke, part of the government-sponsored National Institutes of Health (NIH) in Bethesda, Maryland, in 1984. By then, he realized that in order to understand his receptors, he needed to understand the genes containing the instructions for making them. Those instructions were encoded in the sequence of small molecules called bases that were strung like beads on the long, double-stranded necklace of each DNA molecule. At the time, locating particular genes and determining the sequence of bases in them were extremely slow, laborious processes. Venter, always a man in a hurry, was thrilled to read in 1986 that LEROY HOOD of the California Institute of Technology had invented a machine that could sequence genes automatically. He persuaded his superiors at NIH to let his laboratory test one of the first models.

A second sequencing roadblock lay in the fact that only a small portion of each DNA molecule contains information that cells actually use to make proteins. The rest, which has no known function, has been called "junk DNA." Around 1990, Venter developed a technique for avoiding the junk by synthesizing so-called complementary DNA (cDNA), which contains only sequences that will be "expressed" in making proteins. The cDNA sequences could be determined by machine and matched against known sequences from various organisms. If a cDNA sequence proved similar to part of a known gene, the cDNA's gene would probably have a function similar to that of the known one.

One of Venter's coworkers, Mark Adams, improved the cDNA technique by discovering that only part of an expressed sequence, called an expressed sequence tag (EST), was actually needed for matching. Venter, furthermore, found that he could apply his methods to a whole genome by breaking it into fragments and treating them all at once—what he called a "shotgun" approach. Senior NIH scientists told him his methods would never work, or at least would never produce anything useful, but by 1992, his laboratory had discovered 10,000 genes in human brain and testis cells, of which 2,700 were new. The number of new genes alone almost equaled the total number of human genes found during the previous decade.

Reversing its former lack of interest, NIH applied for patents on 337 of Venter's ESTs in mid-1991. Many scientists were shocked at the idea of patenting parts of human genes and blamed Venter for it, even though, as he repeatedly pointed out, the applications were not in his name. (The U.S. Patent Office rejected NIH's applications in 1993, claiming that the ESTs did not meet the patent requirements of novelty and usefulness.) Some also said that his techniques were not really inventive; JAMES WATSON, for instance, claimed that Venter's sequencing machines could be "run by monkeys." Venter attributed these attacks to envy or to fear that his achievements would weaken Congressional support for the infant Human Genome Project, which used conventional sequencing techniques.

With funding from venture capitalist Wallace Steinberg, Venter left NIH in July 1992 to found The Institute for Genomic Research (TIGR), a nonprofit research institution in Gaithersburg (later Rockville), Maryland. There, he combined his shotgun technique with a new computer program called the TIGR Assembler, which identified overlapping sequences in ESTs and thereby put these pieces of genomic puzzles back in their natural order. In 1995, with the help of Nobel Prize winner Hamilton Smith, TIGR sequenced the genome of a bacterium called *Hemophilus influenzae*, which infects brain, lung, and other tissues. This bacterium, which has about two million base pairs in its genome, was the first cellular organism to have its collection of genes completely sequenced. (FREDERICK SANGER had sequenced the nine-gene genome of a virus in 1977.) In the next few years, the institute sequenced the genomes of several other bacteria.

In 1998, Venter turned over the presidency of TIGR to his wife (he is still chairman of its board of directors) and joined with Michael Hunkapiller, a maker of gene sequencing machines, to form a new company called Celera Genomics. *Celera* comes from the Latin word for "speedy," and Venter clearly intended to live up to the name. He made headlines in May 1998 when he claimed that Celera would sequence the human genome sooner than the Human Genome Project. The government-sponsored program, with a $3 billion budget and laboratories in several countries, expected to finish by around 2003, but Celera proposed to accomplish the feat by 2000 or 2001, at a fraction of the government's cost.

Both Venter and Francis Collins, head of the Human Genome Project at NIH, insisted that their groups were not really in a race, but the public, and probably even most scientists involved, saw it as one. To keep up, the government researchers had to rethink their techniques and even adopt some of Venter's. Meanwhile, as a sort of prelude, Celera announced in 1999 that, in collaboration with scientists at several universities, it had sequenced the genome of the fruit fly (*Drosophila melanogaster*), a staple of genetic research.

The human genome "race" ended in an officially proclaimed tie on June 26, 2000, when Collins and Venter joined President Bill Clinton to announce that both the Human Genome Project and Celera had achieved their aim. Great as this accomplishment was, both men admitted that it was simply the first step in a long process. The sequence information will be truly useful only when scientists find out what the different genes do. Then, they can, for example, develop drugs to fight diseases caused or encouraged by defective genes or print out genetic profiles that forecast individuals' risks of contracting particular illnesses. Ultimately, Venter believes, such advances will revolutionize medicine.

Venter himself continues to take genomic research in new directions. He left Celera in January 2002, and at the end of April he announced that he was forming and heading three new nonprofit organizations: the TIGR Center for the Advancement of Genomics (TCAG), the Institute for Biological Energy Alternatives (IBEA), and the J. Craig Venter Science Foundation. TCAG will explore and seek to help the public and legislators better understand the ethical and social implications of genomic research. IBEA will attempt to use biological pathways and the metabolism of microorganisms to reduce carbon dioxide levels associated with global warming and to produce new, nonpolluting fuels. The Venter Foundation will provide administrative support, coordination, and fund-raising for TIGR, TCAG, and IBEA, as well as encourage science education and scientific innovation. In late 2002, Venter was also planning to build a microorganism "from scratch," adding genes one at a time.

Science writer Ted Anton has said that Craig Venter's achievement is that "he envisioned the relations among computers, sequencers, established [DNA] libraries, and unknown organisms to glimpse each next wave [of genetic technology] before it happened" and then gathered and inspired interdisciplinary teams of top-quality scientists to make his visions real. The biological community has been slow to honor the controversial Venter, but organizations and publications devoted to invention and business have showered him with awards. *R&D Magazine*, for instance, named him its Scientist of the Year in 1998, and *Industry Week* gave him (and Francis Collins) its annual Technology and Innovation Award in 2000. The prestigious London *Times* called him one of the most influential scientists of the century.

Further Reading

Anton, Ted. *Bold Science: Seven Scientists Who Are Changing Our World*. New York: W. H. Freeman, 2000.

Dentzer, Susan. "Breaking the Code." PBS Online NewsHour. Available online. URL: www.pbs.org/newshour/bb/health/jan-june00/extended_venter.html. Posted 2000.

"The Race Is Over." *Time*, July 3, 2000.

Shreeve, James. "The Code Breaker." *Discover*, May 1998.

"Venter, J. Craig." *Current Biography Yearbook 1995*. New York: H. W. Wilson, 1995.

Vesalius, Andreas
(1514–1564)
Flemish/Italian
Anatomist

By producing detailed descriptions of human anatomy based on firsthand dissections rather than on the word of ancient authorities such as GALEN, Andreas Vesalius helped to found the modern study of anatomy. He was born in Brussels, in what is now Belgium but was then called Flanders, or the Lowlands, on December 31, 1514, into a long line of royal physicians. His father, also called Andreas, was the apothecary (pharmacist) to Charles V of the Holy Roman Empire (King Charles I of Spain), a large area centering on present-day Germany. Throughout his adult life, Vesalius used the Latinized form of his family name, van Wesel.

Vesalius first showed his interest in anatomy when he dissected animal corpses in childhood. As a youth, he studied arts and classics at the University of Louvain (now in Belgium) and then, beginning in 1533, medicine in Paris. Like Renaissance artists such as Leonardo da Vinci, he was curious about the internal structure of the human body, and he was disappointed to find that his medical courses included little direct information on this subject. Medical students of his time usually saw only one or two dissections of human corpses each year, during which a professor on a platform read from Galen's works, a barber-surgeon below him did the actual cutting, a demonstrator pointed out the different parts, and hundreds of curious people looked on. The professor never checked to see whether his text matched what the surgeon and demonstrator showed. As Vesalius grumbled later, "less is presented to the spectators [in such dissections] than a butcher in his stall could teach."

Vesalius's Paris studies were cut short when war broke out between France and the Holy Roman Empire in 1536. As a citizen of the enemy country, he had to leave the city. He returned to Louvain, determined to teach himself. For instance, he stole the corpse of an executed criminal from the gallows, dissected it in private, and preserved the skeleton.

After receiving a bachelor's degree in medicine from the University of Louvain in 1537, Vesalius went to the renowned medical school in Padua, Italy, and earned his doctoral degree there later in the same year. Interest in anatomy was high in Padua, and the 22-year-old Vesalius, who had already shown that he knew more about the subject than most of his professors, was asked to join the faculty the day after he took his degree. He was the first person ever specifically paid to teach anatomy (he also taught surgery).

Vesalius broke with tradition by performing his own public dissections, spending three weeks on each one rather than the usual four days. Throngs of students attended them, and both they and Vesalius himself learned much. Like everyone else, Vesalius had assumed that Galen's anatomical descriptions were correct, but his dissections uncovered more and more mistakes—some 200 eventually. Finally, after dissecting a few apes, he concluded that Galen probably had never cut open human corpses at all but instead had made guesses about human anatomy based on dissections of these and other animals. He later wrote that Galen had been "deceived by his monkeys" and scolded other physicians and himself for having accepted these errors rather than checking Galen's statements with their own eyes.

Vesalius began working with artists from the studio of the famous painter Titian, especially a fellow Fleming named Jan van Calcar, to combine his own meticulous notes with equally detailed drawings of human anatomy based on his dissections. He published six large, annotated anatomical drawings, three by van Calcar and three by himself, in 1538. Then, in 1543, when he was only 28 years old, he published his masterpiece, an immense book (almost 700 pages) called *De Humani Corporis Fabrica (On the Workings of the Human Body)*.

Divided into seven sections dealing with different parts of the body, *De Humani Corporis Fabrica* was the first anatomical text based on extensive, firsthand human dissection. Its many large, carefully reproduced drawings also made it a milestone in book production. One drawing, for instance, portrays a standing human figure with the skin removed, showing all the muscles of the body. Others display nerves, blood vessels, and internal organs. Both pictures and text suggest the functions and relationships of body parts as well as showing their structure.

Vesalius's books (he issued a cheaper, abridged text called the *Epitome* at the same time as his larger work) caused a sensation. Many students welcomed them, but professors at Padua and elsewhere attacked them with almost religious fervor. Some critics claimed that Vesalius was insane or accused him of cutting up living human beings. He may have been unprepared for such vituperation. In any case, he publicly burned his remaining manuscripts and notes in early 1544 and abruptly left Padua.

Accepting an invitation from Charles V, Vesalius, like his father and grandfather, became a court physician. He also married later in 1544 and had a daughter a year afterward. Abandoning scientific research, he traveled with the king and, later, with Charles's son and successor, Philip II of Spain. In addition to treating the rulers and their courtiers, he worked as a military surgeon on battlefields that the courts visited. He survived a near shipwreck on the way home from a pilgrimage to Jerusalem, only to die of a sudden illness shortly afterward on the Greek island of Zante (now Zakinthos) on October 15, 1564.

Although Vesalius did not discover all of Galen's errors, *De Humani Corporis Fabrica* ultimately proved as groundbreaking for medicine as Nicolaus Copernicus's *The Revolution of the Heavenly Spheres*, published in the same year, was for astronomy. It pushed physicians to rely on their own eyes rather than on the words of others and to realize that the key to understanding and treating disease lay in understanding the body's structure. In *Doctors: The Biography of Medicine*, surgeon and medical historian Sherwin B. Nuland writes that "anatomy begins with this book, and so does modern scientific medicine."

Further Reading

Nuland, Sherwin B. *Doctors: The Biography of Medicine*. New York: Alfred A. Knopf, 1988.

O'Malley, C. D. *Andreas Vesalius of Brussels, 1514–1564*. Reprint, San Francisco, Calif.: Norman Publishing, 1997.

Vesalius, Andreas. *The Illustrations from the Works of Andreas Vesalius of Brussels*. Reprint, Mineola, N.Y.: Dover Publications, 1973.

Virchow, Rudolf
(1821–1902)
Prussian
Pathologist, Physiologist

Rudolf Virchow tied the cell theory of MATTHIAS JAKOB SCHLEIDEN and THEODOR SCHWANN to the study of bodily functions and the understanding, prevention, and treatment of disease. One of the most eminent biologists of his time, in his later years he was nicknamed "the Pope of German medicine."

Rudolf Virchow, so renowned that he was called "the Pope of German medicine" in his later years, stressed that all disease can be traced to malfunctioning of cells. *(National Library of Medicine)*

Virchow's background was anything but eminent. He was born on October 13, 1821, in the village of Schivelbein, then part of the German state of Prussia and now Swidwin in Poland. His father, Carl Virchow, a farmer with business aspirations, pushed him toward a medical career as a path into middle-class life. The family had no money for education, so Virchow entered the Friedrich-Wilhelms Institute, part of the University of Berlin, which provided a free medical education in return for a term of service as an army surgeon. He earned his M.D. in 1843 and then took his practical training at the Charité Hospital in Berlin.

Drawn to pathology, Virchow began making major discoveries almost immediately. In 1845, he wrote one of the first descriptions of leukemia, which he recognized as being due to excess multiplication of "white cells" in the blood. (John Bennett, a Scottish physician, described this disease, now recognized as a form of cancer, at about the same time.) A year later, he published a paper on blood clotting, coining the terms *thrombus* for a clot that blocks the blood vessel where it is formed and *embolism* for a clot that breaks free and travels through the blood, eventually blocking a vessel in a different spot. No one else had proposed that blood clots could move through the body. Virchow also identified atherosclerosis (hardening of the arteries) as an inflammatory disease that began with irritation of the inner artery wall.

By the time Virchow began teaching pathology at the Charité in 1847, he was already considered an authority on the subject. Unlike previous pathologists such as GIOVANNI BATTISTA MORGAGNI, he held that changes in function (physiology) were more important than changes in form (anatomy) in distinguishing health from disease. Disease, he once wrote, was simply "the process of life under altered conditions."

A tragic event that made headlines in early 1848 introduced Virchow to politics. An epidemic of typhus broke out among famine-stricken peasants in Upper Silesia, and Virchow became the medical officer on a committee that public pressure forced the Prussian government to send to investigate. He concluded that, although the specific cause of the illness might be unknown, the root causes of the epidemic were poverty, miserable living conditions, and the government's oppression and indifference. The only way to prevent future epidemics, he wrote, was to provide "democracy, education, freedom, and prosperity" to all citizens. Throughout his life, he would insist that physicians should be "attorneys for the poor."

Virchow's publicly expressed opinions about the epidemic and his support of a failed revolution in Prussia later in 1848 made the government decide to move the outspoken young physician out of Berlin. It pushed him to accept a professorship of pathologic anatomy (the first for that specialty in Prussia) at the rural, though still highly respected, University of Würzburg. Just before leaving Berlin in 1849, Virchow became engaged to Rose Mayer, the 17-year-old daughter of a friend, and they married in 1850. The marriage, a long and happy one, produced six children.

Virchow did some of his best work at Würzburg. He had become convinced of the correctness of Schleiden and Schwann's theory, which stated that the microscopic, membrane-encased structure called the cell is, as Virchow wrote, "the ultimate irreducible form of every living element." Mirroring his political beliefs, he saw the body as a "democracy of cells" in which the contribution of each semi-independent individual is essential. He proved that muscle, bone, and other tissues contain cells and developed a classification system for different kinds of cells. He also improved the cell theory by correcting a major error that Schwann had made. Schwann had believed that cells could be created out of formless substance inside the body, but Virchow insisted that "every cell arises from another cell" by splitting in two. Extending his earlier stress on physiology, he wrote that normal functioning of cells produces health and abnormal functioning produces disease.

By 1856, as surgeon and medical historian Sherwin B. Nuland writes in *Doctors: The Biography of Medicine*, Virchow had become "the most influential figure in German medicine." The Prussian government dared not keep him in exile any longer, so it invited him back to Berlin as professor of pathology in the city's university. Virchow agreed, but he insisted that an institution be created for his research. The resulting Pathological Institute, part of the University of Berlin, was the first research facility devoted solely to pathology.

Virchow summarized his views on cells in *Cellular Pathology as Based upon Physiological and Pathological Histology*, published in 1858. Although the book's ideas were, as Virchow wrote in the preface to its second edition, "at variance with what is ordinarily taught" and had "found . . . vigorous opponents," it soon became a tremendously respected and influential text, the most important of the more than 2,000 books and papers he penned during his long life. (The next most important was probably *Handbook of Special Pathology and Therapy*, first published in 1854.) Nuland writes that *Cellular Pathology* "enunciate[d] the principles upon which medical research would be based for the next hundred years and more," paving the way for the modern emphasis on cell biochemistry and molecular biology.

If the government hoped that Virchow's quiet years in Würzburg had made him forget politics, it was doomed to disappointment. First, he was elected to the Berlin City Council in 1859, a post he kept for 22 years. His chief concern in city government was public health, a concept just beginning to develop. He installed sewers, built new hospitals, and began keeping statistics on births, deaths, causes of death, housing, and other health-related information. His innovations made Berlin, once known as a disgusting "city built on a sewer," a much healthier place. After his death, the *British Medical Journal* called the improved city his monument.

Virchow did not limit himself to city cleanups. Elected to the Prussian National Assembly in 1861, he became a leader of the liberal Progressive Party. He often argued with the powerful Otto von Bismarck, the country's prime minister and leader of its militarists and conservatives. In the end, Bismarck's views prevailed, uniting the German states into a single country in 1870. Virchow was elected to Germany's new legislature, the Reichstag, in 1880 and remained a member until 1893, but he had little influence there.

Virchow's reputation in science was undiminished, however, and he often spoke and wrote about scientific subjects with political implications. For instance, many Germans of his time, like the later Nazis, believed that the only "true Germans" were blond, light-skinned, and blue-eyed, descendants of a supposed "pure Aryan" race. Virchow surveyed German children in 1876 to determine how often various combinations of skin, hair, and eye color occurred and announced that more than half of the children had mixtures of blond and "brown" coloration. Another survey showed that 11 percent of Jews, a despised group in Germany, had the "Aryan" combination of blond hair, blue eyes, and light skin. Virchow summarized his survey results in 1886 by saying that the pure Aryan or Teutonic race was a myth.

Virchow's research on race was only one aspect of the interest in anthropology and archaeology that dominated his later life. He made a large collection of skulls, organized German anthropology, and financed and took part in archaeological digs around Europe, including Heinrich Schliemann's famous unearthing of the fabled Greek city of Troy in 1879.

By the time Virchow reached his 80th birthday, the onetime rebel had become a deeply revered symbol of German science. That occasion was honored with an international celebration at which scientists and government representatives from all over the world praised his tireless research, writing, speaking, and teaching. The kaiser personally gave him the Gold Medal of Science.

A year later, however, Virchow's energy betrayed him. In a hurry as always, he slipped while trying to leap onto a moving streetcar on a rainy day and fell, fracturing his thigh. He partly recovered, but a second fall that summer brought on heart problems, and he died in Berlin on September 5, 1902. The government gave him an elaborate funeral. One of his biographers wrote that with his death, "Germany . . . lost four great men at once: her leading pathologist, her leading anthropologist, her leading sanitarian [public health advocate], and her leading liberal."

Further Reading

Nuland, Sherwin B. *Doctors: The Biography of Medicine*. New York: Alfred A. Knopf, 1988.

Waksman, Selman A.

(1888–1973)
Russian/American
Microbiologist, Pharmacologist

Selman Abraham Waksman won the 1952 Nobel Prize in physiology or medicine for isolating several antibiotics from soil microorganisms. He was born in Priluka, a village near Kiev, Russia (now the Ukraine), on July 22, 1888, the son of Jacob and Fradia (London) Waksman. Having experienced anti-Semitism in his youth, he was happy to emigrate to the United States in 1910. He earned a B.S. in agriculture from Rutgers College (now Rutgers University) in New Jersey in 1915 and began working at the New Jersey Agricultural Experiment Station immediately afterward. He went on to earn an M.S. from Rutgers in 1916, the same year in which he became a naturalized citizen and married his childhood sweetheart, Bertha (Bobili) Mitnik (they later had one son). He gained a Ph.D. in biochemistry from the University of California, Berkeley, in 1918.

After obtaining his advanced degrees, Waksman began teaching at Rutgers, becoming a professor of soil microbiology in 1930. Throughout his career, beginning in 1921, he did research at the New Jersey agricultural station as well. In the 1920s and 1930s he identified many kinds of soil microbes, including some that increase soil fertility. He found that certain microorganisms make humus and peat, the organic parts of soil, from plant and animal waste. He also studied soil conservation, cultivation of soil microbes, and the making of compost. He advised industrial laboratories and government committees and wrote several books on soil microorganisms, including *The Soil and the Microbe* (with R. L. Starkey, 1931) and an exhaustive text, *Principles of Soil Microbiology* (1927 and 1932). When Rutgers established a new department of microbiology in 1940, he became a professor of that subject and head of the department.

In addition to his soil research, Waksman organized a laboratory of marine bacteriology at the Woods Hole Oceanographic Institution in Massachusetts and headed it until 1942. There, among other things, he developed techniques to keep marine life from damaging ship hulls. The U.S. Navy found these techniques very helpful during World War II.

In 1939, René Dubos, a former student of Waksman's, isolated substances from certain bacteria that killed other types of microbes. These compounds, tyrothricin and gramicidin, proved useful in treating cattle but were too toxic for human beings. Dubos's discovery nonetheless interested Waksman in antibiosis, the phenomenon in which some kinds of microorganisms make substances that kill or halt the growth of others. He knew that members of a large family of soil fungi called actinomycetes, which he had studied extensively, tended to outcompete other soil microbes, and he believed that they probably made such chemicals. He and his coworkers began testing soil samples from all over the world in the hope of finding an antibiotic that could be used in humans.

Waksman's group isolated a number of antibiotics from actinomycetes in the early 1940s, the most important of which were actinomycin (first isolated in 1940), streptomycin (1944), and neomycin (1948). They killed gram-negative bacteria, a large group of bacteria unaffected by penicillin, an antibiotic that HOWARD WALTER FLOREY and others had introduced in the early 1940s. Great excitement was generated when streptomycin, first made available to the public at the end of 1946, was found to be the first effective treatment for tuberculosis. Actinomycin became an anticancer drug, and neomycin was used to fight bacteria that cause wound infections. Waksman's drugs were patented, and most of the income from the patents was used to establish a new Institute of Microbiology at Rutgers, with Waksman as director, in 1949.

In addition to the 1952 Nobel Prize, Waksman won the Star of the Rising Sun from the emperor of Japan (1952), the Lasker Award, and the Trudeau Medal of the National Tuberculosis Association. France made him a Commander of the Legion of Honor, and Brazil made him a Commander of the Order of the Southern Cross. In his later years, he wrote about his work for the public as well as fellow scientists, penning such books as the autobiographical *My Life with the Microbes* (1954) and *The Conquest of Tuberculosis* (1965). Waksman retired in 1958 and died on August 16, 1973.

Further Reading

Foundation for Microbiology. "Selman A. Waksman (1888–1973)." Available online. URL: http://waksman.rutgers.edu/Waks/Waksman/DrWaksman.html. Downloaded 2001.

"Waksman, Selman A(braham)." *Current Biography Yearbook 1946*. New York: H. W. Wilson, 1946.

Wallace, Alfred Russel

(1823–1913)
British/Indonesian
Naturalist, Evolutionary Biologist

Alfred Russel Wallace thought of the theory of evolution by natural selection at about the same time as his more famous compatriot, CHARLES ROBERT DARWIN. He also helped to found biogeography, which studies the geographical distribution of living things. He was born in Usk, Wales, on January 8, 1823, the third son and eighth child of Thomas and Mary Anne (Greenell) Wallace.

Thomas Wallace's unsuccessful business ventures kept his family close to poverty. At age 14, therefore, Alfred had to leave school. He joined the surveying business of his oldest brother, William, to learn the trade. Roaming the hills while doing surveying work interested him in nature.

When loss of business forced his brother to let him go in 1843, Wallace became a teacher at the Collegiate School of Leicester for two years. There, he met Henry Bates, who shared his interest in natural history. (Bates was already on his way to becoming a respected entomologist.) Inspired by CHARLES LYELL's *Principles of Geology* and the exploration narratives of Darwin and ALEXANDER VON HUMBOLDT, the two decided to go to the Amazon River Basin in South America and collect specimens to sell. Wallace also planned to look for evidence supporting the still-controversial idea that species of living things had changed, or evolved, over time.

Wallace and Bates, aged 25 and 23 respectively, arrived in Brazil in May 1848. They separated after about a year. Wallace explored and mapped the Rio Negro, a major Amazon tributary, traveling farther up the river than any other European had done. He collected numerous specimens, but unfortunately he lost most of them, as well as notes, sketches, and nearly his life as well, in a shipwreck and fire on his way home in August 1852. He wrote a book about his journey, *A Narrative of Travels on the Amazon and Rio Negro*, and paid to have it published in 1853. It sold few copies, but it brought him to the attention of other naturalists.

With his thirst for exploration as strong as ever in spite of the disastrous ending of his South American voyage, Wallace obtained backing from the Royal Geographical Society for an expedition to the Far East. He departed in March 1854 to explore the Malay Peninsula, now part of Malaysia, and the islands of the Malay Archipelago, now Indonesia and the Philippines. He traveled some 14,000 miles and collected about 126,000 specimens during his stay, which lasted until 1862. His research during this trip included the first extensive study of orangutans.

In Borneo in 1855, Wallace wrote a paper called "On the Law Which Has Regulated the Introduction of New Species," which stated that all new species had come into existence near to similar species in both time and place. Lyell saw the paper in a natural history journal and told his friend Darwin about it, and Darwin and Wallace began a correspondence. Darwin told Wallace that he, too, was planning to write about "the species question" but did not give any details of his work.

Forced into bed by an attack of malaria in February 1858, Wallace began thinking about economist THOMAS ROBERT MALTHUS's *Essay on the Principle of Population*, which described the limits imposed on population by shortage of food supplies and competition for resources. Malthus had been writing mainly about human society, but Wallace realized that the economist's ideas could apply to other species as well. Then, as he wrote later, "there suddenly flashed upon me the idea of the survival of the fittest."

Wallace concluded, just as Darwin had done when he read Malthus's book, that the stress of competition could help to explain the mechanism of evolution. Like Darwin, he theorized that new species begin when an individual is born with some "little superiority" that makes it better adapted to its environment ("fitter") than others of its kind. This chance variation makes it more likely to survive long enough to raise offspring. The individual's descendants are likely to inherit the useful variation, and if they also survive, they eventually become different enough from the original species to constitute a new species. "Thus, I at once saw, the ever present variability of all living things would furnish the material" of evolution, Wallace wrote.

Wallace wrote a paper describing these ideas, called "On the Tendency of Varieties to Depart Indefinitely from the Original Type," and sent it to Darwin. Darwin was amazed to find that the younger man had independently arrived at almost exactly the same explanation for evolution that he himself had been developing for the past 20 years. He realized that he would have to publish something quickly if he wanted to stake his claim to the theory, but he did not want to deprive Wallace of his share of the credit. After consulting with Lyell and another scientist friend, therefore, he assembled some fragmentary writings into a paper describing his work and arranged for both this paper and Wallace's to be presented at a meeting of the Linnean Society, a prominent scientific society, on July 1, 1858. The presentation produced little stir, and Wallace, still in the Far East, knew nothing of it until after it had happened. Darwin meanwhile began writing *On the Origin of Species*, which was published in 1859.

Although later writers have disagreed about how fairly Darwin treated Wallace, Wallace himself, a modest man, seems never to have complained about the fact that most of the fame for having developed the controversial but extremely influential theory of evolution by natural selection went to Darwin. Indeed, Wallace remained a lifelong admirer and supporter of Darwin's, although the two disagreed about many details of evolution. Most notably, Darwin thought that evolution could account for everything about humanity, including the species's intelligence and moral development, whereas Wallace felt that some "metabiological" agency must account for the unique human psychology. Wallace wrote two books, *Contributions to the Theory of Natural Selection* (1870) and *Darwinism* (1889), that described how his theory differed from Darwin's.

Wallace's role in evolutionary theory may have been downplayed, but the books he wrote about his Malayan explorations made him the greatest living authority on that part of the world. The most important of these writings were *The Malay Archipelago* (1869) and the two-volume *Geographical Distribution of Animals* (1876), still considered classics of biogeography. He showed clearly that biogeography provided excellent evidence for evolution. He also observed that the animals on the large islands of Borneo and Bali resembled animals seen in Asia, whereas those on Celebes and Lombok were like those in Australia. An imaginary line drawn through the archipelago between these islands, he wrote, would have islands with Asia-related animals on the western side of it and islands with Australia-related animals on the eastern side. He discovered that this line, still called Wallace's Line, corresponded to a deepwater channel that would have made that part of the sea difficult for animals to cross during migrations.

During his later life, Wallace found time to marry (in 1866, to Annie Mitten—the 18-year-old daughter of a botanist friend—with whom he later had three children) and pursue a wide variety of interests and sociopolitical causes, ranging from spiritualism to socialism. His prolific lectures and writings included a two-volume autobiography, *My Life* (1905). He was elected to membership in the Royal Society, Britain's top science organization, in 1893 and received the society's Royal Medal (1868), Darwin Medal (1890), and Copley Medal (1908). The Linnean Society, similarly, gave him a gold medal in 1892 and the first Darwin-Wallace Medal in 1898. The British government awarded him the Order of Merit in 1908. Wallace died in Broadstone, Dorset, on November 7, 1913.

Further Reading

"Alfred Russel Wallace: A Capsule Biography." Available online. URL: www.wku.edu/~smithch/index1.htm. Accessed 2001.

Ghiselin, Michael T. "Alfred Russel Wallace and the Birth of Biogeography." *Pacific Discovery*, Spring 1993.

Raby, Peter. *Alfred Russel Wallace: A Life*. Princeton, N.J.: Princeton University Press, 2001.

Wallace, Alfred Russel. *The Alfred Russel Wallace Reader: A Selection of Writings from the Field*. Edited by Jane R. Camerini. Baltimore, Md.: Johns Hopkins University Press, 2001.

Shermer, Michael. *In Darwin's Shadow: The Life and Science of Alfred Russel Wallace*. New York: Oxford University Press, 2002.

Wambugu, Florence

(1953–)
Kenyan
Molecular Biologist

Florence Wambugu has worked to develop genetically engineered crops that resist attacks by viruses and other pests, thereby potentially increasing food supplies and farmers' income in her native Africa. "I support biotechnology because it has the potential to help my people," she says. "Biotechnology is not a silver bullet for all of Africa's problems, but it definitely provides real solutions to our hunger and poverty." She was born on August 23, 1953, in Nyeri, Kenya, near Aberdares Ridges, to a poor farm family. Her mother sold the family's only cow to raise money for Wambugu's secondary education.

Wambugu studied botany and zoology at the University of Nairobi, graduating in 1978. She then took a position at the Kenya Agricultural Research Institute's (KARI) Muguga research station. There, working with scientists from the Centro Internacional de la Papa (CIP), she applied traditional plant breeding methods in an attempt to improve sweet potatoes, a staple Kenyan food crop. These techniques, however, failed to produce a sweet potato variety that could resist viruses.

While working for KARI, Wambugu also learned about newer methods of modifying crops, including tissue culture. She expanded this knowledge by studying plant pathology at the University of North Dakota, Fargo, for two years, obtaining a master's degree in 1984. She then returned to Kenya, where she continued to work with KARI and CIP. She researched diseases of sweet potato plants at the University of Bath in England from 1988 to 1991, when she earned her Ph.D. She found that a complex of seven viruses caused the worst diseases.

Florence Wambugu has helped to create and test genetically altered sweet potatoes and other crops to benefit farmers in her native Kenya and elsewhere in Africa. *(Florence Wambugu)*

Right after she finished her degree work, Wambugu won a three-year fellowship from the United States Agency for International Development (USAID) to study biotechnology at Monsanto Corporation's Life Sciences Research Center in St. Louis, Missouri. She was the first African scientist to obtain such a fellowship. Working with scientists at Monsanto and in Kenya, she adapted techniques developed at Washington University in St. Louis to develop Kenya's first genetically modified sweet potato plants. The plants carry a gene that makes them resistant to the feathery mottle virus, one of the worst pests of this crop. Field tests of the plants began in Kenya in 2000. Tuber yields doubled during the first season, and the amount of foliage, which is used as animal food, also increased. If the tests are fully successful, Wambugu says, sweet potato yields may increase by as much as 80 percent. Monsanto has donated the intellectual property rights for the altered plants to KARI.

Wambugu returned to Kenya in 1994 and became director of the African Centre of the International Service for the Acquisition of Agribiotechnology Applications (ISAAA). After seven years, she left to set up A Harvest Biotech Foundation International (AHBFI), of which she is the executive director. The foundation's vision is to increase agricultural harvests through use of biotechnology tools and to use creative solutions to increase income levels, especially of poor, rural, small-scale farmers.

Wambugu travels and speaks frequently to defend the use of biotechnology in Africa. Countering environmental groups' claims that biotechnology will endanger African ecology or be used to exploit its people, she asserts that, although the technology presents some risks, "there is currently no scientific evidence of any harm, while the benefits are documented and far outweigh the potential risks."

Wambugu has written a book called *Modifying Africa* (2001), which describes the techniques of agricultural biotechnology and its benefits for the continent, especially for poor farmers. Biotechnology, she points out, can reduce the use of chemical pesticides as well as increase crop yields. It can help in growing not only sweet potatoes but also bananas, sugar, cassava, and other staple food and cash crops. She has personally helped to develop genetically altered, disease-resistant forms of some of these crops.

The American Biographical Institute named Wambugu its Woman of the Year in 2001. John Sterling, managing editor of *Genetic Engineering News*, has called her "a voice of reason and logic in the agbiotech controversy."

Further Reading

"About the Author: Dr. Florence Wambugu." Available online. URL: www.modifyingafrica.com/. 2001.

Wambugu, Florence. *Modifying Africa*. India: Pragati Offset Pvt. Ltd., 2001. Available through www.modifyingafrica.com.

———. "Feeding Africa." *New Scientist*, May 2002. Available online. URL: www.ahbfi.org/www/interview.htm.

Watson, James

(1928–)
American
Molecular Biologist

With British coworker FRANCIS CRICK, James (Jim) Dewey Watson discovered the molecular structure of DNA, which many have called the greatest achievement in 20th-century biology. Watson was born on April 6, 1928, in Chicago and grew up there. His father, after whom he is named, was a businessman, and his mother, the former Jean Mitchell, worked in the admissions office of the University of Chicago. Watson's unusual intelligence was apparent from an early age: He appeared on a popular radio show called "Quiz Kids" and entered the University of Chicago at age 15.

At first, Watson planned to focus on ornithology (the study of birds), which had interested him since childhood, but by the time he graduated with a B.S. in zoology in 1947, he had changed his interest to genetics. He did graduate work at the University of Indiana, Bloomington, where prizewinning geneticists and molecular biologists HERMANN JOSEPH MÜLLER and SALVADOR LURIA inspired him. Combining the specialties of both men, he earned his doctorate in 1950 with a thesis on radiation's effects on the genes of viruses that infect bacteria (bacteriophages).

Watson began doing postdoctoral work on bacteriophages at the University of Copenhagen in Denmark, but at a conference in Italy in the spring of 1951 he met MAURICE WILKINS, a New Zealand–born scientist who was studying the structure of DNA at King's College in London. At the time, few biologists believed that DNA was the carrier of genetic material, but Wilkins was one of them, and he soon persuaded Watson as well. "A potential key to the secret of life was impossible to push out of my mind," Watson wrote later.

Watson transferred to the Cavendish Laboratory at Britain's Cambridge University in late 1952 and met Crick there. Crick wrote later, "Jim and I hit it off immediately, partly because our interests were astonishingly similar." One of their shared interests was in DNA. If DNA carries inherited information, they realized, molecules of the substance must be able to reproduce themselves so that each new cell can receive a complete copy of that information. To find out how this happens, they knew they would have to learn the shape of the DNA molecule.

At the time, molecular biologists knew that DNA is a large, chainlike molecule called a polymer, made up of several kinds of smaller molecules: alternating molecules of sugar and phosphate, which form a "backbone," and four kinds of bases (adenine, thymine, cytosine, and guanine). No one was sure how these small molecules were arranged within the larger one, however. To find out, Watson and Crick borrowed two key approaches from other scientists trying to decipher DNA: the X-ray crystallography used by Wilkins's laboratory and the three-dimensional molecular models of American chemist LINUS CARL PAULING, who had already used model building to work out the basic structure of proteins.

Watson and Crick had a vital stroke of luck in January 1953, when Wilkins, who had become a friend of Watson's, showed him a striking X-ray photograph of DNA made by ROSALIND ELSIE FRANKLIN, a chemist and crystallography expert in his laboratory. Watson wrote later in *The Double Helix,* his memoir of the momentous discovery, that when he looked at the photo, "my mouth fell open and my pulse began to race." He realized that the DNA molecule must have the corkscrew shape of a double helix, with two sugar-phosphate backbones twining on the outside and pairs of bases placed between them, like steps on a twisted ladder.

The arrangement of the bases remained to be determined. Experimenting with models made in the Cambridge machine shop, Watson realized that an adenine-thymine pair would have the same overall shape as a cytosine-guanine pair. Furthermore, a biochemist named Erwin Chargaff had shown that the amount of thymine in a molecule of DNA is always the same as the amount of adenine, and the same is true of cytosine and guanine. Watson therefore concluded that adenine must always pair with thymine and cytosine with guanine. Hydrogen bonds could hold each pair together. The official announcement of this discovery, published in the prestigious British science journal *Nature* on April 25, concluded with what *Time* magazine later called "one of the most famous understatements in the history of science": "It has not escaped our notice that the specific pairing we have postulated immediately suggests a possible copying mechanism for the genetic material."

Crick and Watson elaborated on this statement in a second paper published about five weeks later. That paper explained that the hydrogen bonds holding the base pairs together are weak and break apart easily. Just before a cell divides, Watson and Crick theorized, each DNA molecule splits apart lengthwise like a zipper unzipping. Each half of the molecule then attracts the bases it needs to complete its pairs, along with pieces of sugar-phosphate "backbone," from free-floating molecules in the cell nucleus. The result is two DNA molecules identical to the first. Each of the two daughter cells formed by the division then can receive one complete set of DNA molecules. This theory of DNA reproduction was later confirmed by experiment.

Watson returned to the United States in fall 1953. For two years, he studied the structure of a second nucleic acid, RNA, with X-ray crystallography at the California Institute of Technology in Pasadena, then returned briefly to Cambridge to examine the structure of viruses with Crick. He joined the faculty of Harvard University in 1956 and remained there for 20 years, becoming a full professor in 1961. His chief research activity at Harvard was helping to determine how DNA uses RNA as an intermediate in instructing cells how to make proteins.

Along with Crick and Wilkins, Watson received the Nobel Prize in physiology or medicine in 1962 for the discovery of DNA's structure. He also won many other awards, including the Albert Lasker Medical Research

Award of the American Public Health Association (1960), the Medal of Freedom (1977), the Copley Medal of Britain's Royal Society, and the National Medal of Science (1997).

Watson began to turn from research to administration when he became director of Cold Spring Harbor Laboratory on Long Island, New York, in 1968, the same year he married Elizabeth Lewis (they later had two sons). He left Harvard in 1976 to run the laboratory full time. The facility, the first genetics laboratory established in the United States, was in poor condition when he took it over, but he revitalized it and ultimately made it one of the world's most highly regarded institutions for the study of cell biochemistry, molecular biology, genetics, and cancer. He left the directorship, with its day-to-day administrative tasks, in 1994, becoming instead the organization's president.

Watson has always been one of genetic exploration's biggest boosters. When scientists announced the ability to combine genes from different kinds of living things in the early 1970s, he defended the new technology against those who wanted the government to heavily regulate or ban it. In the late 1980s, when scientists began discussing the possibility of working out the sequence of bases in the entire human genome, he supported this proposal as well. The Human Genome Project became a reality in 1988, and Congress chose Watson to head the part of it run by the National Institutes of Health (NIH).

At the same time, Watson has recognized that genetic technologies can be misused. As head of the Human Genome Project, for instance, he earmarked 3 percent (later raised to 5 percent) of the project's $3 billion budget for investigation and debate on the project's ethical, legal, and social implications. He resigned from the project in 1992 because he opposed the patenting of human genes, which Bernardine Healey, then head of the NIH, favored.

In addition to research and administration, Watson has maintained a writing career that began spectacularly with *The Double Helix* (1968), a best-seller praised for its candid, behind-the-scenes look at scientific discovery but also criticized for its opinionated tone and harsh portrait of some fellow researchers, especially Franklin. His other books for a general audience include *A Passion for DNA: Genes, Genomes, and Society* (essays, 2000) and *Genes, Girls, and Gamow* (a second volume of autobiography, 2002). He has also coauthored several textbooks in genetics and molecular biology, including *The Molecular Biology of the Gene* (1965). He makes frequent speaking appearances as well. Recent controversial proposals, such as the suggestion that humans should direct their own evolution by making changes in inheritable (germ line) genes, show that Watson's thinking is as bold as ever.

Further Reading

Edelson, Edward. *Francis Crick and James Watson: And the Building Blocks of Life*. New York: Oxford University Press, 1998.

Watson, James D. *The Double Helix*. New York: Atheneum, 1968.

———. *Genes, Girls, and Gamow: After the Double Helix*. New York: Alfred A. Knopf, 2002.

Wright, Robert. "Molecular Biologists Watson and Crick." *Time*, March 29, 1999.

Weinberg, Robert A.

(1942–)
American
Biochemist, Geneticist

Robert Allan Weinberg and his laboratory discovered the first altered genes shown to cause human cancer. Weinberg was born to Fritz Weinberg, a dentist, and Lore (Reichhardt) Weinberg on November 11, 1942, in Pittsburgh, Pennsylvania. His parents had immigrated to the United States together in 1938, fleeing the persecution of Jews in Nazi Germany.

Weinberg says he felt no interest in science as a youth. Nonetheless, he enrolled in the Massachusetts Institute of Technology (MIT) in Cambridge as a premedical student and, while there, "somehow slipped unconsciously into biology." He earned a B.S. in the subject in 1964 and, although his grades as an undergraduate had not been very good, some of his professors saw his potential and helped him enter the university's graduate school. He earned an M.A. in 1965 and a Ph.D. in 1969, both in biology.

After postdoctoral work at the Weizmann Institute in Rehovoth, Israel, and at the Salk Institute in La Jolla, California, Weinberg returned to MIT in 1972 and has been there ever since. He became a full professor of biology in 1982, an American Cancer Society Research Professor in 1985, and Daniel K. Ludwig Professor for Cancer Research in 1997. He is also a founding member of the Whitehead Institute for Biomedical Research and a member of the Center for Cancer Research, both part of MIT. In 1976, he married Amy Shulman, a teacher, and they have a son and a daughter.

Weinberg's chief research interest has always been cancer. Around the time he obtained his Ph.D., Robert Huebner and George Todaro of the National Cancer Institute theorized that certain viruses cause the disease by inserting genes that Huebner and Todaro called oncogenes (from the Greek word for cancer) into the genomes of cells they infect. Scientists found an apparent oncogene in one cancer-causing virus in the early 1970s. In 1976, however, J. MICHAEL BISHOP and HAROLD E. VARMUS of the University of California, San Francisco, showed that the gene did not come originally from the virus but instead was a cell gene that normally helps to control growth. The gene had become mutated in a way that made it active all the time, causing the cell to reproduce endlessly.

This discovery made cancer researchers, including Weinberg, begin to seek oncogenes in cells rather than in viruses. Weinberg applied chemicals known to cause cancer (carcinogens) to cells in tissue culture to activate their oncogenes, then modified newly developed genetic engineering techniques to transfer suspect genes from these cells to normal cells. He worked with mouse cells at first but later turned to human cells. In 1980, his laboratory isolated the first human gene associated with a particular kind of cancer, a bladder cancer gene that they named *ras*. They went on to find oncogenes associated with leukemia and colon cancer.

Weinberg and his coworkers next analyzed the oncogenes they had discovered. They showed

Massachusetts Institute of Technology scientist Robert A. Weinberg identified several kinds of mutated genes associated with human cancers in the 1980s. *(Donna Coveney/MIT)*

that the genes exist in normal cells but are much less active there than in cancer cells. Working with Edward M. Scolnick and Douglas Lowy of the National Cancer Institute, Weinberg's group found in 1982 that *ras* differs from its normal equivalent by only one base, or nucleotide, out of about 6,000 in the gene. Somehow that tiny change is enough to alter the cell's growth pattern completely. In 1983, Weinberg also helped to show that in rat cells, two or more mutations in different genes are necessary for cancer to develop fully. This explains why most cancers occur in older organisms, which have lived long enough to acquire multiple mutations that may occur years apart.

In the late 1980s, Weinberg's laboratory discovered a second type of cancer-causing gene that is the exact opposite of an oncogene. Oncogenes normally stimulate cell growth, and they cause cancer when mutated in a way that makes them constantly active. This second group of genes, on the other hand, normally stops cells from multiplying, and cancer results when they are mutated in a way that makes them inactive. They have come to be called tumor suppressor genes. Weinberg's group found the first of these genes in a rare eye cancer in 1986 and called it *Rb* for retinoblastoma, the name of the cancer.

Weinberg's work played a leading role in establishing the basic insight that, ultimately, all cancer is caused by defects in genes. Some of these defects are inherited, while others occur during an organism's lifetime, either randomly or as a result of damage caused by environmental agents such as radiation or carcinogenic chemicals. His groundbreaking studies won many awards, including the Bristol-Meyers Award for Distinguished Achievement in Cancer Research (1984), Canada's Gairdner Foundation International Award (1992), and the National Medal of Science (1997). *Discover* magazine chose him as its Scientist of the Year in 1982, and in 1999 he won the Killian Faculty Award from MIT, the greatest honor the faculty can bestow on a member.

In addition to numerous scientific papers, Weinberg has written several popular books describing what he and other scientists have learned about cancer, including *Racing to the Beginning of the Road* (1996) and *One Renegade Cell* (1998). He and Harold Varmus collaborated on *Genetics and the Biology of Cancer* (1992).

Today, Weinberg and his coworkers continue to study the function of oncogenes and tumor suppressor genes in both cancerous and normal cells. They have learned, for instance, that cancer cells contain an activated gene for part of an enzyme named telomerase, which adds DNA to segments called telomeres at the ends of chromosomes. Because this gene is inactive in normal cells, the telomeres become shorter over time, serving as a sort of molecular clock that tells the cells when to stop reproducing and die. If the telomeres are constantly renewed by telomerase, the cells lose this vital signal and go on multiplying. The telomerase gene is not an oncogene, but it works with oncogenes to trigger cancer. Weinberg's group has discovered that blocking this gene in human tumor cells makes the cells die, so either the gene or telomerase might be a target for future anticancer drugs or gene therapy. Weinberg's laboratory is also studying genes involved in breast cancer and normal breast tissue development.

Further Reading

Angier, Natalie. *Natural Obsessions: Striving to Unlock the Deepest Secrets of the Cancer Cell.* Boston: Houghton Mifflin, 1988.

Weinberg, Robert A. *One Renegade Cell: How Cancer Begins*. New York: Basic Books, 1998.

———. *Racing to the Beginning of the Road: The Search for the Origin of Cancer.* New York: Random House/Harmony Books, 1996.

"Weinberg, Robert A(llan)." *Current Biography Yearbook 1983*. New York: H. W. Wilson, 1983.

Wexler, Nancy Sabin
(1945–)
American
Psychologist

Nancy Wexler turned a family tragedy into motivation for understanding a deadly inherited disease. She was born on July 19, 1945, in Washington, D.C., to Milton Wexler, a psychoanalyst, and his wife, Leonore.

In August 1968, when Wexler was 22 years old and her parents had been divorced for four years, her father told her and her older sister, Alice, that he had just learned that their mother was suffering from an inherited brain disorder called Huntington's disease (formerly Huntington's chorea). He explained that the disease, for which there is no cure or treatment, usually does not reveal itself until middle age. It then produces a slow slide into insanity accompanied by uncontrollable twisting or writhing movements. Because a single dominant gene causes the disease, Nancy and Alice each had a 50–50 chance of developing it as well.

The news was devastating, but Milton Wexler said later that Nancy "went from being dismal to . . . wanting to be a knight in shining armor going out to fight the devils." Of a similar mind, Milton contacted the Committee to Combat Huntington's Chorea, an organization led by Marjorie Guthrie, who had been married to famed folk singer Woody Guthrie, the disease's best-known victim. Milton opened a chapter of the group in Los Angeles, and Nancy set up another in Michigan, where she was studying psychology at the University of Michigan, Ann Arbor. (She had obtained an A.B. in social relations and English from Radcliffe College in 1967.) She earned her Ph.D. in psychology in 1974 with research on the psychological effects of being at risk for Huntington's disease.

Guthrie's organization was devoted mainly to improving care for the approximately 40,000 Americans who suffer from Huntington's, but the Wexlers were more interested in sponsoring research. Milton Wexler, with Nancy's help, therefore established the Hereditary Disease Foundation in 1974 to fund research on Huntington's disease. Nancy became the foundation's president in 1983 and still holds this post.

Columbia University psychologist Nancy Sabin Wexler, herself at risk for developing the inherited brain disorder Huntington's disease, helped in identification of the gene that causes the disease by studying a large Venezuelan family in which the illness is widespread; she also gave members of that family, such as this stricken child, "immeasurable love." *(Peter Ginter/ Hereditary Disease Foundation)*

The group agreed that identifying the gene that causes Huntington's offered the best hope for a treatment or cure. At minimum, they could try to develop a simple blood test that would show whether an individual at risk for the disease had inherited the gene and therefore would eventually develop Huntington's. The test could help people at risk make decisions about family, financial, and other matters.

In October 1979, David Housman of the Massachusetts Institute of Technology told the foundation about a new technique that narrowed down the location of unknown genes by using known stretches of DNA called restriction fragment length polymorphisms (RFLPs), which exist in several different forms and therefore could be used as markers. The more often a certain form of RFLP was inherited with a certain form of an unknown gene in a given family, the more likely the unknown gene was to be near the RFLP on the same chromosome.

The only problem was that, at that time, only one human RFLP was known. Finding a RFLP that happened to be near the Huntington's gene might take decades. Still, the Hereditary Disease Foundation gave Housman a grant to try his idea, and Nancy Wexler arranged additional funding through the Congressional Committee for the Control of Huntington's Disease and Its Consequences, of which she had been made executive director in 1976. (She was also a health science administrator at the National Institute of Neurological Diseases and Stroke, part of the National Institutes of Health in Bethesda, Maryland.)

To determine inheritance patterns, the researchers needed a large family in which some members had Huntington's disease. Luckily, Wexler, investigating another aspect of the disease, had learned of such a family in fishing villages on the shore of Lake Maracaibo in Venezuela and had visited them earlier in 1979. In 1981, she and an international research team made the first of what have become yearly trips to collect tissue and blood samples from this family for testing. The Venezuelans cooperated when they learned that Wexler's own family had the disease and she, too, had given samples. In return for the family's help, Wexler's team gives them medical and social aid (the family is very poor), and Wexler personally provides what a Venezuelan team member has called "immeasurable love."

At first, the RFLP project had unusually good luck. James Gusella of Massachusetts General Hospital (part of Harvard University), who was put in charge of the testing, found a RFLP inherited with the Huntington's gene in 1983. It was only the 12th RFLP he had tried. In addition to giving a great boost to the gene hunt, this identification made possible a test that would tell with 96 percent accuracy whether someone would develop Huntington's disease at some point in life.

To improve the chances of locating the Huntington's gene, the Hereditary Disease Foundation, beginning in 1984, persuaded six laboratories in the United States and Britain to collaborate in their research. John Minna, a scientist in the group, told Wexler's sister, Alice, "The person that made everything work . . . was Nancy. . . . It was her acting as . . . glue and go-between, doing whatever was necessary, that was the real key." The group found the Huntington's gene in 1993, and researchers are trying to learn exactly how it damages the brain.

Nancy Wexler's other personal contribution has been tracing the ancestry of the Venezuelan family, which is the largest known family in the world with Huntington's. Their family tree now spans 10 generations, including more than 17,000 members. Wexler has also collected blood samples from more than 4,300 people in the family.

Wexler is Higgins Professor of Neuropsychology in the departments of neurology and psychiatry at Columbia University. She joined the Columbia faculty in 1984 and became a professor of psychology in 1992. She has won numerous awards for her work on behalf of Huntington's and other inherited diseases, including the

Albert Lasker Public Service Award (1993), the J. Allyn Taylor International Prize in Medicine (1994), and several honorary degrees.

Although Wexler's primary interest is still a cure for Huntington's disease, she also gives lectures on the implications of testing for inherited diseases. From 1989 to 1995, she was head of a committee that oversees research on the ethical, social, and legal issues raised by the Human Genome Project, which has worked out the "code" of the complete collection of human genes. Information from this project eventually may make testing possible for any inherited disease or genetic disorder. Such testing could yield medical benefits, but test results could also be used to deny people medical insurance or employment, Wexler points out. She is trying to keep such tragedies from happening.

Further Reading

"A Tale of Pain and Hope on Lake Maracaibo." *Business Week*, June 5, 2000.

Wexler, Alice. *Mapping Fate: A Memoir of Family, Risk, and Genetic Research*. New York: Random House/Times Books, 1995.

"Wexler, Nancy S." *Current Biography Yearbook 1994*. New York: H. W. Wilson, 1994.

Yount, Lisa. *Genetics and Genetic Engineering*. New York: Facts On File, 1997.

Wilkins, Maurice

(1916–)
New Zealander/British
Biophysicist

X-ray photographs of DNA from Maurice Hugh Frederick Wilkins's laboratory helped FRANCIS CRICK and JAMES WATSON work out the molecular structure of this essential biochemical. Wilkins was born on December 15, 1916, in Pongaroa, an isolated community in New Zealand. Both his father, Edgar Wilkins, a physician for the School Medical Service, and his mother, the former Evelyn Whittaker, were Irish immigrants. The family moved to Wellington when Maurice was a baby. He later described his early childhood there as "paradise."

Maurice was six when the Wilkins family moved again, to Birmingham in England, and he grew up there. He attended St. John's College, Cambridge University, as a physics major, graduating in 1938. He earned a Ph.D. from Birmingham University in 1940 and then worked on improving radar screens there as part of the war effort. In 1943, his research group moved to the University of California, Berkeley, to take part in the Manhattan Project, the secret program to develop the atomic bomb. His part of the project was separating uranium isotopes. He later regretted his involvement in the bomb work and became an outspoken foe of nuclear weapons.

Wilkins returned to Britain in 1945 and became a lecturer in physics at St. Andrews University in Edinburgh, Scotland, under John Randall, with whom he had worked in Birmingham. Randall had become interested in applying the techniques of physics to biology, and he persuaded Wilkins to share this interest. Randall's biophysics laboratory, including Wilkins, moved to King's College, London, in 1946, where it became the Medical Research Council Biophysics Research Unit. Wilkins was made assistant director of the facility in 1950 and deputy director in 1955.

Experiments in the United States had provided evidence that DNA was the carrier of inherited information, and in the late 1940s several groups of scientists, including Wilkins's laboratory, began trying to determine the structure of the DNA molecule because they realized that the structure would probably explain how the molecule could reproduce itself and pass on its information. Wilkins began by studying DNA from viruses with a visible-light microscope. After noticing that DNA gel could be stretched into a thin, spiderweblike strand, however, he decided to use instead a relatively new technique called X-ray crystallography, which can provide information about the three-dimensional structure of molecules. Scientists were just starting to use this technique on the complex molecules in living things, and Wilkins thought it would work well on the type of fiber he had seen DNA form.

Wilkins and a coworker, Raymond Gosling, began making X-ray photographs of DNA in 1950. They found that the DNA molecule appeared to have the shape of two intertwined corkscrews, a so-called double helix. Then, in 1951, Randall added British chemist ROSALIND ELSIE FRANKLIN, who had made a specialty of doing X-ray crystallography on biological molecules, to Wilkins's team. Unfortunately, Wilkins and Franklin started with a misunderstanding—Wilkins thought Franklin was supposed to be his assistant, whereas Franklin expected to be an independent member of the team—and they never got along.

Even though Wilkins and Watson, the latter of whom worked with Crick at Cambridge University, were rivals in the search for DNA's structure, they had become friends (indeed, Wilkins had been the one who introduced Watson to DNA research, following their meeting at a scientific conference in 1951), and they often met to talk. During one such meeting in January 1953, Wilkins showed Watson an X-ray photograph that Franklin had made of DNA. This photo gave Watson and Crick crucial information that led to their determination of the molecule's structure. When they won the Nobel Prize in physiology or medicine for that discovery in 1962, Wilkins was included to honor his laboratory's contribution. Franklin could not be considered because she had died in 1958, and Nobel Prizes are never awarded after death.

After Watson and Crick's announcement of DNA's structure in April 1953, Wilkins took further X-ray photographs that confirmed their proposal. He then turned from DNA to the cell's other nucleic acid, RNA. He made the first clear X-ray pictures of RNA molecules in 1962 and showed that they, too, had a double-helix structure. He was a professor of molecular biology at King's College from 1963 to 1970 and then head of its biophysics department until his retirement in 1981. He directed the Medical Research Council's Biophysics Unit from 1970 to 1972 and its Neurobiology Unit from 1974 to 1980. He was also president of a political group, the British Society for Social Responsibility in Science.

In addition to the Nobel Prize, Wilkins (along with Watson and Crick) won the Albert Lasker Medical Research Award from the American Public Health Association in 1960. He was elected a member of the Royal Society, Britain's top scientific group, in 1959 and made a Companion of the British Empire in 1962. He married Patricia Chidgey in 1959, and they have four children.

Further Reading

Judson, Horace Freeland. *The Eighth Day of Creation*. New York: Simon & Schuster, 1979.

"Maurice Wilkins, DNA Enabler." New Zealand Edge. Available online. URL: www.nzedge.com/heroes/wilkins.html. Updated 2002.

"Wilkins, Maurice H(ugh) F(rederick)." *Current Biography Yearbook 1963*. New York: H. W. Wilson, 1963.

Wilmut, Ian

(1944–)
British
Embryologist

Ian Wilmut, a quiet man who prefers walking in the Scottish hills to addressing legislators and reporters, found himself in the center of a media storm when he announced on February 22, 1997, that he and his laboratory at the Roslin Institute had successfully cloned a sheep from a mature adult cell. Wilmut was born in Hampton Lucy,

England, on July 7, 1944, and raised in nearby Coventry. His parents were both teachers. He originally planned to be a farmer, but while studying at the University of Nottingham he became interested in research. He earned a Ph.D. from Darwin College, Cambridge University, in 1971 with a dissertation on freezing boar semen.

After doing postdoctoral work at Cambridge, during which he took part in the production of Frosty, the first calf produced from a frozen embryo, Wilmut began working for the Animal Breeding Research Station, near Edinburgh, Scotland, in 1973. This facility later became the Roslin Institute. Wilmut began thinking about cloning in 1986, when he heard that Steen Willadson, a Danish embryologist with whom he had worked at Cambridge, had cloned calves from cells taken from embryos in late development. Scientists had cloned amphibians, but many researchers had doubted that a mammal could be cloned from such late embryonic stages.

Wilmut and his group produced cloned lambs of their own, called Megan and Morag, from embryonic sheep cells in early 1996. Like other researchers in the field, they used embryonic cells because these cells can mature into many different forms as they multiply. Attempts to make clones from adult cells had always failed. Biologists believed that once a cell differentiated—became a particular type, such as a blood cell or a muscle cell—it could not revert to a state in which it could produce cells of other types. It would therefore be unsuitable for cloning, in which a whole, genetically identical organism is made from the genetic information in a single cell of the "parent" organism.

In the mid-1990s, however, Keith Campbell, a coworker of Wilmut's at the Roslin Institute, found a way to turn back the clock of a mature, differentiated cell. He deprived cultured udder (mammary) cells from adult ewes of nutrients for five days, forcing the cells into a "sleeping" state in which many of their genes shut down. Wilmut and Campbell fused each quiescent mammary cell with a normal sheep egg cell from which the nucleus had been removed, and the cytoplasm in the egg cell somehow reprogrammed the adult cell's genes so that the combined cell could produce offspring cells able to differentiate into many different types. When the fused cell began to develop into an embryo, the scientists implanted it into the uterus of another ewe.

A lamb produced by this method (one of 277 attempts) was born from the mammary cell of a six-year-old Finn-Dorset ewe on July 5,

Ian Wilmut and his coworkers at the Roslin Institute in Scotland made headlines in 1997 when they announced the cloning of a sheep, "Dolly," from a mammary cell of an adult ewe; many commentators feared that cloning of human beings would soon follow, but Wilmut denies any interest in such work. *(Ian Wilmut)*

1996. Wilmut's group named her Dolly, after country-western singer Dolly Parton. They announced her arrival to the scientific community and the public in February 1997. Dolly died of a lung infection in February 2003.

The story of Dolly made headlines in the *New York Times* and other newspapers worldwide. To Wilmut and other biologists, the lamb's importance lay in the fact that she had been created from an adult cell, which showed that the process of cell differentiation could be reversed. This discovery has implications for the study of genetic diseases and the possible production of new tissues and organs to replace damaged ones. The media, public, and legislators of most countries, however, focused on the possibility that the successful cloning of Dolly might lead to the creation of a cloned human child, an idea that many people regarded with horror.

Wilmut has emphasized in many speeches and articles that neither he nor the Roslin Institute sees any reason to clone a human being. He stresses the high failure rate of cloning and says that subjecting humans to such a dangerous process would be immoral. He does, however, favor research with stem cells harvested from human embryos discarded by fertility clinics and then cloned. These cells can develop into many types and could be used in tissue transplants or other treatments. Such cloned embryos would not be allowed to develop beyond the size of a few cells.

From the beginning, Wilmut has said that the institute's aim is to produce clones of genetically altered animals that will be useful in medicine. By implanting selected human genes into sheep or cattle embryos, for instance, they hope to create animals that excrete medically important human proteins in their milk, and Wilmut's team has already had some success in this. Genetic modification of pigs might make their organs suitable for transplantation into humans. Genetic alteration might also benefit agriculture by encouraging greater production or better quality of milk, wool, and meat or creating farm animals that resist disease.

Wilmut, presently head of a department at the Roslin Institute and a scientific adviser to the biotechnology company Geron Bio-Med, continues to improve techniques for genetic alteration and cloning of farm animals by means of nuclear transfer. His work may also help in providing human cells for use in medical treatments. Wilmut is the first to agree that many questions about clones remain to be answered, and he discusses some of them in *The Second Creation: Dolly and the Age of Biological Control* (2000), a book he authored with Keith Campbell and science writer Colin Tudge.

Further Reading

Kolata, Gina. *Clone: The Road to Dolly, and the Path Ahead*. New York: William Morrow, 1999.

Wilmut, Ian, Keith Campbell, and Colin Tudge. *The Second Creation: Dolly and the Age of Biological Control*. New York: Farrar, Straus & Giroux, 2000.

"Wilmut, Ian." *Current Biography Yearbook 1997*. New York: H. W. Wilson, 1997.

Wilson, Edward O.

(1929–)
American
Entomologist, Evolutionary Biologist, Philospher of Science

Edward Osborne Wilson began by describing the society of ants and went on to describe the society of humans and its destructive effects on the environment, producing far-reaching conclusions at every step of his sometimes controversial career. He was born on June 10, 1929, in Birmingham, Alabama, to Edward and Inez (Freeman) Wilson. Because of his father's work as an accountant for the Rural Electrification Administration, the family moved often during Wilson's youth.

A *National Geographic* article that Wilson read when he was nine years old triggered his lifelong fascination with ants, and rural Alabama and Florida provided plenty of study material for what he later called his "childhood bug period." At age 13, for instance, in a vacant lot in Mobile, he discovered the first colony of imported fire ants (an invasive pest insect from Brazil and Argentina) in the United States. By his senior year in high school, he had decided to study insects professionally.

In 1949, the year Wilson obtained his B.S. in biology from the University of Alabama, he wrote the country's first thorough study of fire ants for the Alabama State Department of Conservation. He continued to study these insects while earning a master's degree from the University of Alabama in 1950 and doing additional graduate work at the University of Tennessee. Drawn to Harvard University by its magnificent ant collection, he completed his Ph.D. studies there in 1955 and also married Irene Kelley, a Boston native; they later had one daughter. His thesis, a taxonomic (classification) analysis of an ant genus, was the most detailed work of this kind on any social insect at the time. Wilson has continued to study and classify ants throughout his career and is considered the world's leading authority on these insects.

Edward O. Wilson, a world-renowned expert on ants (one of which is shown here in a much-enlarged model), created controversy by theorizing that human social behavior is partly determined by genes. *(Jon Chase/Harvard University News Office)*

Wilson did postdoctoral work at Harvard and then joined the university's faculty as an assistant professor of biology in 1956. An eager explorer, he traveled through the South Pacific and elsewhere during the 1950s to study ants in wild habitats. His work not only revised the classification of ants but contributed to the new synthesis of taxonomy and evolutionary theory spearheaded by other Harvard professors such as ERNST MAYR. In 1956, for instance, Wilson and coworker William Brown developed the concept of "character displacement," in which evolution increases genetic differences between closely related species when they come into contact. Wilson also expanded the idea of faunal dominance, proposed by other biologists, which claimed that some areas generate unusually large numbers of animal species that colonize and take over other landmasses. He showed that in the case of ants, tropical Asia is the center of faunal dominance for the region stretching from Asia to Australia and the Pacific Islands.

Wilson's studies of faunal dominance and the spread of species led him to formulate what he called the taxon cycle around 1960. In this cycle, some species spread by adapting themselves to marginal environments that encourage travel, such as riverbanks and shorelines. When these species arrive at new destinations, they move inland and split into multiple new species.

Over time, these species decline and new species move in, repeating the cycle.

At about this same time, Wilson discovered that ants communicate by means of chemicals called pheromones. Pheromones can signal sexual attraction, alarm, the presence of a food source, and other messages. Other scientists later found that many kinds of animals, possibly including humans, communicate with pheromones.

After becoming a professor of zoology at Harvard in 1964, Wilson combined his worldwide travel experience with the population biology expertise of ROBERT HELMER MACARTHUR of the University of Pennsylvania to produce a description of the turnover and balance (equilibrium) of species on islands. Among other things, Wilson and MacArthur found that larger islands, and those closer to mainlands, contain more species than smaller or more remote islands. Their 1967 book, *The Theory of Island Biogeography*, turned out to describe not only actual islands but any isolated ecosystem, such as a nature reserve or park surrounded by a "sea" of human settlement. It led conservationists to realize that such reserves must occupy large contiguous or interconnected areas if they are to preserve species successfully.

In the late 1960s, Wilson began to focus on the social organization of ants and the effects of evolution on particular features of ant societies, such as their caste system. He then extended his work to wasps, bees, and termites, which are also social insects. The *Library Journal* called his book on the subject, *The Insect Societies* (1971), "the most masterful synthesis of knowledge of the social insects to appear in the last half-century."

Wilson became the curator of entomology at Harvard's Museum of Comparative Zoology in 1973 and the Frank B. Baird Jr. Professor of Science in 1976. He was later made the Pellegrino University Professor, a member of the highest grade of faculty at Harvard. During the early 1970s, he extended his ideas about social insects to social relationships in other animal groups, founding a new discipline called sociobiology, which he defined as "the systematic study of the biological basis of all social behavior." He described this new discipline in his first controversial book, *Sociobiology: The New Synthesis*, published in 1975. Maintaining that social relationships are shaped by genes and natural selection just as physical characteristics are, Wilson used ideas from population biology and evolutionary theory to explain and predict social behaviors.

Sociobiology aroused debate chiefly because of its last chapter, which stated that human social behavior is shaped by the same genetic forces that control animal behavior. Some critics saw this idea as denying free will or supporting racism or sexism. (One young woman became so angry with Wilson that she poured a pitcher of water over his head.) Surprised and hurt by these attacks, Wilson insisted that his critics had misunderstood or exaggerated his statements. He said that he really saw only "maybe 10 percent of human behavior as genetic and 90 percent [as] environmental [that is, cultural]." Eventually, the controversy died down, and Wilson said in a 1998 interview that sociobiology, which today is often called evolutionary psychology, is "very respectable now."

Far from heeding the critics' implied command to confine his theorizing to animals, Wilson focused on humans throughout his next book, *On Human Nature* (1978). The book repeated and amplified his earlier conclusions, saying that human society is a result of interaction between culture and genetic or "epigenetic" tendencies, among which he includes the drive toward religion. Epigenetic tendencies are genetically determined features of the brain's wiring that make people or animals likely, though not guaranteed, to think or behave in a certain way. In *Consilience: The Unity of Knowledge*, a 1998 book, Wilson continued to reduce social sciences to biology, and biology, in turn, ultimately to physics. These ideas also proved controversial.

Wilson turned to a different subject in *The Diversity of Life* (1992), in which he described the extent of biological diversity (biodiversity for short) on Earth and stressed the importance of that diversity for maintaining the health of ecosystems and keeping the physical environment hospitable to life. He claimed that land clearing and other activities resulting from the rise in human population are producing the worst mass extinction since the one that wiped out the dinosaurs 65 million years ago. If this environmental destruction is not reduced, he wrote, half of the world's species could be extinct by the end of the 21st century. In his most recent book, *The Future of Life* (2002), he proposes to stop the losses while still providing a reasonable standard of living by creating ways in which local populations can profit from preserving rather than destroying their natural environment.

Wilson has received many awards for his work, including the Crafoord Prize, which the Royal Swedish Academy of Science awards to scientists in fields not covered by Nobel Prizes, and the National Medal of Science (1977). He also won the Audubon Society's Audubon Medal for service to conservation in 1995 and the Kistler Prize from the Foundation for the Future in 2000. His writing has been honored as well; both *On Human Nature* (1978) and *The Ants* (1990), the latter of which he wrote with Bert Holldobler, won Pulitzer Prizes.

Wilson is now retired from his teaching duties at Harvard, but he continues to speak and write on loss of biodiversity and other topics. In addition to his other books, he has written an autobiography, *Naturalist* (1994). In 2000, *New York Times* reporter Nicholas Wade called him "perhaps the best-known biologist of his generation."

Further Reading

Wilson, Edward O. *Naturalist*. Washington, D.C.: Island Press, 1994.

———. "The Writing Life." *Washington Post*, June 25, 2000.

"Wilson, Edward O(sborne)." *Current Biography Yearbook 1979*. New York: H. W. Wilson, 1979.

Woese, Carl R.

(1928–)
American
Microbiologist, Molecular Biologist

Carl R. Woese's reclassification of microorganisms led to a complete revision of biologists' understanding of the tree of life. He was born on July 15, 1928, in Syracuse, New York. His father was a consulting engineer and his mother a homemaker.

Woese did not intend to be a biologist at first but rather majored in mathematics and physics at Amherst College in Massachusetts, graduating with a B.A. in 1950. During graduate studies at Yale University, however, he became interested in the physics of living cells, and he earned a Ph.D. in biophysics in 1953. After doing postdoctoral work at Yale, General Electric Research Laboratory, and the Pasteur Institute in France, he came to the Urbana-Champaign campus of the University of Illinois in 1964, and he has spent his entire career there. In 1989, he was appointed to the University of Illinois Center for Advanced Study, the university's highest faculty recognition, and he became holder of the Stanley O. Ikenberry Endowed Chair in 1996.

"The thing that caught me more than anything [else in biology] was evolution," Woese recalls. During his postdoctoral research, he became interested in the evolutionary origin of the genetic code, a puzzle that he realized could be solved only with the aid of a chart of evolutionary relationships, or phylogeny, that covered all organisms. He found that no such chart existed. Biologists generally divided living things into two domains, prokaryotes (cells without nuclei, specifically bacteria) and eukaryotes (cells with nuclei, which included all other living things). A great deal was known about evolutionary relationships among eukaryotes, but biologists had learned almost nothing about relationships among prokaryotes. To Woese, "it was as if you went to a zoo and had no way of telling the lions from the elephants from the orangutans—or any of these from the trees."

Unlike most biologists of the time, Woese felt that classification of bacteria and understanding of their evolutionary development were both possible and important. Around 1966, therefore, while most molecular biologists were studying DNA or proteins, he began working with the RNA that makes up the ribosomes, the cell's protein-manufacturing organelles. Ribosomal RNA is much more similar from one organism to another than DNA is, which suggests that it arose earlier in evolution. It is also easy to extract from cells in relatively large quantities. Woese planned to work out evolutionary relationships among bacteria by comparing the sequence of bases in their ribosomal RNA.

During the next decade, Woese used a painstaking technique developed by FREDERICK SANGER to analyze ribosomal RNA from about 60 kinds of bacteria. The result was hundreds of sheets of film displaying patterns of blurry dots, which he clipped to individual light boxes or to his "luminescent wall," a giant sheet of plastic with lights behind it, and compared for days on end. He was practically the only person who could read the films—or wanted to. Most other microbiologists thought his technique would not work or at least would not produce any useful information. His only support from outside the university was a small grant from the National Aeronautics and Space Administration (NASA), which thought his research might shed light on possible extraterrestrial life-forms.

In spite of colleagues' skepticism, Woese began to fill in the blanks on his bacterial chart. Then, in 1976, on the advice of Ralph Wolfe, a close friend and fellow professor at the university, he focused on methanogens, unusual microorganisms that excrete a gas called methane and live in inhospitable habitats such as hot springs. To his amazement, Woese found that methanogens' ribosomal RNA lacked sequences that he had found in all the bacteria he had studied. "These things aren't bacteria," he told Wolfe.

After further work confirmed this startling conclusion, Woese announced in the November 3, 1977, *Proceedings of the National Academy of Sciences* that the tree of life should have three main branches rather than two. He claimed that methanogens and other microorganisms that he now calls archaea, meaning "ancient ones," comprise a third domain, separate from prokaryotes and eukaryotes and equal to them in importance. (At first, he called the members of this new domain archaebacteria, but he later dropped the "bacteria.")

Most American microbiologists, as well as prominent evolutionary biologists such as ERNST MAYR of Harvard University, greeted Woese's revolutionary claim with disbelief. In Germany, on the other hand, his work gained the support of well-known microbiologist Otto Kandler, and some German scientists began trying to verify it. Woese and others accumulated more evidence in favor of Woese's theory during the 1980s, and most microbiology textbooks began showing the archaea as separate from bacteria and eukaryotes. Botanists and zoologists, however, have been slow to accept the new classification scheme.

For many microbiologists, the most convincing support for Woese's claim came from the sequencing of the genome of an archaean microorganism by Woese's group and The Institute for Genome Research, then headed by controversial gene sequencing entrepreneur J. CRAIG VENTER, in 1996. The archaean genome showed substantial differences from those of bacteria and similarities to those of eukaryotes. More than half of its genes were unlike any seen in any other organism. "It's like something out of science fiction," Venter said.

Other scientists have refined Woese's techniques for analyzing ribosomal RNA and expanded them to include other molecules "conserved" relatively unchanged during evolution, and they and Woese have gone far toward establishing a phylogeny of archaea and bacteria by comparing these molecules in different microorganisms. Their findings have shed light on the beginnings of evolution, for instance supporting the theory, propounded by University of Massachusetts, Amherst, microbiologist LYNN ALEXANDER MARGULIS and others, that cell organelles such as mitochondria and chloroplasts were once free-living microorganisms that came to live symbiotically inside other cells. In 1989, Wolfe called Woese's contributions to the understanding of early evolution "among the most significant since Darwin."

Woese said in 2000 that "the central task of biology in the new century will be to lay out and elaborate this overarching framework of relationships among living organisms," especially microorganisms, which he calls "the underpinnings of everything." Such knowledge has practical implications, he emphasizes: "We have to understand how the biosphere works at the microbial level if we're going to be able to cope with man's stressing of it." His work also underlines the radical and humbling philosophical insight that, as *Science* reporter Virginia Morell put it in 1997, "most life is one-celled, and all Eukarya are but a twig on what amounts to a great microbial tree of life."

As ridicule faded into acceptance, Woese began to be honored for his work. He received a "genius" grant from the John D. and Catherine T. MacArthur Foundation in 1984, the Leeuwenhoek Medal of the Dutch Royal Academy of Science (microbiology's top honor, awarded only once a decade) in 1992, the National Medal of Science in 2000, and the Royal Swedish Academy of Sciences' Crafoord Prize in 2003. Today, with computers replacing films and light boxes, he continues to investigate the archaea in particular and microbial evolution and diversity in general. He also studies the structure and evolution of ribosomal RNAs and explores the idea that, as he put it in 1989, "processes (evolution, development, mind) somehow underlie genes, cells, brains, etc., not the reverse."

Further Reading

Anton, Ted. *Bold Science: Seven Scientists Who Are Changing Our World*. New York: W. H. Freeman, 2000.

Morell, Virginia. "Microbiology's Scarred Revolutionary." *Science*, May 2, 1997.

Woese, Carl R. "Archaebacteria." *Scientific American*, June 1981.

Yalow, Rosalyn Sussman
(1921–)
American
Biophysicist

Rosalyn Yalow and her research partner, Solomon Berson, invented a technique called radioimmunoassay that measures substances in body fluids so accurately that reporters have said it could detect a lump of sugar dropped into Lake Erie. For this advance, Yalow won a share of the Nobel Prize in physiology or medicine in 1977.

Yalow was born Rosalyn Sussman on July 19, 1921, in the South Bronx area of New York City. Her parents, Simon and Clara (Zipper) Sussman, had grown up in the city's immigrant community. Simon Sussman owned a small paper and twine business, which made just enough money for his family to live on. Nonetheless, the Sussmans planned to make sure that their two children somehow obtained a college education.

By the time Rosalyn was eight, she had decided that she was going to be a "big deal" scientist—and marry and have a family as well. As a young woman, she attended Hunter College (now part of the City University of New York), which charged no tuition to New York City residents. Her first scientific choices had been mathematics and chemistry, but at Hunter she turned to physics because, she wrote in her Nobel Foundation autobiography, "in the late thirties . . . nuclear physics was the most exciting field in the world." She graduated with high honors in January 1941.

As a Jewish woman with little money, Rosalyn Sussman had three strikes against her in trying to enter a graduate or medical school. At first, she thought her only hope was to do secretarial work for a professor at Columbia University Medical School, which would allow her to take classes there for free. Impending war drained universities of men and created new openings for women, however, and Sussman obtained a teaching assistantship in physics at the University of Illinois, Urbana-Champaign. (The engineering dean told her that she was the first woman admitted to the department's faculty since 1917.) On her first day of classes in fall 1941, she met another Jewish New Yorker, a rabbi's son from Syracuse named Aaron Yalow. They married in 1943 and later had two children.

After Rosalyn Yalow obtained her Ph.D. in 1945—only the second woman ever to earn a physics doctorate from Illinois—she returned to New York (her husband joined her shortly afterward) and became the first woman assistant engineer in International Telephone and Telegraph's Federal Telecommunications Laboratory. When the laboratory moved away a year later, she began teaching physics at Hunter. Hunter had no research facilities, however, and she wanted to do research. Aaron, who had entered the new field of medical physics, suggested that Rosalyn do so as well. Research in medical physics focused on radioactive forms of certain elements, or radioisotopes, and she was already an expert in working with radioactive substances.

Rosalyn Yalow consulted Edith Quimby, a pioneer researcher in medical physics at Columbia, and Quimby in turn introduced Yalow to her chief, Gioacchino Failla. On Failla's recommendation, the Bronx Veterans Administration (VA) Hospital hired Yalow as a part-time consultant in December 1947. Her laboratory, one of the first radioisotope laboratories in the United States, began in what had been a janitor's closet, and she had to design and build most of her own equipment.

Yalow stopped teaching at Hunter in January 1950 and joined the hospital full time. A few months later, she found her ideal professional partner in a young physician named Solomon Berson. One coworker told science writer Sharon McGrayne that Yalow and Berson

In the late 1950s, biophysicist Rosalyn Sussman Yalow, shown here, and Solomon Berson coinvented the radioimmunoassay, an extremely sensitive method for detecting biochemicals in body fluids. *(National Library of Medicine)*

had "a kind of eerie extrasensory perception. Each knew what the other was thinking. . . . Each had complete trust and confidence in the other." Their collaboration lasted 22 years.

The VA thought of radioisotopes mainly as a cheaper substitute for radium in the treatment of cancer, but Yalow and Berson learned that these substances could also be attached to molecules and used to track chemicals through reactions in the body or in test tubes. In one of their first studies, published in 1956, they used radioisotope tagging to show that the immune systems of diabetics, who must take daily injections of the hormone insulin to make up for their body's lack of it, formed antibodies in response to the insulin they took, which came from cows or pigs and thus was slightly different from human insulin. The antibodies kept the insulin from being removed from the blood as fast as it was in normal people, who lacked such antibodies.

This finding was startling enough—researchers had believed that insulin molecules were too small to produce an immune response—but, even more importantly, Berson and Yalow realized that they could turn their discovery on its head to create a very sensitive way of measuring insulin or almost any other biological substance in body fluids. They injected the substance they wanted to test for into laboratory animals, making the animals produce antibodies to it. They then mixed a known amount of these antibodies with a known amount of the substance to which radioactive atoms had been added and a sample of the fluid to be tested.

Antibodies attach to molecules of the substance that caused their formation. The nonradioactive substance in the sample attached to some of the antibodies, keeping the radioactive substance from doing so. After a certain amount of time, Yalow and Berson measured the amount of radioactive material that was not attached to the antibodies. The more substance had been in the sample, the more radioactive material would be left over. This test, called the radioimmunoassay, can detect as little as a billionth of a gram of material. In 1978, *Current Biography Yearbook* termed it "one of the most important postwar applications of basic research to clinical medicine."

Berson and Yalow first described the radioimmunoassay in 1959. They spent the 1960s perfecting the test and persuading researchers to use it, a difficult task at first. Scientists eventually applied their technique to make a host of discoveries about the way both the immune system and biochemicals such as hormones function in health and disease. Radioimmunoassays have also revealed illegal drugs, helped doctors work out the best doses of medicines, and detected dangerous viruses in donated blood.

Yalow and Berson worked together less often after 1968, when Berson became chairman of the department of medicine at Mount Sinai School of Medicine and Yalow became acting chief of the Bronx hospital's radioisotope service. Still, they remained close until Berson died suddenly of a heart attack in 1972, at age 54. His death devastated Yalow both personally and professionally. She found she had to prove her worth all over again as a solo researcher, which she did by making discoveries about a variety of hormones. In the early 1970s, when her hospital became affiliated with the the Mount Sinai School of Medicine, she became a Distinguished Service Professor at the medical school. She also headed the hospital's nuclear medicine service from 1970 to 1980.

Yalow accumulated many honors for her work, including the American Medical Association's Scientific Achievement Award and election to the National Academy of Sciences in 1975. In 1976, she became the first woman to win the Albert Lasker Medical Research Award, often considered a prelude to the medical Nobel Prize. A year later, she won the Nobel itself, sharing it with two researchers who had made discoveries about hormones in the brain. This was the first time that the surviving member of a research partnership had been honored for work done by both. Yalow was only the second woman (after GERTY THERESA RADNITZ CORI) to win a Nobel Prize in physiology or medicine and was the first American-born woman to win any science Nobel. In 1988, she also won the National Medal of Science, the highest science award in the United States.

After Yalow's retirement from the Bronx hospital in 1991, she spent much of her time giving lectures on such subjects as nuclear power, which she feels is unjustly feared; the need for better science education in the United States; and the need for more women scientists. As she said in her Nobel Prize acceptance speech, "The world cannot afford the loss of the talents of half its people."

Further Reading

McGrayne, Sharon Bertsch. *Nobel Prize Women in Science: Their Lives, Struggles, and Momentous Discoveries*. New York: Birch Lane Press, 1993.

Pizzi, Richard A. "Rosalyn Yalow: Assaying the Unknown." *Modern Drug Discovery*, September 2001. Available online. URL: http://pubs.acs.org/subscribe/journals/mdd/v04/i09/html/09timeline.html. Posted 2001.

Straus, Eugene. *Rosalyn Yalow, Nobel Laureate: Her Life and Work in Medicine*. Reprint, Cambridge, Mass.: Perseus Books, 1998.

Yalow, Rosalyn S. "Rosalyn Yalow—Autobiography." *Les Prix Nobel 1977*. Nobel Foundation. Available online. URL: www.nobel.se/medicine/laureates/1977/yalow-autobio.html. Last updated 2001.

Entries by Country of Birth

Argentina
Milstein, César

Australia
Burnet, Frank Macfarlane
Florey, Howard Walter

Austria
Frisch, Karl von
Landsteiner, Karl
Lorenz, Konrad

Austria-Hungary
Cori, Carl Ferdinand
Cori, Gerty Theresa Radnitz
Mendel, Gregor
Purkinje, Jan Evangelista
Semmelweis, Ignaz Phillipp
Szent-Györgyi, Albert

Canada
Avery, Oswald Theodore
Banting, Frederick Grant
Galdikas, Biruté
MacArthur, Robert Helmer

China
Li, Choh Hao
Tsui, Lap-Chee

Estonia
Baer, Karl Ernst von

Flanders (later Belgium)
Vesalius, Andreas

France
Bernard, Claude
Boussingault, Jean-Baptiste
Buffon, Georges-Louis, comte de
Carrel, Alexis
Fabre, Jean-Henri
Lamarck, Jean-Baptiste, chevalier de
Lavoisier, Antoine-Laurent
Monod, Jacques
Montagnier, Luc
Pasteur, Louis

Germany
Chain, Ernst Boris
Delbrück, Max
Domagk, Gerhard
Funk, Casimir
Katz, Bernhard
Krebs, Hans Adolf
Mayr, Ernst
Meyerhof, Otto Fritz
Nüsslein-Volhard, Christiane
Schleiden, Matthias Jakob
Schwann, Theodor

Hanover
Koch, Robert

Hesse
Ludwig, Karl Friedrich Wilhelm

Prussia
Behring, Emil von
Ehrlich, Paul
Helmholtz, Hermann von
Humboldt, Alexander von
Röntgen, Wilhelm Conrad
Virchow, Rudolf

Württemburg
Cuvier, Georges, Baron

Great Britain
Adrian, Edgar Douglas
Bateson, William
Black, James Whyte
Crick, Francis
Dale, Henry Hallett
Darwin, Charles Robert
Doll, Richard
Edwards, Robert
Fisher, Ronald Aylmer

Fleming, Alexander
Franklin, Rosalind Elsie
Goodall, Jane
Haldane, J. B. S.
Hales, Stephen
Harvey, William
Hodgkin, Alan Lloyd
Hodgkin, Dorothy Crowfoot
Hooke, Robert
Jenner, Edward
Leakey, Mary
Lind, James
Lister, Joseph
Lovelock, James
Lyell, Charles
Malthus, Thomas Robert
Medawar, Peter Brian
Mitchell, Peter Dennis
Ray, John
Sanger, Frederick
Sherrington, Charles Scott
Snow, John
Starling, Ernest Henry
Wallace, Alfred Russel
Wilmut, Ian

GREECE

Aristotle
Galen
Hippocrates
Theophrastus

HOLLAND/NETHERLANDS

Beijerinck, Martinus Willem
De Vries, Hugo
Kolff, Willem Johan
Leeuwenhoek, Antoni van
Swammerdam, Jan
Tinbergen, Niko

INDIA

Khorana, Har Gobind
Ross, Ronald

ITALY

Golgi, Camillo
Levi-Montalcini, Rita
Luria, Salvador
Malpighi, Marcello
Morgagni, Giovanni Battista
Spallanzani, Lazzaro

Papal States

Galvani, Luigi

JAPAN

Kimura, Motoo
Kitasato, Shibasaburo
Ohta, Tomoko
Takabe, Tetsuko
Takamine, Jokichi
Tonegawa, Susumu

KENYA

Dawkins, Richard
Leakey, Louis S. B.
Leakey, Richard
Wambugu, Florence

NEW ZEALAND

Wilkins, Maurice

RUSSIA/SOVIET UNION

Mechnikov, Ilya Ilyich
Pavlov, Ivan Petrovich
Sabin, Albert Bruce
Waksman, Selman A.

SPAIN

Ochoa, Severo
Ramón y Cajal, Santiago

SWEDEN

Linnaeus, Carolus

SWITZERLAND

Haller, Albrecht von
Hess, Walter Rudolf
Paracelsus

TAIWAN

Ho, David

UNITED STATES

Anderson, W. French
Baltimore, David
Beadle, George Wells
Beaumont, William
Berg, Paul
Bishop, J. Michael
Boyer, Herbert Wayne
Burkholder, JoAnn Marie
Calvin, Melvin
Carson, Rachel Louise
Cohen, Stanley
Cohen, Stanley N.
Colborn, Theo E.
Cushing, Harvey Williams
Earle, Sylvia Alice
Elion, Gertrude Belle
Enders, John Franklin
Folkman, Moses Judah
Fossey, Dian
Gallo, Robert
Gilbert, Walter
Gould, Stephen Jay
Hershey, Alfred Day
Hitchings, George Herbert
Hood, Leroy
Horner, John R.
Johanson, Donald C.
King, Mary-Claire
Kornberg, Arthur
Lederberg, Joshua
Margulis, Lynn Alexander
McClintock, Barbara
Miller, Stanley Lloyd
Morgan, Thomas Hunt
Morton, William Thomas Green

Entries by Country of Major Scientific Activity

Prusiner, Stanley B.
Rosenberg, Steven A.
Rous, Peyton
Sabin, Albert Bruce
Salk, Jonas
Simpson, George Gaylord
Skinner, B. F.
Sperry, Roger Wolcott
Stevens, Nettie Maria
Sutherland, Earl Wilbur, Jr.
Takamine, Jokichi
Temin, Howard Martin
Varmus, Harold E.
Venter, J. Craig
Waksman, Selman A.
Watson, James Dewey
Weinberg, Robert A(llan)
Wexler, Nancy Sabin
Wilson, Edward O.
Woese, Carl R.
Yalow, Rosalyn Sussman

Entries by Year of Birth

500–451 B.C.
Hippocrates

400 B.C.–351 B.C.
Aristotle
Theophrastus

100–150
Galen

1450–1499
Paracelsus

1500–1549
Vesalius, Andreas

1550–1599
Harvey, William

1600–1649
Hooke, Robert
Leeuwenhoek, Antoni van
Malpighi, Marcello
Ray, John
Swammerdam, Jan

1650–1699
Hales, Stephen
Morgagni, Giovanni Battista

1700–1709
Buffon, Georges-Louis, comte de
Haller, Albrecht von
Linnaeus, Carolus

1710–1719
Lind, James

1720–1729
Spallanzani, Lazzaro

1730–1739
Galvani, Luigi

1740–1749
Jenner, Edward
Lamarck, Jean-Baptiste, chevalier de
Lavoisier, Antoine-Laurent

1760–1769
Cuvier, Georges, Baron
Humboldt, Alexander von
Malthus, Thomas Robert

1780–1789
Beaumont, William
Purkinje, Jan Evangelista

1790–1799
Baer, Karl Ernst von
Lyell, Charles

1800–1809
Boussingault, Jean-Baptiste
Darwin, Charles Robert
Schleiden, Matthias Jakob

1810–1819
Bernard, Claude
Ludwig, Karl Friedrich Wilhelm
Morton, William Thomas Green
Schwann, Theodor
Semmelweis, Ignaz Phillipp
Snow, John

1820–1829
Fabre, Jean-Henri
Helmholtz, Hermann von
Lister, Joseph
Mendel, Gregor
Pasteur, Louis
Virchow, Rudolf
Wallace, Alfred Russel

1840–1849
De Vries, Hugo
Golgi, Camillo
Koch, Robert
Mechnikov, Ilya Ilyich
Pavlov, Ivan Petrovich
Röntgen, Wilhelm Conrad

1850–1859
Behring, Emil von
Beijerinck, Martinus Willem
Ehrlich, Paul
Kitasato, Shibasaburo
Ramón y Cajal, Santiago
Ross, Ronald
Sherrington, Charles Scott
Takamine, Jokichi

1860–1869
Bateson, William
Cushing, Harvey Williams
Landsteiner, Karl
Morgan, Thomas Hunt
Starling, Ernest Henry
Stevens, Nettie Maria

1870–1879
Avery, Oswald Theodore
Carrel, Alexis
Dale, Henry Hallett
Rous, Peyton

1880–1889
Adrian, Edgar Douglas
Fleming, Alexander
Frisch, Karl von
Funk, Casimir
Hess, Walter Rudolf
Meyerhof, Otto Fritz
Waksman, Selman A.

1890–1899
Banting, Frederick Grant
Burnet, Frank Macfarlane
Cori, Carl Ferdinand
Cori, Gerty Theresa Radnitz
Domagk, Gerhard
Enders, John Franklin
Fisher, Ronald Aylmer
Florey, Howard Walter
Haldane, J. B. S.
Müller, Hermann Joseph
Szent-Györgyi, Albert

1900–1909
Beadle, George Wells
Carson, Rachel Louise
Chain, Ernst Boris
Delbrück, Max
Hershey, Alfred Day
Hitchings, George Herbert
Krebs, Hans Adolf
Leakey, Louis S. B.
Levi-Montalcini, Rita
Lorenz, Konrad
Mayr, Ernst
McClintock, Barbara
Ochoa, Severo
Patrick, Ruth
Pauling, Linus Carl
Pincus, Gregory Goodwin
Sabin, Albert Bruce
Simpson, George Gaylord
Skinner, B. F.
Tinbergen, Niko

1910–1919
Calvin, Melvin
Crick, Francis
Doll, Richard
Elion, Gertrude Belle
Hodgkin, Alan Lloyd
Hodgkin, Dorothy Crowfoot
Katz, Bernhard
Kolff, Willem Johan
Kornberg, Arthur
Leakey, Mary Douglas Nicol
Li, Choh Hao
Lovelock, James
Luria, Salvador
Medawar, Peter Brian
Monod, Jacques
Salk, Jonas
Sanger, Frederick
Sperry, Roger Wolcott
Sutherland, Earl Wilbur, Jr.
Wilkins, Maurice

1920–1929
Berg, Paul
Black, James Whyte
Cohen, Stanley
Colborn, Theo E.
Edwards, Robert
Franklin, Rosalind Elsie
Khorana, Har Gobind
Kimura, Motoo
Lederberg, Joshua
Milstein, César
Mitchell, Peter Dennis
Nirenberg, Marshall W.
Watson, James Dewey
Wilson, Edward O.
Woese, Carl R.
Yalow, Rosalyn Sussman

1930–1939
Anderson, W. French
Baltimore, David
Bishop, J. Michael
Boyer, Herbert Wayne
Cohen, Stanley N.
Earle, Sylvia Alice
Folkman, Moses Judah
Fossey, Dian
Gallo, Robert
Gilbert, Walter
Goodall, Jane
Hood, Leroy
MacArthur, Robert Helmer
Margulis, Lynn Alexander
Miller, Stanley Lloyd
Montagnier, Luc
Ohta, Tomoko
Temin, Howard Martin
Tonegawa, Susumu
Varmus, Harold E.

1940–1949

Dawkins, Richard
Galdikas, Biruté
Gould, Stephen Jay
Horner, John R.
Johanson, Donald C.
King, Mary-Claire
Leakey, Richard
Mullis, Kary B.
Nüsslein-Volhard, Christiane
Prusiner, Stanley B.
Rosenberg, Steven A.
Takabe, Tetsuko
Venter, J. Craig
Weinberg, Robert A.
Wexler, Nancy Sabin
Wilmut, Ian

1950–1959

Burkholder, JoAnn Marie
Ho, David
Tsui, Lap-Chee
Wambugu, Florence

CHRONOLOGY

ca. 400 B.C.	Hippocrates stresses natural causes of illness and close observation by physicians.
ca. 335–323 B.C.	Aristotle describes and classifies more than 500 animals.
320s B.C.	Theophrastus describes and classifies more than 500 types of plants.
A.D. 162	Galen moves to Rome and begins writing books on medical subjects.
1520s	Paracelsus introduces new drugs and chemical concepts into medicine.
1543	Andreas Vesalius issues first detailed, accurate book on human anatomy.
1628	William Harvey describes circulation of blood.
1658	Jan Swammerdam sees red cells in blood of frog.
1661	Marcello Malpighi observes capillaries.
1665	Robert Hooke names and describes cells.
1670s	Jan Swammerdam works out life cycles of insects and shows that they have organs.
1674	Antoni van Leeuwenhoek sees first microorganisms.
1676	Antoni van Leeuwenhoek sees bacteria.
1677	Antoni van Leeuwenhoek sees spermatazoa.
1682	John Ray improves classification of plants.
1693	John Ray improves classification of animals.
1727	Stephen Hales uses physics to study plants.
early 1730s	Stephen Hales measures animal blood pressure.
1747	James Lind shows that citrus fruits cure scurvy.
1749	First volume of Georges-Louis Buffon's *Natural History* published.
1757	Albrecht von Haller begins publishing encyclopedia of human physiology.
1758	Carolus Linnaeus describes two-name system of classifying plants and animals.
1760s	Lazzaro Spallanzani shows that gastric juice dissolves food; artificially inseminates animals; shows that microorganisms cannot be created from nonliving matter.
1761	Giovanni Battista Morgagni ties disease to damage in specific organs.
1790	Antoine-Laurent Lavoisier measures gases involved in human respiration.
1791	Luigi Galvani shows that electricity makes muscles from dead frogs contract.
1796	Edward Jenner tests vaccination against smallpox.
1798	Thomas Robert Malthus publishes *Essay on the Principle of Population*.
1805	Alexander von Humboldt begins publishing account of South American voyage.
1809	Jean-Baptiste Lamarck describes theory of evolution based on use and disuse of body parts.
1812	Georges Cuvier establishes paleontology.

1815 Jean-Baptiste Lamarck improves classification of invertebrates.

1825 William Beaumont begins digestion experiments on man with opening in his stomach.

1826 Karl Ernst von Baer finds eggs in ovary of a mammal.

1830 Charles Lyell says changes in Earth are slow and gradual.

1830s Karl Ernst von Baer shows that embryos begin with little form and develop generalized features before specialized ones.

1836 Theodor Schwann extracts first animal enzyme.

1838 Matthias Jakob Schleiden suggests that cells are basic unit of structure in plants.

1839 Jan Evangelista Purkinje identifies fibers in heart that coordinate heartbeat.

Theodor Schwann extends Schleiden's cell theory to animals.

1846 Karl Friedrich Wilhelm Ludwig invents the kymograph.

William Thomas Green Morton popularizes ether anesthesia.

1847 Ignaz Phillipp Semmelweis reduces deaths from puerperal fever by insisting that physicians wash hands in disinfectant solution.

1850s Claude Bernard shows that birds and mammals can control body temperature and internal environment.

Jean-Baptiste Boussingault discovers nitrogen cycle.

early 1850s Hermann von Helmholtz measures speed of nerve transmission.

1851 Hermann von Helmholtz invents ophthalmoscope.

1854 John Snow shows that cholera is spread by contaminated water.

1856 Hermann von Helmholtz begins publishing book that describes human vision.

1858 Rudolf Virchow stresses that disease is caused by abnormal functions of cells.

Alfred Russel Wallace develops theory of evolution by natural selection independently of Charles Robert Darwin.

1859 Charles Robert Darwin publishes *On the Origin of Species by Means of Natural Selection*.

1860s Louis Pasteur proves that fermentation is carried out by microorganisms.

1865 Karl Friedrich Wilhelm Ludwig invents perfusion.

1866 Gregor Mendel describes laws of heredity.

Louis Pasteur invents pasteurization.

1867 Joseph Lister describes antiseptic surgery.

1870s Louis Pasteur propounds germ theory of disease and develops techniques to create vaccines.

1873 Camillo Golgi describes stain that shows nerve cells clearly.

1876 Robert Koch proves that certain bacteria cause anthrax.

Alfred Russel Wallace describes biogeography of Malay Archipelago.

1879 Jean-Henri Fabre publishes first volume of *Entomological Memories*.

1880s Charles Scott Sherrington shows that nerve cells are separate, yet communicate as network.

1882 Robert Koch identifies bacteria that cause tuberculosis.

Ilya Ilyich Mechnikov discovers phagocytes.

1884 Robert Koch identifies bacteria that cause cholera.

1885 Louis Pasteur successfully tests rabies vaccine on injured boy.

1886 Rudolf Virchow shows that "pure" German race is myth.

1890s Charles Scott Sherrington shows that muscles contain sensory nerve endings.

1891 Emil von Behring demonstrates diphtheria antitoxin.

1895 Wilhelm Conrad Röntgen discovers X rays.

1897 Ivan Petrovich Pavlov shows that nervous system controls digestion.

1898 Martinus Willem Beijerinck describes first virus.

Ronald Ross and Giovanni Battista Grassi independently prove that malaria is transmitted by mosquitoes.

1900	Hugo De Vries and two German scientists independently rediscover Mendel's paper.
	Karl Landsteiner discovers blood types.
early 1900s	Paul Ehrlich describes how antisera affect immune system.
1901	Hugo De Vries describes mutations.
	Jokichi Takamine isolates adrenalin.
1902	Alexis Carrel improves techniques for reconnecting blood vessels in surgery.
1904	Shibasaburo Kitasato and Alexandre Yersin independently identify bacteria that cause bubonic plague.
	Santiago Ramón y Cajal shows that neurons do not touch.
1905	Ernest Henry Starling and William Bayliss establish concept of hormones.
	Nettie Maria Stevens and Edmund B. Wilson independently show that gender is determined by Y chromosome.
1906	William Bateson suggests calling the new science of biological inheritance *genetics*.
1906–1910	Alexis Carrel and Charles Guthrie transplant organs in dogs.
1909	Paul Ehrlich creates first drug that kills specific disease-causing microorganism inside body.
1910	Thomas Hunt Morgan shows that eye color gene in fruit flies is on X chromosome.
1911	Thomas Hunt Morgan's laboratory creates first chromosome maps.
1912	Casimir Funk describes vitamins.
	Peyton Rous suggests that a virus causes a chicken cancer.
19-teens	Ivan Petrovich Pavlov develops concept of conditioned reflex.
1914	Henry Hallett Dale isolates first neurotransmitter.
1915	Harvey Williams Cushing reduces brain surgery mortality from 90 percent to 8 percent.
	Thomas Hunt Morgan and coworkers tie breeding experiments to activities in cells.
1919	Karl von Frisch shows that bees communicate by "dances."
	Ernest Henry Starling shows that the more the heart fills during relaxation, the more strongly it contracts when pumping blood.
early 1920s	Henry Hallett Dale and Otto Loewi prove that nerves use chemicals to send signals.
1921	Frederick Grant Banting and Charles Best isolate insulin and show that it controls diabetes.
1924	J. B. S. Haldane shows that enzymes obey laws of thermodynamics.
1925	Ronald Aylmer Fisher publishes *Statistical Methods for Research Workers*.
	Walter Rudolf Hess begins experiments that show functions of different parts of brain.
late 1920s	Otto Fritz Meyerhof and Gustav Embden independently describe details of glycolysis.
1926	Hermann Joseph Müller shows that X rays increase rate of mutation.
1928	Alexander Fleming discovers penicillin.
1929	Carl Ferdinand Cori and Gerty Theresa Radnitz Cori describe basic cycle of carbohydrate use in mammals.
	Otto Fritz Meyerhof's laboratory discovers energy donor ATP.
1930	Ronald Aylmer Fisher uses statistics to reconcile Mendel's and Darwin's theories.
1930s	Edgar Douglas Adrian shows that increasing stimuli to nerves makes the nerves fire more often rather than more strongly.
early 1930s	Alexis Carrel and Charles Lindbergh develop pumps that keep organs alive in the laboratory.
	J. B. S. Haldane, Ronald Aylmer Fisher, and Sewall Wright found theoretical population genetics.
1932	Gerhard Domagk isolates first general-purpose antibacterial drug for internal use.
	Albert Szent-Györgyi and Charles C. King identify vitamin C as ascorbic acid.
1935	Konrad Lorenz discovers imprinting.

late 1930s Konrad Lorenz and Niko Tinbergen found ethology.

Peyton Rous shows that development of cancer can involve two stages.

B. F. Skinner invents Skinner box.

Roger Wolcott Sperry shows that nerves' functions are innate.

1937 Hans Adolf Krebs describes cycle by which body breaks down and builds up carbohydrates.

1940 Karl Landsteiner discovers Rh antigen.

Ernst Mayr redefines species.

Roger Wolcott Sperry proposes that brain and nerve connections are determined by gene-controlled movement of chemicals in embryos.

Albert Szent-Györgyi shows that muscle fibers are made of two proteins.

1940s Ruth Patrick develops way to determine pollution effect on streams.

early 1940s George Wells Beadle and Edward L. Tatum show that a single gene usually controls the making of a single protein (enzyme).

Albert Szent-Györgyi and coworkers create artificial muscle fibers and make them contract.

1941 Howard Walter Florey tests penicillin on first sick human.

1943 Ernst Boris Chain proposes structure for penicillin molecule.

Salvador Luria shows that bacteria have genes that can mutate.

1944 Oswald Theodore Avery shows that nucleic acids can change genetics of bacteria.

Barbara McClintock shows that some genes can change position on chromosomes.

Selman A. Waksman creates streptomycin, first drug effective against tuberculosis.

1945 Willem Johan Kolff saves first human life with artificial kidney (dialysis machine).

late 1940s John Franklin Enders and coworkers show that viruses can be grown in cultured cells.

Alan Lloyd Hodgkin, Andrew Huxley, and Bernhard Katz show how changes in electrical activity conduct messages in nerves.

1946 Max Delbrück and Alfred Day Hershey discover independently that viruses can exchange or combine genes.

Dorothy Crowfoot Hodgkin determines molecular structure of penicillin.

1947 George Herbert Hitchings begins to look for compounds that will kill cancer cells by interfering with DNA.

Joshua Lederberg shows that bacteria can exchange and recombine genes.

1948 John Franklin Enders, Frederick Robbins, and Thomas Weller develop method for growing large amounts of poliovirus in culture.

Linus Carl Pauling discovers alpha helix structure of proteins.

1949 Linus Carl Pauling and coworkers show that sickle-cell anemia is caused by defective gene that makes abnormal hemoglobin molecule.

1950 Richard Doll and Austin Hill show smokers' increased risk of lung cancer.

Gertrude Belle Elion creates drug that fights cancer by interfering with cancer cells' nucleic acid.

Rita Levi-Montalcini discovers nerve growth factor.

Frederick Sanger works out sequence of amino acids in a protein.

Maurice H. F. Wilkins and Raymond Gosling begin making X-ray crystallography photos of DNA.

1950s George Gaylord Simpson revises evolutionary history of mammals.

early 1950s Frank Macfarlane Burnet proposes that the immune system's ability to identify foreign antigens develops before birth.

Gerty Theresa Radnitz Cori shows that inherited diseases can be caused by lack of particular enzymes.

Bernhard Katz shows how neurotransmitters convey nerve messages across synapses.

1952 Rosalind Elsie Franklin makes key X-ray photograph of DNA.

	Alfred Day Hershey and Martha Chase show that nucleic acids carry genetic information.
	Joshua Lederberg shows that viruses can exchange genes with cells they infect.
	Salvador Luria discovers restriction enzymes.
1953	Francis Crick and James Watson work out structure of DNA molecule and determine how DNA reproduces.
	Peter Brian Medawar confirms that immune system develops tolerance for certain antigens before birth.
	Stanley Lloyd Miller describes production of amino acids under possible primitive Earth conditions.
	Niko Tinbergen describes life of herring gulls.
1955	Severo Ochoa synthesizes RNA.
	Jonas Salk's injectable polio vaccine approved in United States.
late 1950s	Stanley N. Cohen finds that Nerve Growth Factor is protein.
	Earl Wilbur Sutherland Jr. identifies "second messenger" chemical by which hormones act.
1956	Choh Hao Li and coworkers determine composition and structure of ACTH.
1957	Melvin Calvin works out steps in photosynthesis.
	Francis Crick and Sydney Brenner propose that "letter" of genetic code is set of three bases in DNA molecule.
	Arthur Kornberg synthesizes DNA outside cells.
	B. F. Skinner describes relationships between behavior and reinforcement.
1959	Louis S. B. Leakey and Mary Leakey find skull of *Zinjanthropus boisei*.
	Rosalyn Sussman Yalow and Solomon Berson develop radioimmunoassay.
late 1950s/early 1960s	Francis Crick suggests mechanism through which cells manufacture proteins following DNA instructions.
1960	Drug developed by Gertrude Belle Elion makes possible first successful kidney transplant between unrelated humans.
	Jane Goodall begins research on chimpanzees.
	Contraceptive pills developed by Gregory Goodwin Pincus and others are approved by FDA.
	Albert Bruce Sabin's oral polio vaccine approved in United States.
	Edward O. Wilson shows that ants communicate by means of pheromones.
1960s	Roger Wolcott Sperry and Michael S. Gazzaniga show that brain hemispheres have different functions and communicate through corpus callosum.
early 1960s	James Whyte Black develops drug for angina and high blood pressure.
	Marshall W. Nirenberg, Har Gobind Khorana, Robert W. Holley, and others decipher genetic code.
	Louis S. B. Leakey and Mary Leakey find partial skeleton of *Homo habilis*.
	Jacques Monod and François Jacob propose that operators and repressors control action of structural genes.
1961	Peter Dennis Mitchell proposes theory describing how cells generate energy.
	Jacques Monod and François Jacob propose that messenger RNA is intermediate between DNA instructions and protein manufacture.
	Marshall W. Nirenberg and J. Heinrich Matthaei decipher first "letter" of genetic code.
1962	Rachel Louise Carson warns of pesticides' harm to environment.
	Stanley Cohen purifies Epidermal Growth Factor.
	Maurice H. F. Wilkins shows that RNA molecules have double-helix structure.
1963	Ernst Mayr describes geographic factors affecting formation of new species.
1966	Walter Gilbert and Benno Müller-Hill identify first genetic control element.
	Lynn Alexander Margulis says cell organelles were once free-living microorganisms.
1967	Dian Fossey begins research on mountain gorillas.

	Arthur Kornberg synthesizes biologically active viral DNA.
	Richard Leakey discovers hominid fossil site in Kenya.
	Robert Helmer MacArthur and Edward O. Wilson describe theory of island biogeography.
1968	Motoo Kimura proposes neutral theory of molecular evolution.
1969	Dorothy Crowfoot Hodgkin determines molecular structure of insulin.
	James Lovelock proposes Gaia theory.
1970	David Baltimore and Howard Martin Temin independently discover enzyme that lets RNA be copied into DNA.
	Choh Hao Li and coworkers synthesize human growth hormone.
early 1970s	James Whyte Black develops drug for ulcers and heartburn.
1971	Moses Judah Folkman suggests that cancers create own blood supply.
	Biruté Galdikas begins studies of orangutans.
1972	Paul Berg combines DNA from two kinds of living things.
	Stephen Jay Gould and Niles Eldredge propose evolutionary theory of punctuated equilibrium.
1973	Herbert Wayne Boyer and Stanley N. Cohen move DNA from one kind of living thing to another and show that it functions in new location.
1974	Donald C. Johanson finds skeleton of hominid "Lucy."
1975	Paul Berg organizes conference to consider safety of genetic engineering.
	Mary-Claire King shows that humans and chimpanzees have more than 99 percent of genes in common.
	César Milstein and Georges Köhler invent monoclonal antibodies.
	Edward O. Wilson claims that social behavior, including that of humans, is determined partly by genes.
1976	J. Michael Bishop, Harold E. Varmus, and coworkers show that cancer-causing genes began as normal cell genes.
	Herbert Wayne Boyer and Robert Swanson found first company based on genetic engineering technology.
	Richard Dawkins publishes *The Selfish Gene*.
	Har Gobind Khorana synthesizes gene and shows that it can make a protein.
	Susumu Tonegawa and coworkers show that multiple, movable genes explain diversity of antibodies.
1977	Frederick Sanger works out sequence of bases in genome of a virus.
	Carl R. Woese announces existence of archaean microorganisms and claims that they represent third domain of life.
1978	Robert Edwards and Patrick Steptoe produce first "test tube" baby.
	Robert Gallo isolates first virus shown to cause human cancer.
	John R. Horner finds first dinosaur nest.
	Mary Leakey finds 3.6-million-year-old footprints of hominids walking upright.
1980	Walter Gilbert devises way to find sequence of bases in DNA.
	Robert A. Weinberg isolates first oncogene involved in human cancer.
1980s	Christiane Nüsslein-Volhard and Eric Wieschaus identify genes that control embryonic development.
early 1980s	Tomoko Ohta develops nearly neutral hypothesis of molecular evolution.
1981	Nancy Sabin Wexler begins studying Venezuelan family with Huntington's disease.
1982	Herbert Wayne Boyer's company sells first genetically engineered commercial product.
	Stanley B. Prusiner suggests that certain proteins can reproduce and cause disease.
	Robert A. Weinberg and others show that human *ras* oncogene differs from normal gene by only one base.
1983	Luc Montagnier isolates virus later known as HIV.

	Kary B. Mullis invents polymerase chain reaction.
	Robert A. Weinberg shows that two or more mutations in different genes may be necessary to trigger cancer.
1984	Robert Gallo announces finding virus that causes AIDS.
	Richard Leakey's team finds almost-complete *Homo erectus* skeleton.
1985	Steven A. Rosenberg temporarily controls cancers with stimulated immune cells.
late 1980s	Leroy Hood develops automatic gene and protein sequencers and synthesizers.
	John R. Horner suggests that some dinosaurs cared for young and were warm-blooded.
	Robert A. Weinberg and others discover first tumor suppressor gene.
1988	Human Genome Project, headed by James Watson, begins.
1989	Steven A. Rosenberg and W. French Anderson give genetically altered cells to human being.
	Lap-Chee Tsui and Francis Collins find cystic fibrosis gene.
1990	W. French Anderson oversees first human gene therapy.
	Mary-Claire King's laboratory locates breast cancer gene on lower arm of chromosome 17.
	J. Craig Venter and coworkers develop methods for isolating and sequencing only DNA that will be expressed in proteins.
1990s	Tetsuko Takabe makes genetically engineered, salt-tolerant rice.
	Robert A. Weinberg and others show that abnormalities in telomerase gene may be involved in cancer.
1991	JoAnn Marie Burkholder and coworkers identify *Pfiesteria* microorganisms as cause of massive fish kills.
	Theo E. Colborn warns of health and environmental dangers from pollutants that affect hormones.
1992	Edward O. Wilson claims that human activities are causing major mass extinction.
1993	With assistance of Nancy Sabin Wexler, researchers find Huntington's disease gene.
1995	J. Craig Venter's organization sequences first cellular genome.
1996	David Ho suggests that HIV may be controllable by early administration of drug combination.
	Sequencing of genome of archaean microorganism confirms Carl R. Woese's claim that archaea are not bacteria.
1997	Moses Judah Folkman and coworkers stop growth of tumors in mice with compounds that block new blood vessels.
	Ian Wilmut announces cloning of sheep from adult cell.
1998	Sylvia Alice Earle begins study of U.S. National Marine Sanctuaries.
2000	Human Genome Project and company led by J. Craig Venter independently sequence human genome.
	Testing of genetically modified, virus-resistant sweet potato developed by Florence Wambugu begins in Kenya.

GLOSSARY

AIDS Acquired immunodeficiency syndrome, a virus-caused disease that suppresses the immune system and is usually fatal.

amino acid One of 20 kinds of small molecules combined to make proteins.

anatomy The structure of the body and its parts, or the study of that structure.

angiogenesis Growth of new blood vessels.

anthrax An epidemic disease caused by a bacterium, usually fatal; it chiefly affects cattle and sheep but can affect humans.

anthropology The study of humans, including their physical characteristics and evolution.

antibiosis Production by one kind of microorganism of a chemical that kills or stops the growth of other kinds.

antibiotic A substance made by one kind of microorganism to kill or halt the growth of another kind; a drug made from such a substance.

antibody A protein made by certain cells in the immune system; each antibody fits onto a particular antigen and, when attached to a cell carrying the antigen, marks that cell for destruction by other cells in the immune system.

antigen A protein on the surface of a cell; the immune system recognizes antigens not belonging to the body and makes antibodies to match them.

antiseptic A substance that prevents infection by killing microorganisms outside the body.

antiserum Liquid part of the blood containing antibodies made in response to injection of a particular bacterium or toxin, given as a treatment for disease; also called antitoxin.

autopsy Dissection of a dead body for the purpose of determining cause of death.

bacteriophages A group of viruses that attack bacteria (the name means "bacteria eaters").

base One of four kinds of small molecules combined to make DNA or RNA; the bases in DNA are adenine, guanine, cytosine, and thymine, but in RNA, uracil substitutes for thymine.

biogeography The study of the geographical distribution of living things.

botany The study of plants.

carbohydrates Sugars, starches, and celluloses; most living things break them down to obtain energy.

cell A microscopic unit of living matter, surrounded by a membrane; the basic unit of which all living things are composed.

chemotherapy Drug treatment, especially of cancer.

cholera A disease caused by bacteria and usually spread by drinking water contaminated by waste from people with the disease; it produces severe diarrhea and vomiting and often kills quickly by dehydration.

chromosomes Threadlike bodies in the nucleus of the cell, made primarily of DNA and carrying inherited information.

clone A duplicate of a cell or living thing carrying exactly the same genes as the original.

culture A group of cells or bacteria grown in a nourishing substance in the laboratory.

cytoplasm The jellylike material that fills cells; formerly called protoplasm.

DDT Dichlorodiphenyltrichloroethane, a powerful insecticide now banned in the United States because it is harmful to the environment.

diabetes A disease caused by lack of the hormone insulin or failure to respond to the hormone, in which sugar accumulates in the blood and damages tissues.

dialysis A process in which small molecules are separated from larger ones by being filtered through a membrane; a form of it purifies the blood of people with kidney failure.

diphtheria A disease caused by bacteria, often fatal, especially to children, in which breathing becomes difficult and nerves are damaged by a poison.

dissection Cutting apart a body, usually for purposes of learning.

DNA Deoxyribonucleic acid, the substance that carries inherited (genetic) information in most living things.

ecology The study of relationships among living things and between living things and their environment.

embryo A living thing undergoing development before birth, especially in its early stages.

embryology The study of development before birth.

entomology The study of insects.

enzyme A protein that speeds up a chemical reaction; enzymes make possible most reactions in the bodies of living things.

epidemiology The study of the way diseases spread in and affect populations.

Escherichia coli (E. coli) A common bacterium that lives, usually harmlessly, in the human intestine; it has been used often in genetic experiments.

ethology The study of animal behavior, especially in the wild.

evolution The slow change of types of organisms over time.

exobiology The study of possible extraterrestrial life.

fermentation The breakdown of substances such as sugars, carried out by living things or enzymes taken from them.

fertilized egg A single cell resulting from combination of an egg (female sex cell) and a sperm (male sex cell), from which a complete new organism can arise.

fossil Remains or imprints of organisms that lived long ago, usually turned into stone.

gene A segment of a nucleic acid molecule that contains information for a specific activity, usually making a protein or controlling another gene.

genetic code The pattern of transmission of inherited information through the sequence of bases in DNA or RNA; each "letter" of the code (codon) is a group of three bases.

genetic engineering The technology of changing genes or combining genes from different kinds of living things.

genome An organism's complete collection of genes.

genus A group of species that share a close evolutionary relationship and many common features; plural *genera*.

geology The study of the physical nature and history of the Earth.

gerontology The study of aging and medical conditions of old people.

histology The study of the structure of tissues or organisms through a microscope.

HIV Human immunodeficiency virus, the virus that most scientists believe is the cause of AIDS.

hominid Member of a family (group of related genera) of two-legged primates, including humans.

hormone A substance made by an organ that travels through body fluids and affects a different organ or tissue.

immune system The body's defense system, consisting of cells and biochemicals that attack foreign substances (those not belonging to the body), such as bacteria and viruses.

infectious disease A disease caused by microorganisms or other parasites.

insulin A hormone produced by special cells (islets of Langerhans) in the pancreas that controls the body's use of sugar and other carbohydrates.

invertebrate An animal without a backbone.

ion An electrically charged atom.

larva The immature form of an animal that changes substantially in structure when it becomes an adult; plural *larvae*.

leukemia A cancer of white blood cells (immune system cells).

lysozyme A weak antibacterial substance found in mucus, tears, and other secretions.

malaria A disease caused by a microscopic parasite that infects red blood cells; the disease is spread by mosquitoes and causes fever, chills, and sometimes death.

marine Pertaining to the sea.

metastases Secondary growths of a cancer at sites distant from the primary tumor.

microbe Microorganism; also sometimes called *germ*.

mitochondria Organelles in the cytoplasm of cells in which the reactions that let the cell release energy are carried out.

molecular biology The study of the chemical and physical principles related to the properties, composition, and activities of molecules in cells.

mutation A change in a gene or a sudden change in an inheritable characteristic; the act of changing a gene or characteristic.

National Institutes of Health (NIH) A large group of institutions in Bethesda, Maryland, devoted to medical research and sponsored by the U.S. government.

natural history Studies of the living and nonliving parts of the Earth, including botany, zoology, geology, and so on.

naturalist Old term for a person who studies all of nature or all living things.

natural selection The process by which, over many generations, living things become better adapted to their environment because those that possess more adaptive features are more likely to reproduce and pass on their genes than those that are less well adapted.

nucleic acids DNA and RNA, substances that carry inherited information in living things.

nucleotide A unit within DNA or RNA, made up of a base and an attached piece of phosphate-sugar "backbone."

nucleus An organelle within most cells (except bacteria and archaea) that contains the bulk of the cell's hereditary material.

oncologist A physician who specializes in treating cancer.

organelle A small body within a cell, separated from the rest of the cell by a membrane and carrying out a specialized function.

paleontology The study of fossils.

pathology The study of diseases and the changes they cause in organs and tissues.

PCBs Polychlorinated biphenyls, formerly used as insulators in electrical equipment but now banned because of potential damage to environment and health.

pharmacology The study of drugs and their effects.

physiology The functions of parts of the body, or the study of these functions.

plague Any epidemic disease, or, specifically, bubonic or pneumonic plague, an often-fatal disease caused by certain bacteria and spread by fleas.

polio Short for poliomyelitis (formerly infantile paralysis), a disease caused by a virus that frequently causes permanent paralysis or death.

primatology The study of primates, an order of mammals that includes monkeys, apes, and hominids (animals closely related to humans).

proteins A large class of biochemicals that do most of the work in cells; they are complex molecules made up of amino acids, manufactured by cells according to instructions carried in DNA and RNA.

respiration Breathing; also, the chemical reactions in living things that use oxygen and release carbon dioxide or other products.

retina The light-sensitive tissue in the back of the eye that makes vision possible.

retrovirus A virus that has genetic material made of RNA and uses an enzyme called reverse transcriptase to copy its genome into the genome of a cell so that the cell will reproduce the virus along with its own genes; some retroviruses cause cancer, and one causes AIDS.

RNA Ribonucleic acid, a type of nucleic acid that "translates" instructions from DNA into protein in most cells; it carries genetic information in certain viruses.

Royal Society In full, the Royal Society of London, an organization of scientists formed in the 17th century; considered to be Britain's most prestigious scientific organization.

scurvy A condition caused by a lack of vitamin C (ascorbic acid) in the diet, producing weakness, bleeding gums, and sometimes death.

serum The liquid part of the blood.

side effect An undesirable effect of a medical treatment.

species A group of similar living things that normally interbreed only among themselves.

statistics Numerical facts or data, especially those pertaining to large groups.

strain A subtype within a species, usually descendants of a common ancestor.

symptom A sign of disease.

synthesize To make complex chemicals from combinations of simpler ones.

taxonomy A system of biological classification or the study of such systems.

tetanus A disease resulting when certain bacteria enter wounds, producing severe muscle spasms and often death.

toxin A poison.

tuberculosis A disease caused by bacteria, which chiefly affects the lungs and frequently causes death if untreated.

typhus An often fatal disease caused by microorganisms and transmitted by lice and fleas.

vaccine A preparation of killed or weakened disease-causing microorganisms put into the body in order to activate the immune system so it can fight off attacks by full-strength microbes of the same type.

vertebrate An animal with a backbone.

virus A microorganism consisting of genetic material in a protein coat, able to reproduce only within living cells; considered to be on the borderline between living and nonliving things.

X-ray crystallography A procedure in which X rays are shone through a crystal or other solid and strike film on the other side, producing a pattern of dots that provides information about the three-dimensional structure of molecules in the solid.

zoology The study of animals.

INDEX

Note: Page numbers in **boldface** indicate main topics. Page numbers in *italic* refer to illustrations.

C

D

E

J

K

M

O

Q

R

T